THE AGE OF SUBTLETY

The Early Modern Exchange

SELECTED TITLES

Redreaming the Renaissance: Essays on History and Literature in Honor of Guido Ruggiero, edited by Mary Lindemann and Deanna Shemek

Ordering Customs: Ethnographic Thought in Early Modern Venice, Kathryn Taylor

The Waxing of the Middle Ages: Revisiting Late Medieval France, edited by Tracy Adams and Charles-Louis Morand-Métivier

Gendering the Renaissance: Text and Context in Early Modern Italy, edited by Meredith K. Ray and Lynn Lara Westwater

Storytelling in Sixteenth-Century France: Negotiating Shifting Forms, edited by Emily E. Thompson

England's Asian Renaissance, edited by Su Fang Ng and Carmen Nocentelli

Performative Polemic: Anti-Absolutist Pamphlets and Their Readers in Late Seventeenth-Century France, Kathrina Ann LaPorta

The Age of Subtlety

Nature and Rhetorical Conceits in Early Modern Europe

Javier Patiño Loira

University of Delaware

University of Delaware Press

Newark

Library of Congress Cataloging-in-Publication Data

Names: Patiño Loira, Javier, author.
Title: The age of subtlety : nature and rhetorical conceits in early modern Europe / Javier Patiño Loira.
Description: Newark, DE : University of Delaware Press, 2024. | Series: The early modern exchange | Includes bibliographical references and index.
Identifiers: LCCN 2023047797 | ISBN 9781644533444 (paperback) | ISBN 9781644533451 (hardcover) | ISBN 9781644533468 (epub) | ISBN 9781644533475 (pdf)
Subjects: LCSH: Rhetoric—Europe—History—17th century. | Rhetoric—Italy—History—17th century. | Rhetoric—Spain—History—17th century. | Europe—Intellectual life—17th century. | Italy—Intellectual life—1559–1789. | Spain—Intellectual life—1516–1700. | Metaphor.
Classification: LCC P301.3.E85 P38 2024 | DDC 808.0094/09032—dc23/eng/20240208
LC record available at https://lccn.loc.gov/2023047797

A British Cataloging-in-Publication record for this book is available from the British Library.

udpress.udel.edu

Distributed worldwide by Rutgers University Press
Printed and bound by CPI Group (UK) Ltd, Croydon, CR0 4YY

To my parents, Nieves and José Manuel

Contents

Acknowledgments

In the past three years, I have incurred numerous debts. The book in its final form exists thanks to the support of institutions and people who believed in it at different stages of its conception and growth. I am grateful to the American Council of Learned Societies, which made it possible for me to work full-time on the manuscript during the darkest hours of the COVID-19 pandemic. It was then, in October 2020, that I started writing the book in the same place where I now finish it, at a table overlooking the river in Florence, when its shores and bridges, now overflowing with people, were all but empty. I thank the Herzog August Bibliothek for two memorable months in Wolfenbüttel during the spring of 2021; the Folger Institute in Washington, D.C., for providing time for writing and resources in the summer of 2022; and the Linda Hall Library in Kansas City, Missouri, where Ben Gross and a wonderful and dedicated team made my life extremely easy as I revised the manuscript when fireflies started to spark at dusk in June 2023. I am grateful to my institution, the University of California, Los Angeles (UCLA), and especially to the Dean of the Humanities and the Spanish and Portuguese Department for supporting my leave in 2020–2021. I thank the UCLA Council of Research, Academic Senate, for funding research stays in Rome in the summers of 2022 and 2023.

At UCLA, Barbara Fuchs has provided unwavering support and valuable advice throughout the years. I am beyond grateful for that. I also thank Teo Ruiz for being always willing to share his experience and help. In a different and almost unrecognizable guise, *The Age of Subtlety* began to exist much earlier, during my years as a graduate student at Princeton. There, I found guidance and support in Marina Brownlee and Claudia Brodsky, who first inspired my interest in rhetorical conceits eleven years ago. I am grateful for that, and for much more. From the years at Princeton, I also thank Christina Lee, the late Ron Surtz, Leonard Barkan, and Kathy Eden. I especially thank Enrique García Santo-Tomás, who always offered valuable mentorship, and Lia Markey, who facilitated my first contact with the early modern cabinets of curiosity and a number of topics and scholarship that ended up playing a prominent role in my research.

I am grateful to Matthias Roick for inviting me to spend ten days at the Herzog August Bibliothek in 2019. When I went back there in the spring of 2021, Elizabeth Harding and a cohort of fellows including Justin Begley, Jonathan Nathan, and Tomás Valle provided feedback on what later became chapter 2 of the book. Charlotte Colding Smith, Jeremy Thompson, and their daughter Clara filled the deserted streets of Wolfenbüttel under COVID-19 with company and warmth. Jeremy was the first person to read the manuscript and he saved me from more than a few inaccuracies in matters of medieval philosophical and theological references. It is exciting to write a book at the crossroads of several disciplines; it is also a minefield for mistakes. Thus, I thank Zrinka Stahuljak at the UCLA CMRS Center for Early Global Studies for organizing a book workshop in which Eileen Reeves and Alexander Marr were instrumental in helping me transform a previous draft of *The Age of Subtlety* into the book in its final form. I am also grateful to the colleagues and friends who participated in the event: Raphaëlle Burns, Cesar Favila, Barbara Fuchs, Susana Hernández Araico, Robin Kello, Fabien Montcher, Laura Muñoz, Rhonda Sharrah, and Cheché Silveyra. They not only helped me write a better book but also convinced me that I had written something valuable. I thank the members of the Medieval & Early Modern Iberian Cultures group (MEMIC): Marta Albalá Pelegrín, David Horacio Colmenares, Lexie Cook, Noel Blanco Mourelle, Nicole Hughes, Miguel Ibáñez Aristondo, Michele Lodone, Juan Carlos Mantilla, Mariana Meneses Muñoz, and Leonardo Velloso-Lyons provided important feedback on chapter 4 of the book in the spring of 2022. I am also grateful to the scholars and friends with whom I have shared time and ideas during the years that I researched and wrote *The Age of Subtlety*: Fausta Antonucci, Patricia Arroyo Calderón, Helen Deutsch, Esther Fernández, Eloy García Domínguez, Claire Gilbert, Stefano Gulizia, Efraín Kristal, Patrícia Lino, Valeria López Fadul, Miguel Martínez, Patrícia Martins Marcos, James Nelson Novoa, Xavier Tubau Moreu, Lorena Uribe Bracho, Felipe Valencia, Miguel Valerio, Juan Vitulli, and many others.

Throughout the years, librarians, archivists, and staff in Italy, Spain, Germany, and the United States have helped me access early modern manuscripts and printed books. I thank Irene Pedretti and Martín Morales (APUG), Patrizia Ricca (Biblioteca Lancisiana), Vanessa Pintado (Hispanic Society of America), and also the personnel at the UCLA Library, the Biblioteca

Apostolica Vaticana, the Biblioteca Nazionale Centrale di Roma, the Biblioteca Vallicelliana, the Biblioteca Magliabecchiana, the Biblioteca Medicea Laurenziana, the Biblioteca Moreniana, the Biblioteca Nazionale Centrale in Florence, the Museo Galileo, the Herzog August Bibliothek, and the Linda Hall Library.

I am grateful to Julia Oestreich at the University of Delaware Press for her help throughout the process and to the two anonymous reviewers for their insightful comments.

I thank the graduate students at UCLA who enthusiastically discussed dissimulation, Tacitism, and nature's transformative and generative powers in seminars on early modern studies and, of course, the undergraduate students who asked time and again, "When is the book coming out?"

I leave for the end those who have accompanied me for the longest time: my family, and especially Marta, my brother, my grandmother Carmen—she is 101 as I write—and my parents, who have been behind me every single step and have made everything possible. I dedicate *The Age of Subtlety* to them.

Introduction

The Early Modern Culture of Ingenuity

It was well before dawn when Christoph Scheiner (1573–1650) set off with his instruments toward the fields outside the city of Ingolstadt. Having learned from his disciples that large stains on the sun had become visible to the naked eye, Scheiner hastened back from a trip to Munich just in time to accomplish a strange and unexpected discovery. It was September of 1612 and he, a professor of mathematics and Hebrew at the Jesuit University of Ingolstadt in Bavaria, was still immersed in a fierce and notorious dispute with Galileo Galilei (1564–1642) regarding the presence of freckles disfiguring the face of the sun, which both claimed they had been the first to detect with the aid of a telescope during the previous year.

Now, as the sun started to rise, Scheiner could see for himself what his disciples mentioned in their letter to him. When he subsequently peered through his telescope, however, something else, wondrous and new, absorbed all of Scheiner's attention. Floating above the horizon, the sun appeared to be oblong rather than round. Filled with curiosity, he decided that the matter was worth investigating.[1] A diagram illustrating the phenomenon decorated the thesis that Scheiner's student Johann Georg Locher (1592–ca. 1633) published in 1614, where a note stated its cause with exacting brevity: refraction motivated by the density of the atmosphere (Figure I.1).[2] For a detailed account, the reader was advised to consult *Sol ellipticus* (*Elliptical Sun*), the first of two works that Scheiner devoted to the topic. Scheiner signed the dedication of *Sol ellipticus* on

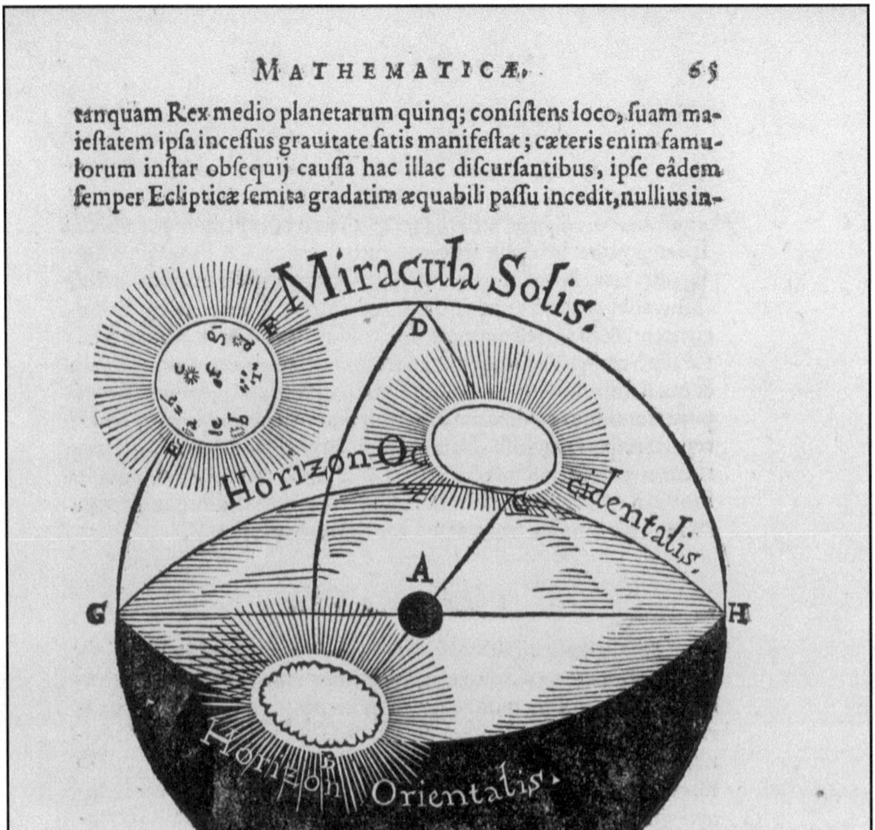
MATHEMATICÆ. 65

tanquam Rex medio planetarum quinq; conſiſtens loco, ſuam maieſtatem ipſa inceſſus grauitate ſatis manifeſtat; cæteris enim famulorum inſtar obſequij cauſſa hac illac diſcurſantibus, ipſe eâdem ſemper Eclipticæ ſemita gradatim æquabili paſſu incedit, nullius in-

digus, omnibus aliorum obtutibus ad ſe conuerſis, omnibus ipſe intendens, & de ſua abundantia copioſè adminiſtrans. Ipſe enim ſua face omnia colluſtrat, & abſq; ipſo foret, tenebris ipſa ſidera inuoluta iacerent; nulláq; lux quæ ſolis non ſit, à ſuperiore mundo in hunc deſcendit. Magnitudo eius à ſtella nuperâ in ſidere Caſſiopææ anno 1572. viſâ, ſuperata eſt bis. Terram centies & quadragies minimùm excedit ſecũdum omnes probatiores Aſtronomos;

I 2 Sphæri-

Figure I.1. The sun appearing oblong near the horizon. Christoph Scheiner and Johann Georg Locher, *Disquisitiones mathematicae de controversiis et novitatibus astronomicis* (Ingolstadt: Elisabeth Angermaier, 1614), 65. Courtesy of The Linda Hall Library of Science, Engineering, and Technology.

December 28, 1614. In 1615, the lean volume was in the hands of the community of astronomers. Before the year was over, Scheiner had sent Galileo a copy.[3]

Given its date of publication, Scheiner made a point that *Sol ellipticus* was meant to serve as "a New Year's present" or "strena" for Archduke Maximilian of Austria. Almost no one among the book's potential readers would have missed the reference to *Strena, seu de Nive Sexangula (New Year' Gift, or the Six-Angled Snowflake)*, a publication in which astronomer Johannes Kepler (1571–1630) presented diplomat Johannes Matthaeus Wacker von Wackenfels with a crystal of snow to celebrate the new year at the turn of 1611.[4] Patrons and friends had grown accustomed to extraordinary gifts. In March 1610, Galileo traded the four moons that he had discovered orbiting Jupiter for a court position in Florence.[5] What was Scheiner's gift, though, and how was he supposed to present it to Maximilian?

The contribution of *Sol ellipticus* was the link between refraction and what Scheiner later called the sun's metamorphosis or transformation.[6] Refraction was a topic that a professor of mathematics who pursued optics and astronomy with comparable enthusiasm was well prepared to examine. It would have been entirely out of place, however, to discuss it in the dedication. Even if Maximilian might have been able to understand Scheiner's technical account—and the conversations they had in Innsbruck earlier in 1614 suggest that the archduke was at least interested in astronomy—etiquette should prevail.[7] The dedication was a space in which Scheiner had to captivate his patron, and indirectly a learned and sophisticated readership, without resorting to any form of specialized knowledge. Preferably, he should have reminded whoever read *Sol ellipticus* that they, along with the author, belonged to a community of cultivated and quick-witted individuals. Instead of discussing refraction, Scheiner had to produce a different kind of argument to explain why the sun appeared oval-shaped. For instance, he could and did say that a fearful and modest sun, cowering at the prospect that it might itself be a gift for such a powerful prince, had adopted the shape of an egg in order to dissimulate its presence among other foodstuffs on the table, lest it distract Maximilian from more critical and urgent matters.[8]

What might strike us today as a curious, even eccentric way of approaching a patron—let alone of opening a book on optics and astronomy—was a standard practice for Scheiner's contemporaries. He had used a conceit, a rhetorical device that connected diverse facets of reality in a playful and irreverent way through figurative language. First, Scheiner pretended that Maximilian's gift was not the book *Sol ellipticus* but rather the egg-shaped sun that constituted the book's object of study. Second, he endowed the egg-shaped sun with feelings and agency. In what seems to be a form of impostor syndrome, Scheiner's unwilling protagonist perceives itself as lacking and unworthy of its task. The dedication departs from sound logic in a third and final trope when the egg-shaped sun morphs into a real egg placed on the table and ready for Maximilian to relish. Any individual trained in the early modern arts curriculum knew that Aristotle had expelled equivocal terms from the province of logic. English jurist and natural philosopher Francis Bacon (1521–1626) carried the idea even further, claiming there is a component of ambiguity that language is never entirely able to avoid, which stands on the path between the mind and the advancement of natural philosophy.[9]

To take a book about the sun for the sun itself, as Scheiner did, reeked of sophistry. It implied the purpose of misleading both Maximilian and the public, except for the fact that Scheiner's conceit was self-evidently humorous. With its flagrant disrespect for the rules of sound reasoning, it prevented even the most naïve of readers from taking it at face value. Indeed, a conceit used one or more tropes to alert readers that they should adopt a lenient and accommodating approach to its logic and truth value.[10] They ought to react with the same blend of belief and dispassionate detachment that enabled them to engage with theatrical plays and epic or chivalric tales without believing them to be true.[11] Scheiner's conceit catered to a mind in the mood for fiction, ready to wonder at the disconcerting, volatile link that the three tropes established between otherwise unrelated terms: between an egg and the sun and between the sky and Maximilian's table.

Conceits are harmless. They make readers aware of the trick as they rejoice in their innocent and temporary inadvertence. As polymath Emanuele Tesauro (1592–1675) once put it, conceits act like someone who surreptitiously removes the chair in which you are about to sit. You fall and then get up, after which, making a show of the virtue that Aristotle called *eutrapelia,* you let everyone see that you can take a joke.[12] There is between sophistry and conceits the same difference that exists between "a true viper, which suddenly bites and poisons you, and a painted viper, which seems to be willing to bite you, and yet you delight in it."[13] Similar to the automatons that captivated the attention of early modern audiences even when the ropes and gears that put them in motion were visible, conceits aroused a form of pleasure linked to their artificial or artful quality. The three tropes in Scheiner's example were the ropes and gears that transformed his conceit's argument into a coherent, working whole, a form of trickery that announced itself openly and begged for praise at the artifice.

Unlike the plausible truths of dialectic and the certain truths of mathematics, conceits sought to arouse admiration at their author's ability and audacity. Despised by those who deplored seeing intellectual and creative abilities wasted in trifles that failed to produce solid knowledge, conceits received more favorable attention from those who saw their increasing proliferation and refinement as a sign of epochal transformations in the ability of humans to compete with nature. Conceits may have been absent from the most prominent works celebrating the feats of early modern discovery and invention, such as the series of *Nova reperta* (*New Discoveries*, ca. 1600) engraved after late sixteenth-century designs by Johannes Stradanus (Jan van der Straet, 1523–1605), yet they appeared to the advocates of the new fashion as a testament to the present's superiority over ancient Greek and Roman culture. Like Scheiner's discovery of the causes that made the sun appear oblong and the invention of the telescope that made that discovery possible in the first place, conceits were evidence of *ingenium* or talent.

The same individual could, and in most cases did, cultivate different manifestations of *ingenium*, complicating any distinction between "scientific" and "literary" talent. For starters, Scheiner both treasured and mastered conceits. In December 1611, he resorted to one when he mistakenly believed Venus to have failed to pass before the sun. Venus, he wrote, "evidently blushed and ran away to keep her nuptials secret."[14] This was a meandering, clever way of saying that the planet did not show up to the meeting as expected. As he would do with the egg-shaped sun and his book on it, Scheiner conflated the planet Venus with the goddess of the same name, endowing the planet with agency and feelings and transforming its passage in front of the sun into a wedding.

Scheiner could have a hard time letting go of a conceit. Following his claim that the sun had adopted the shape of an egg in order to pass unnoticed on Maximilian's table, he went on to add that the warmth of the archduke's generosity would keep the egg-shaped sun safe until a mighty and stately bird came out of its shell. Not without humor, he even assured Maximilian that the egg was not rotten.[15] Two years later, Scheiner opened *Refractiones Coelestes* (*Celestial Refractions*, 1617) by recalling that he had left Maximilian's eagle of empire "incubating a certain egg" in Innsbruck.[16] A poem signed by Bernardus Renanus prefacing the volume and arguing that the sun had adopted the shape of a shield, not an egg, challenged Scheiner to a friendly, lighthearted competition in the name of inventiveness.[17] Conceits were evidence of talent, and a look at the publications on astronomy and optics that Scheiner brought forth in the years that followed demonstrate that there was more to them than mere exhibition and display. Conceits allowed for different forms of sociability and emulation. By resorting to conceits, individuals built on one another's work but also defied and confronted one another, all while asking the reader to be part of the game.

Around the same years, pockets of heated discussion targeting the use of conceits set minds ablaze with questions: What made a statement qualify as a conceit? What did it take to invent one? How did a conceit touch the intellect and the emotions of the audience? Jesuit scholars active in Rome in the 1610s and 1620s rushed to weigh in on the topic. Among the disputants were the famous historian of the wars of Flanders Famiano Strada (1572–1649), a professor of Hebrew, soon-to-be botanist, called Giovanni Battista Ferrari (ca. 1584–1655), and the poet of Polish origins Maciej Kazimierz Sarbiewski (Sarbievius, 1595–1640), who authored the earliest extant course exclusively devoted to conceits. The archives brim with echoes of a lively, not always peaceful, exchange carried out face-to-face and through correspondence among Jesuits writing in different locales on the continent. It was a matter of time before a stinging and unforgiving invective against the new fashion, printed in 1636 by former Jesuit Agostino Mascardi (1590–1640), and the ensuing publication of *Delle acutezze* (*On Conceits*, 1639), the long and meticulous response by Matteo Peregrini (ca. 1595–1652), relocated the debate to the printed page.

The controversy was henceforth open to a broader and more diversified community that comprised individuals who were neither Jesuits nor academics. Freelance writers à la mode Pier Francesco Minozzi (1611–1672) and Giovanni Battista Manzini (1599–1664) published works in 1641 and 1652, respectively, that were, at least in part, dedicated to conceits. Jesuits Baltasar Gracián (1601–1658) and Pietro Sforza Pallavicino (1607–1667) published how-to manuals for would-be inventors of conceits in 1642 and 1646, respectively, which were clothed in the language of Aristotelian logic and rhetoric and were destined to undergo several editions. In 1654, culminating three decades of reflection about conceits, former Jesuit Emanuele Tesauro published *Il cannocchiale aristotelico* (*The Aristotelian Telescope*), undoubtedly the most detailed and most widely reprinted work on the topic.

Despite the success and, more often than not, the polemic that accompanied writings on conceits, they were unlikely to make a scholar's career. The majority of authors dedicated at most a single publication to the topic, and often less than that. Sarbiewski became renowned as a poet and earned the title of "the Polish Horace," whereas Ferrari went on to pioneer work on citrus fruits and the cultivation of flowers. Peregrini and Gracián became names of reference on life at court and its perils, while Pallavicino, who only wrote on conceits in a treatise on the style appropriate for didactic writing, went on to write a history of the Council of Trent, in addition to works on logic, ethics, and metaphysics. Minozzi inserted two digressions on conceits in the narration of his travels, partly in reaction to what he disliked about Peregrini's publication on the topic, whereas Manzini, a famous and best-selling writer of fiction, discussed conceits only to justify why—and how often—he used them.

Only Tesauro remained determined to develop a method for the invention of conceits throughout his life, both in his days in the Society of Jesus and after he left the order in 1635. Around 1625, when he was still relatively young, Tesauro wrote two tracts on the topic, but they remained in manuscript form.[18] Not until three decades later did he finally publish *Il cannocchiale aristotelico*, a work that he kept updating for the rest of his life. Conversely, for other characters in the story that is about to unfold, conceits furnished only a sporadic occupation. Yet even though the debate over them oscillated between conversation and cacophony, this is not to say that it lacked a center. Gracián and Peregrini exchanged untenable accusations of plagiarism, and Pallavicino acknowledged—and criticized—Peregrini's efforts.[19] Tesauro failed to acknowledge his debt with his contemporaries but borrowed freely from Peregrini. Some of the authors wrote in direct response to a previous publication on the topic: Minozzi replied to Peregrini, who to some extent wrote in reply to Mascardi, as did Manzini. For all of them, there was more at stake than dissenting views on language. From different milieus and backgrounds, the scholars involved in the controversy voiced conflicting ideas on what made the invention of conceits both similar to and distinct from other forms of intellectual and creative pursuit. I will argue that, whether they stood for or against conceits, seventeenth-century scholars writing on the topic constantly mobilized conflicting ideas on nature and the way that humans relate to it through the invention of the rhetorical and poetic devices.

I use "conceit" to translate a wide and heterogeneous range of terms featured in both Latin and vernacular (mostly Italian and Castilian) writings on conceits dating from around 1619 to 1654. Words such as *acutum*, *acutezza*, and *agudeza* suggested the sharpness that rendered conceits apt at making a deep and lasting impression on the reader's mind. At times, however, "acutum" referred to the author's intellectual and creative penetration of the topic about which he or she was speaking or writing.[20] In a different vein, terms connected with life (*spiriti*, *vivezze*, *vivezas*) elaborated on the claim that conceits provided language and reasoning with soul, liveliness, and motion. *Conceptus*, *concetto*, and *concepto* focused on a widespread analogy between the conception of living and intellectual "creatures." Each of these nouns told a different but complementary story about how conceits worked and how they impacted the reader, and all of them will be the object of analysis in the present book.

English-speaking scholars writing on seventeenth-century metaphysical poetry favor the term "conceit."[21] For the sake of clarity, I always use "conceit," but indicate the original term in parentheses when I translate a passage. Although *concetto* and *concepto* share the floor with *acutezza* and *agudeza* in the Spanish and Italian traditions, critics unanimously favor the terms *concettismo* and *conceptismo* to designate the seventeenth-century craze for ingenious or witty language.[22] In addition to maintaining continuity with the work of previous scholars, "conceit" also has the advantage that its seventeenth-century equivalents were at the center of word families that were household concepts for Gracián's contemporaries, forming part of the vocabulary of theory and criticism in use by specialist and nonspecialist writers. Adjectives such as *concettoso* and *conceptuoso*, for instance, conveyed unambiguously to Italian and Spanish audiences the idea that a piece of writing or a speaker was abundant in conceits.[23] These adjectives' existence and the fact that they have disappeared from usage are testaments to the vitality of the world they once evoked, but also to the strenuous and meticulous effort that is required to bring it back to life again.

Engaging with Nature: Ingenuity as Problem-Solving

The Age of Subtlety approaches seventeenth-century conceits as sites of engagement with nature. By analyzing how-to manuals on the invention of conceits in parallel with writings on natural philosophy, I argue that conceits appeared to Scheiner's contemporaries as embodiments of a fluid but intricate correspondence between natural and human creativity in which knowing and making were two faces of the same process.

In recent years, art historians have devoted keen attention to the relationship between the production of knowledge by the artist and that by the natural philosopher.[24] Matter—whether a colony of living bugs crushed into cochineal pigment or stone blocks extracted from the depth of the earth's bowels—served as a point of encounter in a creative and transformative process implicating nature in the artist's work and presupposing the artist's familiarity with nature's behavior. The materials placed in the hands of early modern artists and artisans carried hidden stories within them, which told of natural processes and change. Matter could comply or resist, and in doing so opened both challenges to and possibilities for human understanding and creativity. Artists learned about nature in the act of manipulating and experimenting with it.[25]

How is this relevant to conceits? Indeed, artists and artisans interacted physically with nature. Casting silver figures from life or meddling with pigments entailed a direct, hands-on relationship with matter that is difficult to fully compare with the manipulation of language.[26] Yet language can be pliable or resistant; it too brims with lifelike stubbornness and initiative, never passive to the operations of tongue or pen, always ready to offer some possibilities and reject others. Instead of brush and scalpel, the inventor of conceits used tropes as tools. It was metonymy that allowed Scheiner to transform a book about the sun into the sun itself and prosopopoeia that enabled him to claim that the sun felt shame and reserve; more conspicuously, he resorted to metaphor to transform the egg-shaped sun into an egg. Tropes shaped the matter of language into new forms with a degree of plasticity and "materiality" that, from the perspective of logic, represented a dangerous but alluring threat.[27] Scheiner's conceit revealed a set of relationships—between an egg and the sun, and between the sky and Maximilian's table—that did not exist prior to the act of discovery. They were products of metaphor, prosopopoeia, and metonymy, but could not have been created if they were not latent in language—i.e., in the words "sun" and "egg"—in the first place.[28]

This, at least, is how Gracián and Tesauro perceived the invention of conceits, which lay dormant in the matter of language and waited for the author's ingenuity in the same way that nature, without itself ever producing a clock, provides artisans and engineers with all the clues to fabricate one. Conceits existed halfway between discovery and creation, the two meanings of the rhetorical notion of *inventio*, not unlike the work of artists, artisans, and engineers who learned about nature's processes by manipulating it in the production of artifacts. For Gracián, and even more for Tesauro, the poet's use of language resembled a process of experimentation in the mind's "workshop." It was about putting "sun" and "egg" together and seeing what happened. To brace against the multiplicity of results, the poet had to employ his or her insight to understand the properties of each term (what lies within "sun" and "egg"?) and mix them in correct proportion. Like machines and other products of engineering, conceits required taking notice of nature's behavior—here, language, with its vain aspiration to be denotative and univocal—in order to trick it appropriately.[29]

Seventeenth-century scholars—those who favored them, at least—praised the invention of conceits as the result of intimacy with nature, often expressed by the twin properties of sharpness (or acuity) and subtlety. A sharp mind, as Jesuit professor Baltasar López (d. 1650) told a crowd of students and civic authorities gathered in Mexico City's Jesuit College of Saint Peter and Saint Paul in November of 1644, "enters in what is deepest and inmost, and becomes immersed in the marrow of things."[30] López, for one, was skeptical that conceits accomplished that and was resolved to debate with dissenters, such as Peregrini, Gracián, or Tesauro. Like acuity, subtlety denoted the ability to access a realm of intellection and experience reserved to those gifted enough to see beyond appearances. In a work first published in 1550, physician Girolamo Cardano (1501–1576) defined "subtlety" (from the Latin word *subtilitas*, and connected with the adjective *subtilis*, meaning "thin" or "delicate") as a property of objects. Objects were subtle when they were difficult to perceive through the senses or to understand through the intellect. Interestingly, subtlety was a reflexive or reciprocal quality that also applied to the senses and the intellect capable of perceiving or understanding difficult—that is, "subtle"—objects. Subtlety was objective as much as subjective and material as well as intellectual: it could at once describe theorems, hair-splitting arguments, and the wings of a moth, but also a person capable of understanding or seeing each of those.[31] As with sharpness or acuity, the notion of subtlety was metaphoric in origin. It was as if subtlety rendered one's *ingenium* thin enough to insinuate itself into the cracks of things and see distinctions where others could not.

The characterization of conceits as "sharp" or "subtle" placed them in a suggestive but dangerous proximity to a variety of seemingly unrelated phenomena. Girolamo Fracastoro (1478–1553), also a physician, claimed in 1525 that only extremely sharp and subtle particles can travel far enough to enter bodies and propagate illness. Conceits also traveled like the plague, if we are to believe Jesuit authorities determined to stymie the success with which the new fashion was met among preachers of the order across continents and oceans.[32] Conceits were creations of a sharp and subtle mind. They effectively penetrated reality, while insinuating themselves into the reader's intellectual and creative yearnings. Yet the term "mind" as I have used it here, is consciously vague. From what or where did conceits come into being, according to seventeenth-century scholars?

Conceits represented a material, embodied alternative to abstract thinking, one that muddied the operations of the intellect with a transgressive infusion of sensory activity, as when the similarity between an egg and the sun's shape suggested itself to Scheiner. Embodiment and materiality were also characteristics associated with the faculty to which early modern scholars ascribed the invention of conceits. Inventing conceits earned the author the epithet "ingenious" (Latin *ingeniosus*, Italian *ingegnoso*, and Spanish *ingenioso*) because they were the product of *ingenium*. Aligning with recent scholarship, I favor "ingenuity" over "wit" as a more faithful and comprehensive translation of *ingenium* and its vernacular counterparts (Italian *ingegno* and Spanish *ingenio*).[33] Leaving aside the suggestion of "superficial cleverness" associated with wit, the fact that *ingenium* came from the verb *gignere*, to "produce" or "generate," confirms the link between natural and intellectual generation that is the topic of much of what follows.[34] "Ingenuity," a more comprehensive and all-encompassing term, also preserves the connection with both engineering and cunning, two aspects that were inseparable from medieval and early modern accounts of *ingenium*. Engineering and cunning, I will show, were central to understanding the place of conceits in early modern debates about how humans placed themselves in relation to and in competition with nature.

Ingenium has been described as "a capacious umbrella term" that reunited a set of different but interrelated meanings.[35] It could refer to inborn talent, a meaning that lexicographers often explained through the etymology *in-genitum*, "in-born." The idea of individual difference, for instance, was at the root of the phrase *ingenium soli* ("ingenuity of the soil"), which evoked the properties that made a soil fertile in a way that was different from other terrains.[36] Educators took notice of the analogy between earth and intellect and encouraged paying attention to the particular traits of each student in order to "cultivate" his or her mind accordingly, as indicated in *Examen de ingenios para las ciencias (The Examination of Men's Wits*, 1575), a successful but controversial treatise published by Spanish physician Juan Huarte de San Juan (1529–1588), following both ancient and early modern humanistic precedents.[37] Conceived as that which is innate to the individual, *ingenium* stood opposite to *ars*, a set of general rules that can be taught in order to complement and perfect one's nature.[38]

However, when, in 1642, Gracián entitled his treatise on conceits *Arte de ingenio* (*Art of Ingenuity*), "ingenuity" referred no longer to inborn talent understood as individual difference but to imagination, one of the three faculties that, together with memory and the intellect, formed reason, according to Huarte.[39] In 1652, determined to solve the ambiguity surrounding the term *concetto*—which in Italian could mean "conceit" as much as "concept"—Manzini explained that humans can mentally conceive and bear offspring in one of two ways: either by abstracting intellectually what they gather from the senses, thus creating a concept (*concetto, notizia*), or by elaborating it further through imagination into a conceit. Only the second, Manzini claimed, qualified as the mind's offspring, which acquired its artificial or artful quality through the involvement of the author's ingenuity, identified with imagination and the use of tropes.[40]

In early modern Aristotelian philosophy, imagination was one of the internal senses, the others being common sense, the estimative faculty, and memory.[41] The internal senses were embodied faculties that mediated between the stimuli conveyed by the external senses and the immortal and disembodied intellect. Specifically, imagination furnished the intellect with "phantasms," or *phantasmata,* images elaborated from the stimuli that common sense collected from sight, hearing, smell, taste, and touch. Even though imagination was sense-oriented, its products were indispensable for the intellect, which, according to Aristotle's dictum, "never thinks without a phantasm."[42] With its liminal, mediating role between body and mind, imagination posed a threat to the dogma of the immortality of the intellect: if the intellect alone was expected to survive the death of the individual, how would it remain capable of thought without the aid of imagination?[43]

In Blaise Pascal's (1623–1662) formulation, however, imagination was also a mistress of "error and falsity" that chased certainty out of reach because of its tendency to rarely limit itself to evoking objects that were absent, preferring instead to compound phantasms with one another and bring forth new beings that lacked a counterpart in nature, such as mountains of gold and centaurs that were part-human and part-horse.[44] Accordingly, some scholars distinguished between imagination, which produced phantasms from the stimuli gathered by common sense, and fantasy, its unruly and creative counterpart. Others, however, conflated the two faculties.[45] Among them was Juan Luis Vives (1493–1540), who, in a widely reprinted and influential treatise first published in 1531, described fantasy as "marvelously free and disengaged. [Fantasy] invents, reproduces, combines, and dissolves everything at will. It conjoins the most distant things and separates the most united. Therefore, if it is not controlled and bridled by reason, it shakes up and disturbs the mind as a storm stirs up the sea."[46] Such unrestrained joining and disjoining brings to mind Samuel Johnson's oft-quoted appraisal of conceits in his biography of the seventeenth-century English poet Abraham Cowley, published in 1781. Johnson (1709–1784) described ingenuity or wit, the force responsible for the invention of conceits, as causing "a combination of dissimilar images, or discovery of occult resemblances in things apparently unlike." In a process made possible by the poet's imagination or fantasy, Johnson wrote, "the most heterogeneous ideas are yoked by violence together."[47]

Although imagination's ability to pair far-fetched realities was its most conspicuous and spectacular side, it was its material, sensory aspect that disrupted syllogistic logic, with its aim at a direct, unambiguous language in which single words represent single ideas. A conceit's reliance on imagination meant that its terms were trapped in a helpless plurality of interconnected reference networks. Its tropes tied words and things together based on sensory relationships, even similarity in sound, positioning the inventor of conceits closer to the artist and the artisan, in their attentiveness to matter's metamorphic and partially unpredictable properties, than to the philosopher.

To twenty-first-century readers, the operations of imagination described by Vives suggest a form of freedom and defiance that for over two centuries has become the province of artists and poets. However, the purview of imagination used to extend further. As Lorraine Daston argues, the views on the role of imagination in what we today call science underwent rapid change starting in the late eighteenth century. Daston claims that Immanuel Kant's 1790 *Critique of Judgment* (*Kritik der Urteilskraft*) played a crucial role in such transformations by restricting genius—Kant's *Genie*, which reunited early modern notions of genius, or individual difference, and *ingenium*, understood as creative and inventive power—to the artist and the poet, while denying it to mathematicians and natural philosophers, even Isaac Newton! As imagination and genius became increasingly celebrated as the powers behind creative pursuits, they also lost the relevance they once had in the search of truths about nature, now restricted to experiment and facts. A few decades into the nineteenth century, Daston explains, the forces supposedly driving scientific and technological innovation were no longer the same as those that guided the poet and the artist.[48]

To interpret early modern writings on conceits in light of a modern, post-Kantian understanding of creativity implies severing the ties that connected the likes of Gracián or Tesauro to views on ingenuity and imagination that predated Vives and had started to wane by the time of Johnson's assessment. It means effacing the links that connected the invention of conceits with the craft of the artisan, the ploys of the statesman, the dissimulation of the courtier, and the machines of the engineer, all aspects of creativity more deeply and intricately entangled than has been thus far acknowledged.[49] Early modern scholars and practitioners attributed to ingenuity a vast and heterogeneous range of activities, so much so that there were conscious efforts to curb the labeling of any achievement as "ingenious." For instance, the 1623 edition of the Accademia della Crusca dictionary credited to Francesco da Buti's fourteenth-century commentary on

Dante's *Commedia* the idea that "we speak only in a figurative way of the ingenuity of the hands, which one should call aptitude."[50] The authors enlisted a venerable but archaic source to remind the reader that it was wrong to conflate dexterity of mind and hand. Usage went down a different path, however, ascribing ingenuity to conceits as much as to the creations of painters, sculptors, and artisans, but also to the acts of jugglers and mountebanks performing in taverns and public squares.[51]

Ingenuity not only straddled the boundaries between hand and mind, and between "high" and "low" culture. It also encompassed both right and wrong, linking the feats of the engineer to deceptive schemes in a way that eventually cast a negative, or at least ambivalent, light on conceits. Take, for instance, the dome designed by architect and sculptor Filippo Brunelleschi (1377–1446) and completed in 1436 for Florence cathedral, celebrated as one of the period's greatest achievements, and described by Brunelleschi's biographer Antonio Manetti (1423–1497) as a feat of ingenuity. However, Manetti also praised Brunelleschi's ingenuity in a short story recounting a strange anecdote dating to the winter of 1409, when Brunelleschi successfully choreographed a group of Florentine aristocrats, artisans, and officials with such dexterity and coordination that they were able to persuade the woodworker Grasso that he was, in fact, another person. Manetti thus brought home the point that the greatest engineer of fifteenth-century Florence was also its greatest trickster. The dome embodied Brunelleschi's ingenuity as much as did his ability to deceive a friend by altering the appearances around him to "transform" him into another person.[52]

It has been argued that, in the early modern period, "most vernaculars gave cunning and machines a prominent place within their vocabularies of ingenuity," a legacy that carried on from the medieval period, when two prominent meanings of *ingenium* were those of "machination, legal subterfuge" and "machine."[53] Both "ingenuity" and "machine" formed word families that evoked resourcefulness in practical matters, including war, hunting, and politics.[54] Resourcefulness, however, was rarely taken in neutral or indifferent terms. Thus, when in 1609 Girolamo Vittori translated *maquinar* ("to plot") as "to imagine and find out some way or scheme," he merely reinforced a link between imagination, craftiness, and deceit that was already centuries old.[55] "*Ingenium*," as Patricia Ingham claims, "raised ethical questions: cleverness and wit were virtues; yet one could be too clever by half, deploying ingenuity in the service of deceit or fraud."[56] In 1575, Huarte justified the problem in physiological terms, arguing that imagination inclined hot-tempered individuals to do harm and to use cunning and prudence.[57]

More generally, however, imagination made it possible to apply theory to practice. For Huarte, imagination was useful not only for poets but also for rulers and physicians, who need to apply a set of principles to ever-different, changing circumstances.[58] In fact, writings on early modern architecture and engineering, two domains often conflated in early modern sources in the wake of Vitruvius, praised ingenuity in terms of problem-solving. The Greek word for machine, μηχανή, much like its Latin and vernacular derivations (*machinari, macchinare, maquinar*), often implied attack, deception, and ill-will, but could simply mean "device."[59] Ancient and early modern machines were "ingenious" because they used indirect, nonobvious means to solve a problem. The archetype of ingenuity as machination reduced to its basic, essential form was Odysseus, whom Homer described as being full of μηχανή. He was also gifted with μῆτις or *sollertia*, a form of quick and practical intelligence that enabled the subject to find a solution where others were at a loss.[60] In fact, Luca Simeoni has argued that, for Vitruvius, *sollertia* was the ability to "come up with a device, a tool, a means that allows to overcome a specific practical difficulty and to accomplish operations that would otherwise be impossible."[61]

In 1611, lexicographer Sebastián de Covarrubias (1539–1613) built into his definition of *ingenio* the wealth of meanings discussed above: "We usually call ingenuity (*ingenio*) a natural force of the intellect that investigates anything that can be attained through reason and discourse in all kinds of sciences, disciplines, liberal and mechanical arts, subtleties, inventions, and deceits. Thus, we call engineer one who creates machines for defense and attack against the enemy; and we call one of subtle and delicate ingenuity ingenious."[62] Ensconced in the notion of ingenuity, Covarrubias included natural philosophy (knowledge or *scientia*), rhetoric and poetics (liberal arts), art and engineering (mechanical arts), as well as the component of deceit that Brunelleschi's biographer Manetti connected with the cunning of the engineer's machines and that early modern authors often invoked in relation to Odysseus. Covarrubias concluded with a reminder that the term *ingenio* could apply to any machine, like those in use for sugar and silk, and "anything created with one's understanding that makes it easier to do something that was difficult and costly with one's strength alone."[63]

The dome in Florence and the trick played on the woodworker Grasso were different but comparable examples of "machine" in the sense that each was a product of Brunelleschi's ingenuity, his ability to find out and implement nonobvious means to accomplish a goal. In a certain way, the same was true of conceits. In *Arte de ingenio*, Gracián claimed that conceits linked different aspects of reality in what could be understood as a form of problem solving. Inventors of conceits found themselves in a predicament similar to the one of Odysseus when he and his companions hid under sheep in order to escape the cave of the cyclops Polyphemus.[64] Odysseus used the resources available to him in a way that was less than straightforward or conventional. Admittedly, sheep are not a means of transportation, and the idea that humans could be carried by them while lying underneath rather than riding on them is even stranger. Inventors of conceits twisted the terms involved in an argument in order to make them fit with one another in new and unheard-of ways. Like the feat accomplished by Odysseus, conceits remained simply out of reach for those who were reluctant to think outside the box.

Even more than Odysseus, however, the character in ancient myth who embodied Gracián's approach to ingenuity as problem solving was Theseus, the prince of Athens who, thanks to Ariadne's thread, found a way out of King Minos's labyrinth. "There are labyrinths of discourse," Gracián asserted, "that the mental Theseus measures and conquers with the precious skein of a fitting statement."[65] Inventing a conceit was, for Gracián, a trick that achieved what was otherwise impossible. What machines accomplished for engineers, tropes did for inventors of conceits by transforming terms through imagination. As mentioned above, tropes served as ropes and gears in a conceit's "machine."

To situate conceits in the early modern culture of ingenuity, defined as it was as a virtue of problem-solving, suggests placing writers in a weak, imperiled position—acting on the ground, so to speak, rather than from the philosopher's disengaged, authoritative position. The expression *hallar un medio*, which for Gracián came to epitomize the invention of conceits, is significantly polysemous, as it could mean "to find a middle term" (for a syllogism) but also "to find some means" (to solve a problem or get out of trouble).[66] Gracián clearly enjoyed picturing to himself the speaker's ingenuity when, "besieged by perplexity and with all the roads already taken by the enemy, it finds a sudden way out with the aid of promptness."[67] For Gracián, the archetype of promptness was Alexander the Great, who in the city of Gordion in Phrygia proceeded to simply cut the knot he was expected to loosen.[68] Alexander's bold but unorthodox way of achieving a goal provided a parallel in the realm of conceits, where tropes acted as tools that constrained otherwise incompatible terms to fit with one another.

As a matter of fact, ingenuity and knots were not strangers. They had shared center stage in Lodovico Castelvetro's (ca. 1505–1571) commentary on Aristotle's *Poetics* (1570), a work that often stressed the role of ingenuity in the poet's activity. Prefiguring Gracián's view of conceits as acts of problem solving, Castelvetro claimed that poets bring pleasure to the audience in proportion to the difficulties they overcome and the degree of ingenuity they show in devising the plot.[69] Tragic and epic poets demonstrate greater ingenuity than those who write comedies because, unlike the latter, tragic and epic poets build on well-known stories and thus are "restricted within certain boundaries that it is not licit to abandon."[70] Tragic and epic poets work within constraints: they are not allowed to prevent Oedipus from committing incest but can decide on the circumstances of the crime.

According to Castelvetro, poets should introduce novelty into traditional and well-known stories, while ensuring that each part fits the whole. When faced with a problem, poets solve it by means of ingenuity. Castelvetro illustrated the point through a story centered on none other than Michelangelo Buonarroti (1475–1564), the artist who, perhaps more than anyone else in the early modern period, embodied the ideal of prompt, even impetuous creativity and talent. All started when a statue representing the personification of a river was unearthed in Rome. It was large and made entirely of marble, but "its beard was in pieces and part of it was missing."[71] In addition to that, there was a mystery that only Michelangelo was able to solve:

> The part of the beard that stuck out from the chin had been sculpted in a way that suggested that, when the beard was still whole, it would have reached the navel according to proportion. And yet the extremity of the beard had also survived, but one could see it attached to the chest. Everyone wondered at this but no one was able to imagine how the beard could have been arranged before it broke to pieces. Only Michelangelo Buonarroti, a sculptor of the rarest ingenuity who was there present, after he pondered for some time finally understood what it all was about. "Bring me clay," he said. With it, he shaped the part of the beard that was missing, long in proportion to the part that survived, and brought it down to the navel. Then he twisted the beard in a knot, so that everyone might be witnesses that the beard he had shaped hit the statue's chest exactly on the spot in which one still could see the extremity of the beard that had been broken.[72]

Michelangelo "twisted the beard in a knot" just as the inventor of conceits resorted to tropes to bend the terms of the argument, making them fit in otherwise impossible ways. The onlookers agreed that Michelangelo had demonstrated subtle ingenuity ("sottilità d'ingegno"), more so than if he had shaped a beard from beginning to end without consideration for preexisting constraints.[73] The recourse to new, indirect means in the act of problem-solving was central to early modern understandings of ingenuity.

With its broad, almost all-encompassing range, embracing the liberal and the mechanical arts and mental as much as material processes, the culture of ingenuity evoked by Covarrubias's definition, the example of Brunelleschi, and the views of Castelvetro and Gracián offers a rich and meaningful background to study the invention of conceits, understood as products of imagination and embodied acts of making and knowing, akin to the operations of engineers and artists. Conceits offer a vantage point to explore how a focus on early modern rhetorical and poetic thought might contribute to the recent and ongoing reassessment of the relation between knowing and making and between mind and hand promoted by Pamela H. Smith, Lissa Roberts, Simon Schaffer, and others.[74] Smith has contrasted the manner in which artists approached nature with that of natural philosophers: "If scholars conceived of problems, or indeed of reality, primarily in terms of words and the manipulation of words, artisans might see reality as intimately related to material objects and the manipulation of material."[75] However, I claim that inventors of conceits occupied a liminal position in the alternative between materiality and words. Even if they certainly manipulated words, they did so as artificers plunged into the intricacies of sense- and imagination-based forms of association, which were a far cry from the detached and disembodied character of truth-oriented forms of reasoning.

By locating conceits within the early modern culture of ingenuity, I engage with recent claims that the notion of ingenuity helped seventeenth-century scholars and artists embrace and explore the continuity between human and natural, nonhuman creative processes. Early modern individuals ascribed ingenuity to nature in highly specific and distinct ways. Richard J. Oosterhoff argues that "matter's powers to conceive, generate, nurture, and birth complex things made it ingenious, in the early modern vocabulary of invention," whereas, according to Jennifer Rampling, Latin and vernacular terms designating ingenuity were used "as a euphemism for inner, material processes that could not be directly observed," a derivation from *ingenium*'s etymology as "in-genitum" or "inborn." Alchemists partnered their ingenuity with that of nature by understanding and manipulating nature correctly.[76]

Within the early modern culture of ingenuity, conceits engaged with nature because nature was also ingenious. Nature conceived and transformed creatures just as, in Manzini's above-mentioned definition, the intellect conceived and transformed conceits. In fact, Manzini likened the invention of conceits to spontaneous generation, the process widely believed to lead to the emergence of insects, reptiles, fungi, and other creatures from mud and decaying matter, a paradigm of creativity in which nature generated something from something else that was unlike it. Likewise, trickery permeated nature's processes as much as it did the invention of conceits. Doina-Cristina Rusu argues that Bacon saw a form of ingenuity in the way vapors penetrate hostile or unsympathetic bodies by means of deception and the creation of false appearances. Bacon's example suggests that the link between ingenuity and deceit or indirectness—of the kind that tropes lent to conceits—applied to nonhuman and human ingenuity alike.[77]

For Scheiner's contemporaries, there was a clear parallel between what they saw as natural and human *ingenium*. Scheiner, for one, did not sit on his hands in the face of nature's ingenuity, which transformed the sun's appearance in a way that struck him as remarkable and highly unexpected. To expose nature's feat, he used the telescope, a scheme that allowed him to peer into the secrets of the sky thanks to the principles of optics that humans had learned from nature in the first place. But Scheiner was eager to demonstrate that he too could perform tricks of transformation, competing for the epithet "ingenious" with nature itself. With its power to transform objects by sheer sleight-of-mind, and to alter relations of pertinence, cause, and effect, a conceit acted as a form of alchemy that, in order to work, required from its inventor some knowledge and experience of the topic being discussed.

The claim that conceits engaged with nature's ingenuity failed to convince everyone. As one might expect, the disagreement grew only larger with time, laying the grounds for a controversy that soon would transcend the sphere of rhetorical and poetic criticism, giving birth to a culture war with long-lasting consequences.

A Culture War: To Define Art, Define Nature First

In *The Age of Subtlety*, I argue against a prevalent, oft-repeated assessment of conceits and other instances of seventeenth-century so-called Baroque taste as products of a fatal ebb in the confidence that language could accurately and effectively represent reality. Scholars have claimed that the religious and epistemological crisis tied to the early modern rediscovery of ancient skepticism encouraged a form of escapism, with the pursuit of wonder in both poetry and the arts as a substitute for comprehension of the world.[78] There is certainly merit to the view that conceits were a symptom of the appetite for fiction and playfulness that permeated seventeenth-century European culture as a result of widespread intellectual and moral disenchantment. Yet contrasting playfulness and fiction with a deep and sustained form of engagement with nature means imposing on the past ideas from the present. Acknowledging that nature—considered as the object of imitation of both art and poetry—was never a plain, unproblematic notion is a necessary step to examining without prejudice the ways in which conceits related to it. In the face of blanket distinctions between the preference for natural versus unnatural representations and forms of art, I contend that the debate over art's relation to nature should start with a discussion as to what the term "nature" exactly stood for in the seventeenth century. *The Age of Subtlety* elucidates how a set of debates that have been studied so far as controversies on literary fashion and taste were also part of a broader reassessment of the relationship between natural and human ingenuity.

The notion that conceits represent a departure away from nature peaked in the eighteenth century, at a time when they became the focus of a culture war involving the position of European nations and empires vis-à-vis one another in the race for intellectual and aesthetic advancement. Conceits became the villains, if not the scapegoats, on a scale that today might seem disproportionate, in a tale that opposed the spirit of enlightenment and science against tradition and obscurantism. That was never more the case than in Italy in the last third of the eighteenth century, when conceits took center stage in the controversy ignited by the influential *Storia della letteratura italiana* (*History of Italian Literature*), published by Jesuit Girolamo Tiraboschi (1731–1794) beginning in 1772.

Elaborating on views expressed by fellow Jesuit Saverio Bettinelli (1718–1808) in *Dell'entusiasmo delle belle arti* (*The Enthusiasm of Fine Arts*, 1769), Tiraboschi asserted that Spanish rule in Italy had been directly responsible not once, but twice, for the dissemination of a contagious and objectionable taste for extravagance—one that, he had to admit, Italians had ended up embracing: "Just as subjects tend to adopt their lords' inclinations and habits, Italians became, so-to-speak, Spaniards."[79] Tiraboschi depicted Spaniards as stubbornly committed to bringing ruin upon the taste of Italians past and present. They did so in the first century CE, initiating the decadence that terminated the golden age of letters after the death of Augustus. In the seventeenth century, they again managed to impose their irrational and capricious ways in Italy, especially through the theater of Lope de Vega (1562–1635), with its proud disrespect for the precepts of art, and through the embrace of conceits.

To a certain extent, Tiraboschi was repeating accusations advanced earlier by the French Jesuit René Rapin (1621–1687), who in two works published in

1671 and 1674 made Spaniards responsible for a similar perversion of French taste.[80] However, the century separating the remarks of Rapin and Tiraboschi meant that they rested on different assumptions and ideals. Rapin worshipped a form of classicism best exemplified by Roman literature in the age of Augustus (27 BCE–14 CE), while Tiraboschi wrote in the genre of literary histories then in vogue, self-styled philosophical works devoted to analyzing, through either a national or a comparative scope, the "origins" and "progress" of human creative and scholarly pursuits in order to carry them to "perfection."[81] Tiraboschi sought to understand what had gone amiss in specific periods as a means to extirpate the problem and prevent it from reemerging. At a time in which the "Black Legend" had transformed all things Spanish into embodiments of backwardness and obscurantism, conceits played a role in the controversy over the place of each nation in the narrative of progress, listed among such examples of Spanish wrongdoing as inquisitorial excess and the mistreatment of American native peoples.

The opinions of Bettinelli and Tiraboschi stirred a strong and immediate reaction within the community of Jesuits transplanted to Italy after the expulsion of the order from American and Iberian Spanish territories in 1767.[82] At once, the group of Jesuit exiles became implicated in a dispute characterized by outbursts of national pride and rivalry. In *Saggio storico-apologetico* (*Historical and Apologetic Essay*, 1778–1781), Jesuit exile Xavier Lampillas (1731–1810) vindicated Lope de Vega's theater. After all, even Lope's critics acknowledged his talent, even though they often relegated him with some condescension to the status of a tasteless and artless prodigy of nature.[83]

Conversely, no one came to the defense of conceits. In 1781, writing from Spain in response to Tiraboschi, the brothers Rafael (1725–1787) and Pedro Rodríguez Mohedano (1722–1773) agreed that conceits were a calamitous evil, and concentrated on disproving that Spain was in any way responsible for their introduction and use.[84] In doing so, they echoed the opinion that the Spanish Jesuits in Italy had been circulating for years.

In Italy, the work by Lampillas was followed by a letter *Sopra una pretesa cagione del corrompimento del gusto* (*About an Alleged Cause of the Corruption of Taste*, 1776) authored by Jesuit exile Joan Andrés i Morell (1740–1817).[85] Similar to Lampillas, Andrés agreed with Bettinelli and Tiraboschi that conceits were like plague, except he believed that Italy was as much at fault as Spain for the devices. The decay of good literature, according to Andrés, started at the same time in the two territories. In fact, if conceits were only Spain's fault, as Bettinelli and Tiraboschi proclaimed, how was it that seventeenth-century Italians, at the height of the craze for the rhetorical and poetic devices, consistently identified Neapolitan poet Giovanni Battista Marino (1569–1625) and the school of Marino's followers as the instigators of the revolution in taste?[86] Andrés believed that Tiraboschi's argument was not only flawed but also malicious. He believed it was ridiculous to maintain that Spanish taste dominated Italians in the seventeenth century, when they were lords, but also in the first century CE, when they formed a community of immigrants in Rome.[87]

The Age of Enlightenment viewed conceits as a tool of darkness, a motif that became literal when Andrés imputed them with "a certain subtlety, a mysterious sublimity that departs from nature and renders writing obscure."[88] Obscurity—understood as the difficulty to convey reason and meaning—appeared to Andrés as a departure from nature, the consequence of "a certain subtlety" that, for him, was synonymous with bad taste and irrationality. For Tiraboschi, in turn, subtlety was central to the legacy of Spain, a nation then in the process of being construed by its political and cultural rivals in opposition to triumphant reason. Subtlety enabled the intellect to see distinctions that others were unable to detect; it was a virtue until carried to excess, when it became a vice. Subtlety was the reason why Spain had contributed important names to scholasticism but almost none to the so-called New Science. Using subtlety, philosophers focused on hair-splitting quibbles that detracted from important matters. Tiraboschi associated subtlety of mind with scholasticism; likewise, Andrés described it as the production of "whims, and subtleties without substance."[89]

Subtlety could be seen as the ability to distract audiences with minutiae, which afforded the opportunity to deceive them. Tiraboschi and Andrés considered the difference between the subtlety of the sophist, aimed at instilling false opinion, and the playful and harmless subtlety of conceits to be nonexistent or irrelevant. To them, both appeared equally removed from the clear, straightforward ways of reason. Conversely, for seventeenth-century scholars writing on conceits, the distinction between sophistry and conceits—between deception and play—made all the difference. Useless for deception, conceits were subtle not because they departed away from nature, as maintained by Andrés, but rather because they had a close and intense relation to it. Seventeenth-century writings suggest that conceits used obscurity because nature also sought to obscure its actions and processes from everyone's sight. Conceits used indirect, meandering ways of expression because, to seventeenth-century individuals, nature also appeared as elusive, always intent on escaping both the senses and reason. *The Age of Subtlety* argues that conceits were subtle because they were one of the multiple and varied tools through which individuals immersed in the early modern culture of ingenuity undertook to imitate, and compete with, nature's often tortuous and opaque ways. Conceits were full of trickery and eager to arouse surprise because that is how writers and orators, as well as readers and audiences, saw nature, alternately donning the hat of the trickster and the magician.

The contrast between seventeenth- and eighteenth-century discourses on conceits indicates that, whenever we speak of art as imitating nature or competing with it, it is critical that we historicize the term "nature." Conceits struck Andrés as unnatural because he took nature to be a stable and unproblematic notion. The issue of how humans use art—whether devising the plot of a tragedy, a pulley to lift weights, or a string of conceits—to relate to nature depends directly on how they understand nature. Eighteenth-century critics of conceits reified nature just as they turned their backs on the previous century's efforts at making sense of nature's creative powers. It comes as little surprise that the contemporaries of Andrés, for whom nature was "uniform, regular, and simple," did not recognize the views on nature that inspired seventeenth-century inventors and theorists of conceits.[90]

Conversely, there was latitude in what natural philosophers and artists understood nature to be at the height of the craze for conceits. Nature was then seen as a set of norms governing that which happened "always or for the most part," following Aristotle's formulation, rather than as a set of necessary and unbreakable laws, leaving ample room for phenomena and objects that were possible but rare enough to arouse curiosity and pleasure in the observer.[91] In 1633, for instance, Jesuit natural philosopher Juan Eusebio Nieremberg (1595–1658) argued that nature was never more agreeable than when it departed from the norm and explored all that of which it was capable.[92] When nature did that, its ways became "subtle" and "obscure," the behavior that, more than a century later, Andrés would associate with conceits. Conceits resembled nature when, as Bacon once suggested, it invited humans to play hide-and-seek in search of its secrets.[93]

When seventeenth-century sources did describe conceits as being unnatural, they did so by drawing a distinction between two kinds of behavior in nature: nature "for the most part" and nature in all its power and playfulness, at the height of its acuity and subtlety. Conceits were unnatural only insofar as they imitated nature's delight in playful operations, never truer to itself than when it obscured its processes and thrived in a precarious and unstable equilibrium. That was also the approach to nature that justified the period's cabinets of curiosity being furnished with objects that, because of their presumed rarity and novelty, warranted attention and study in natural-philosophical and mixed-mathematical publications. Comets, animals preserved in amber, and creatures that glowed in the dark despite the absence of fire provided a model for the invention of conceits, being, as they were, examples of nature's ability to arouse surprise by means of possible but rare causes and processes.

The notion that conceits take from nature's wildest and most creative impulses received confirmation in Tesauro's claim, in 1654, that God, nature, and humans were all capable, albeit to varying degrees, of inventing conceits.[94] In fact, when Lope de Vega commended freedom, he did not seek to break away from nature—unless we understand nature in its narrower sense, as what happens "for the most part." Lope clarified that his plays and poems were monstrous and departed from what earlier generations considered natural because he preferred to imitate nature's ability to be varied and, like Proteus in ancient myth, ever different from itself.[95] Lope's argument, with its half-ironic, half-passionate endorsement of monstrosity, reflected contemporary approaches to monsters not as un- or antinatural, but—as Bacon famously pointed out—as instances of nature's creativity at its peak.[96] Similarly, when Lope's contemporaries identified him as "a monster of nature," any negative connotations that they may have wanted to attach to the label coexisted with sincere fascination toward nature's fertility gone wild, but not awry.

As natural philosophy increasingly focused on particular individuals with rare and wondrous characteristics, scholars and writers consistently viewed conceits through the lens of a more capacious and inclusive approach to natural objects and phenomena.[97] Freedom was not the opposite of nature: freedom was the mark of the inventor's nobility, which was identified with nature's ability to leave the norm aside and revel in the full use of its inventive and creative powers. Conceits mimicked nature as it took time out of the duty of keeping the world running, pausing to paint oblong rather than circular dots on a beetle, to decorate flower petals with infinite colors without apparent purpose, or to sculpt bizarre creatures in fossils. Seventeenth-century writings on conceits thus expanded the literature about nature's playfulness, which Paula Findlen has studied in depth.[98] As a form of art, conceits modeled themselves on nature, but a version of nature that liked to play the artist.

Conceits may have appeared transgressive for similar reasons to the ones that in 1686 led physicist Robert Boyle (1627–1691) to claim that the exaltation of nature's freedom touted by some of his contemporaries was blasphemous and diminished God's authority.[99] The debate on the ingenuity of conceits was tightly bound to the ingenuity of nature: the rights and limits of one impacted the rights and limits of the other. The link between nature and conceits and the lack of agreement about what "nature" means explain why conceits became the center of a culture war in eighteenth-century Italy in which all the parties involved claimed to be fighting the same enemy. The conflict that played out between Jesuit and lay scholars, and between Italians and Spaniards, concerned the way in which one should understand nature, and its moral authority to determine what is right and what is wrong.[100] Transformed into a token for traditionalism and irrationality, or clothed as a form of rebellious but—in the eyes of detractors, at least—childish freedom, conceits were the tip of a larger and epochal disagreement on what nature was supposed to be. In turn, the question as to what nature was and how conceits related to it was symptomatic of a shift in the conceptualization of human intellectual and creative abilities. The lack of understanding that Tiraboschi and Andrés displayed toward conceits was not only the product of a new approach to nature in which secrecy, playfulness, and surprise had lost currency but also the consequence of the decline of the early modern culture of ingenuity, with its conflation of the liberal and the mechanical arts, of engineering and deceit, and of the role of imagination as a tool of both making and discovery.

The Age of Subtlety reacts to twentieth- and twenty-first-century scholars who, like their eighteenth-century Italian and Spanish predecessors, have also failed to historize nature. Whether condemning or praising conceits, modern scholars have done so on the grounds that the rhetorical and poetic devices sought to explicitly depart away from what is both "natural" and "reasonable." In 1899, Italian philosopher and critic Benedetto Croce delivered a scathing judgment of the taste promoted by Gracián and some of his contemporaries: "The literary production and the aesthetic theory of a given period are not connected by any sort of necessity," Croce argued. "They may find themselves in intimate agreement or disagreement: they may be both good or both bad, or the one good and the other bad. In the period that we call 'secentistico' [i.e., the seventeenth century], both theory and practice were bad."[101] Croce's view was far from exceptional. In 1913, Gracián's *Arte de ingenio* and its revised and augmented sequel *Agudeza y arte de ingenio* (*Acuity and Art of Ingenuity*) were dismissed by the author's biographer, Adolphe Coster, as works devoted to "lesser manifestations of intelligence."[102] A few years later, they struck a historian of the Society of Jesus in Spain as "a manual on how to become mad."[103]

Since the days of Croce, the perception of seventeenth-century rhetorical and poetic taste has become more nuanced. What remains, however, is the view that conceits stemmed from a desire to break from nature. The avant-garde cult of novelty awakened sympathies toward the seventeenth-century rupture with traditional views on imitation. Formalist and structuralist approaches to language reveled in the possibility of inventing and replicating conceits in infinite variation. The linguistic turn also cast a positive light on conceits, now perceived as language games without real referents. If that was, perhaps, too fast a conclusion, it was also merely one stage in a protracted, decades-long process that saw a variety of ideological, scholarly, poetic, and artistic trends appropriate conceits and imbue them with different and conflicting values. For better or for worse, however, the result is that conceits are now firmly on the table after two centuries of neglect.

The idea of conceits as language games became increasingly accepted in the 1950s and 1960s, when scholars such as Guido Morpurgo Tagliabue, Eugenio Donato, and Franco Croce reacted against the views of Joseph A. Mazzeo, who maintained that conceits should be understood as the expression of occult, underlying correspondences in nature.[104] In fact, the two positions were incompatible only in appearance, since even the fictitious and playful nature of conceits was self-consciously modeled on nature's ability to "play jokes." Taken individually, conceits did not express hidden links found in nature, as Mazzeo would have it, but the ingenuity responsible for conceits was believed to resemble the ingenuity of nature.

Ingenuity has been the focus of recent scholarship interested in early modern perceptions of the relationship between natural and human transformative forces. Andrés Vélez-Posada remarks that, for early modern individuals, "the idea of ingenuity proposed an ecological link between human action and its environment."[105] If conceits pointed to "an ecological link" between natural and human ingenuity and helped bridge the divide between nonhuman and human creativity, they did so by imitating not nature's order but its tantalizing ability to seemingly escape from order through playfulness, monstrosity, and creative freedom. Conceits engaged with nature in ways that were variously understood through the tropes of imitation, competition, and counterfeiting. With ingenuity as a bridge and a common language, writings on conceits became a fecund and suggestive battleground in which the fraught relationship between natural and human creativity was constantly the object of negotiation.

The Age of Subtlety

In their conspicuous, self-evident use of deception, conceits imitated nature's ability to behave in ways that appeared less than straightforward. Conceits and deception went hand in hand from a historical and anthropological perspective as well. Seventeenth-century scholars remarked that, in the present, humans had become increasingly more apt not only at inventing conceits but also at deceiving and manipulating others. One might argue that conceits were central to seventeenth-century culture because the features that contemporaries appreciated in nature had become second nature to humans as inventors of conceits, and as creators more generally. It was as if the transformation in what interested humans about nature had brought about a change in their abilities as imitators and rivals of nature.

The awareness that human ingenuity had gone through a process of refinement left traces in contemporary poetic and artistic criticism. Spanish scholar and critic Jusepe Antonio González de Salas (1588–1654) opened *Nueva idea de la tragedia antigua* (*New Idea of Ancient Tragedy*, 1633) by pondering on the gulf in taste that alienated him and his contemporaries from Greco-Roman theorists. He recalled that rhetorician Hermogenes of Tarsus (161–180 CE) once objected against a certain conceit that referred to vultures as "animated tombs" or "tombs with soul," based on the abyss that, at least for Hermogenes, separated tombs and birds.[106] González de Salas believed that seventeenth-century readers were unlikely to find anything odd or noteworthy about calling vultures "animated tombs." Vultures eat dead bodies and then carry them during digestion, just as if they were tombs. Met with a lesser challenge, he claimed, seventeenth-century readers would not even notice that it was a conceit.[107] Salas initiated a plea for updating the principles of ancient Greek and Roman art in view of a world that bore little resemblance to that of Hermogenes. Art is a corrective of nature in the sense of the individual's inborn talent, he explained, but the two entertain a dynamic and reciprocal relationship: if art is to function appropriately, then it needs to be codified according to nature.[108] The taste for and the ability to create increasingly elusive and sophisticated conceits revealed that humans had changed in nature. Now, it was art's turn to follow suit and to adapt to the subjects it was supposed to serve.

According to Alessandra Russo, early modern Europeans often construed the ingenuity of New World artifacts as evidence of the humanity of those who created them.[109] Likewise, seventeenth-century scholars interpreted the growing relevance and refinement of the theory and practice of conceits as evidence that humans had reached a new stage of intellectual and creative development, defined by a superior ability to imitate nature and compete with it. Conceits were thus elevated to equivalents of the scientific discoveries and technological inventions traditionally identified with the belabored idea of "modernity," but also with its pitfalls. In *Arte de ingenio,* Gracián claimed that the ability to invent conceits transformed humans into angels, a paradigm of disembodied knowledge that it is difficult to reconcile with the embodied and mechanical aspects that he imagined as being crucial to conceits.[110] More cynically, five years later, Gracián noted in *Oráculo manual* (*A Pocket Oracle,* 1647) that the art of being "a person" had now reached perfection. Gracián was referring to the skills required to survive and thrive in the hostile, competitive environment of seventeenth-century courts, a talent epitomized by the ability to conceal one's thoughts and emotions by putting on a mask (Latin *persona*) and, in turn, to penetrate other people's interiority. "It takes a greater effort to deal with a single individual in our days than with a whole population in the past," Gracián pithily concluded.[111] For better or for worse, humans had entered the age of subtlety.

The scholars who between 1619 and 1654 produced a body of writings on conceits agreed on at least two points. First, conceits represented a form of intellectual and cultural refinement without a precise equivalent in the past. Greco-Roman orators and writers used devices that Gracián would not have hesitated to call conceits. What was missing was a broad, all-encompassing notion that would have allowed them to reflect on conceits as a coherent intellectual and creative phenomenon the way that seventeenth-century supporters and detractors of conceits did. The awareness that these were a distinct and culturally significant phenomenon turned them into a topic that could inspire entire books and scholarly disputes. The second aspect on which there was consensus was that the ancients mostly restricted the use of conceits to specific rhetorical or poetic genres and situations. While several of Gracián's contemporaries advocated similar restraint, conceits mushroomed in use to religious sermons, histories, and all forms of fiction writing.

Among seventeenth-century scholars, it became common wisdom that the ancients had failed to produce a theory and a method of conceits. Peregrini excavated the works of the Greek and Roman rhetoricians, but all he was able to find was a partial overlap between Aristotle's notion of "smart sayings" or *asteia* (ἀστεῖα) and conceits. Complicating the issue even further was the problem that the ancients had placed in the same group statements that certainly qualified as conceits and others that lacked any acuity or subtlety.[112] To make things even worse, Aristotle had failed to isolate the component that would enable one to distinguish conceits from statements that were not conceits. Peregrini believed that conceits were a source of pleasure because they allowed readers to notice the author's acuity or subtlety and, as a consequence, to feel ingenious themselves. He was surprised, and even disappointed, that the ancients seemed to have had little awareness that conceits were all about self-love and peer recognition.[113] Following in Peregrini's footsteps, Pallavicino noticed other instances of discontinuity between ancient and modern ingenuity, and concluded that conceits had a partial predecessor in Cicero's notion of *facetiae* or witty sayings but had been hardly in use in the days of Aristotle, three centuries earlier.[114]

Others saw the matter differently. Tesauro, for one, claimed that Aristotle had bequeathed a theory of conceits to posterity in concealed form, which he was the first to uncover. Tesauro's version of Aristotle's method appeared shrouded by the opacity of guild secrets and new inventions.[115] To the controversies that involved Mascardi, Peregrini, Minozzi, and Manzini between 1636 and 1639, one could add Sarbiewski's exchanges on the topic with a long list of Jesuit interlocutors around 1623 and, especially, the instances of theft and plagiarism that for several decades besieged the manuscript containing *De acuto et arguto*, Sarbiewski's course on conceits.[116]

Sarbiewski illustrated his method of conceits through poet Marcus Valerius Martial (ca. 40–ca. 104 CE) and philosopher Lucius Annaeus Seneca (4 BCE–65 CE), two of the authors that Rapin and Tiraboschi considered directly responsible for the ruin of taste in Ancient Rome. Sarbiewski did not consider the ancients to have been any less capable in the invention of conceits than the moderns. The problem was that the ancients had failed to lay down a theory that could account for conceits philosophically. This is how Gracián saw it too. He called Martial "the first child of ingenuity" and considered other periods in history to have been fertile in conceits.[117] For instance, in the second, augmented edition of *Arte de ingenio*, Gracián argued that late medieval Castilian poetry could and should serve as a model to the present for the invention of conceits.[118] Yet even as conceits flourished in different periods of history, Gracián contended that they had been left orphaned, deprived of a discipline that might put them under scrutiny the way that logic and rhetoric taught the rules of syllogisms and tropes. Gracián sought to remedy what he perceived to be a missing link in the disciplines of language and reason.[119] In his view, to replace conceits as a form of invention rooted in practice with theoretical awareness would bestow them with increased legitimacy, just as contemporary inventions—say, Galileo's telescope—gained dignity by being reframed as products of theory rather than of trial and error.[120]

Others believed that seventeenth-century writers were simply more gifted at invention. When it comes to conceits, Minozzi claimed, "modern minds . . . leave the ancients far behind."[121] In *A Letter in Self-Defense* (*Lettera apologetica*, 1637), Manzini acknowledged that the use of conceits had earned him criticism from the few but was also the reason why his books circulated widely across the continent, translated into numerous languages. Manzini's works saw reprints not once or twice but "twice ten times, and with such applause that it has almost made the present century change its way of writing."[122] Conceits conquered a market in which reputation could be measured based on the number of books a writer sold.[123] Manuals on conceits also enjoyed considerable success with the public. Some of them went through several editions and even appeared in translation.[124]

The claim that the seventeenth century was the age of subtlety was met with enthusiasm and ridicule alike. Minozzi wrote of conceits that "such a noble seasoning becomes food for whoever has a stomach . . . adapted to digest the substance of sugar and honey."[125] Dismissing a trend that likened conceits to honey or sugar—treats for children that, like spices, should be used only sparingly—, he saw it as a mark of the present's subtlety that individuals had developed the ability to feed on what those in less delicate ages had used merely as a condiment.[126] Behind Minozzi's observation lay the belief in a transformation wrought at the level of culture and taste through continuous exposure to stimuli that, to continue the metaphor, had strengthened people's stomachs. The notion that the present was, first and foremost, "a delicate century" was popular with detractors of conceits, who mocked the claim that humans had reached a new stage in the cultivation of the mind. The underlying idea was that the spirits that traveled in the blood and served as vehicles between body and mind had become exceptionally rarefied, almost immaterial thanks to increasingly sustained exposure to the works of imagination.[127] Salas's reflection on the conceit about the vultures proved that the shift was real. Pondering on the rise of subtlety as a cultural and physiological condition of seventeenth-century society, Manzini concluded, "The world becomes increasingly refined, especially in the works of ingenuity. Many things once taken for rare are now ordinary."[128]

Conceits and "Science"

Even though the view of conceits as signifiers of a literary break between past and present was far from universal, those who shared it saw themselves inaugurating a new stage in human creativity and inventiveness, modeled on a view of nature that lured both eye and intellect with the same characteristics of acuity, subtlety, and artifice that had become central to social and political life. Conceits eventually lost the currency that, for several decades, had turned them into trailblazers of a new stage of ingenuity. Like the body of writings that sought to account for them, conceits have been cast away from the epoch of seventeenth-century invention and discovery. Now the time has come to place them again in dialogue with the ideas and concerns with which they once shared space.

In *Baroque Science* (2013), Ofer Gal and Raz Chen-Morris argue that the terms paired in the title of their book are less antithetical than usually believed. The authors criticize the belief that the art of the period that was once called the Scientific Revolution was the expression of a violent, emotionally overloaded impulse that stood in stark contrast to the rational and rigorous approach to nature that became prevalent in the age of Galileo. Conversely, they argue that the period's artistic and scientific endeavors are the expression of similar interests and concerns. Artists, natural philosophers, and mathematicians were similarly haunted by a blend of fascination and uncertainty regarding the role of artifice in the study of nature.[129] Scheiner's playful conceit might seem far removed from the arguments that, mobilizing knowledge of optics and natural philosophy, explained why the sun appeared to be egg-shaped, yet the two disciplines, which Scheiner might have seen as distinct but comparable acts of ingenuity, were similarly capable of making a case for the author's talent and sophistication, and were similarly apt at problematizing the access to truths about nature. Scheiner's *Sol ellipticus* was a work of optics, a discipline of mixed mathematics that applied quantitative methods from pure mathematics (arithmetic or geometry) to the study of nature.[130] Mathematics was on the rise, and geometry had started to replace logic as the discipline that lay at the root of every form of knowledge, prompting Galileo's claim in *Il saggiatore* (*The Assayer*, 1623) that the book of nature was "written in the language of mathematics, and its characters are triangles, circles, and other geometrical figures."[131] Geometry and mathematics also mattered in the controversy over conceits. In 1639, Peregrini argued that Martial's conceits and Euclid's theorems offered different but comparable evidence of the mind's ability to penetrate reality. It is not a coincidence that Peregrini introduced *Delle acutezze* as a reply to mathematician Vincenzo Renieri (1606–1647), a disciple of Galileo.[132] The scholars who wrote on conceits lived in a world in which current developments in the study of nature took place in proximity to debates on rhetorical and poetic invention.

The Age of Subtlety unfolds along several points of encounter between seventeenth-century writings on conceits and natural-philosophical and mathematical works. By establishing a bridge between intellectual and creative universes that later disciplinary boundaries have managed to keep separate, I join recent attempts at understanding how early modern fiction and art intersected with what we today call science.[133] While I am not the first to connect writings on conceits with early modern notions of nature and technology, existing studies remain few and, for the most part, are restricted to Tesauro's views on metaphor and the relationship between metaphor and optics. In 2000, Zakiyah Hanafi devoted a chapter to monstrosity in Tesauro's *Il cannocchiale aristotelico,* while more recently Jon R. Snyder has led the way in analyzing the role of optical devices in Tesauro's theory of conceits.[134] I expand on such approaches and analyze writings on conceits that circulated on the Italian and Iberian peninsulas between the 1620s and 1650s in parallel with works concerned with the study of nature.[135]

Outline of Chapters

Two instances of nature's creativity—comets and fireflies—served as bywords for conceits in seventeenth-century controversies on style. Part I of the book centers on comets, part II on fireflies.

Part I, "Comets," starts off with a lecture by Jesuit mathematician Orazio Grassi delivered in the winter of 1619. Humans are so made, Grassi maintained, that they feel excitement at new and rare phenomena that seem to defy nature's usual course. Take comets, for instance. They alter the existence of a world that would be perhaps more orderly without them but undeniably less thrilling. One would say that nature introduces them artificially merely to avoid appearing repetitive. As one of Grassi's contemporaries aptly put it, conceits were "poetry's comets," which likewise catered to the hunger for novelty and artifice that is congenital to humans. In chapter 1, I argue that the lavish and inventive display of conceits that adorned Grassi's lecture pointed in the same direction as his reflection on comets. The craze for conceits and the trend that led natural philosophers to indulge, even focus on, rare and new phenomena were two symptoms of a culture that had grown comfortable with psychological features, such as the appetite for novelty and the preference for the artificial over the natural, that not long before had been considered sinful or flawed. Together, the taste for comets and conceits signaled a wish to adapt both natural philosophy and the pursuit of beauty to humans as they actually were—with their defects and limitations—rather than to a prelapsarian or ideal notion of how they ought to be.

Chapter 2 examines conceits as a site of reflection on the relationship between natural and human ingenuity. It does so by focusing on the analogy between optical devices and conceits, widely established among seventeenth-century scholars shaken by the discovery that lenses revealed things to be different than they appeared to the naked eye. Just as the microscope transformed lice into elephants, so conceits made or transformed the reality that they claimed to merely show from within. I argue that knowing and making became two conflicting yet simultaneous aspects of engagement with nature because seventeenth-century scholars imagined both lenses and conceits in competition with nature's ingenuity. Put simply, conceits and lenses transformed reality in order to know it better, ostensibly because they both obtained knowledge by undoing nature's artifice, which had veiled reality in the first place. Characteristically, seventeenth-century scholars portrayed the competition between natural and human artifice as affording the intellect a pleasure that would be impossible to find in a world defined by transparency and openness.

Part II, "Fireflies," draws on a comparison between conceits and fireflies that became widespread in Italian controversies on rhetorical and poetic style between 1622 and 1641. Just as fireflies pale in contrast with the sun and the stars, the analogy went, so conceits appear inconsequential when compared to sound arguments. I demonstrate that, in 1639, Minozzi mounted a defense of conceits and fireflies that turned the tables of the debate and transformed any shortcomings of both fireflies and conceits into strengths. Chapter 3 argues that Minozzi's enthusiasm for both fireflies and conceits relied on a new sensibility toward nature's smallest creatures, reinforced by the advent of the microscope. The brevity of conceits aligned with a cult of smallness that viewed with fascination the ability of both nature and humans to work at small scale, whether in the structure of a flea or in Homer's *Iliad* written inside a nutshell. In a bounteous and atomized discourse, the reader enjoyed a succession of brief sparks of ingenuity in which multiplicity and variety became embodiments of the author's freedom of invention, assimilated to a form of nobility that humans first learned from nature's ability to "play."

Fireflies also outdid the sun and the stars in other ways. They were living creatures, with a dim and flickering light that gave them a mysterious and enchanting allure. Chapter 4 examines the claim, repeatedly found in period writings and often grounded on natural-philosophical beliefs, that conceits were also endowed with life and even a soul. Manzini claimed that conceits, like insects, were endowed with life because the intellect created them through "spontaneous generation," by analogy with the process believed to give rise to insects, toads, and a wealth of allegedly imperfect creatures, as opposed to through sexual reproduction, which enabled the propagation of mammals, and humans specifically. Other scholars and writers, however, accounted for the life of conceits by drawing on disciplines such as alchemy, which for centuries had claimed the ability to create life artificially, natural magic, and mechanics. In the last part of chapter 4, I move on to the feature that originated the analogy between fireflies and conceits in the first place: the parallel between the precarious but captivating light of one and the problematic but attractive logic of the other. By analyzing a range of seventeenth-century attempts at explaining the cause of fireflies' light, I demonstrate that the involvement of the imagination in both the invention of conceits and the pursuit of natural-philosophical truths rendered the divide between them extremely porous.

The focus of what follows is not on the use of conceits in works of natural philosophy or mathematics from the period that was once called the Scientific Revolution—although Scheiner's example makes obvious that they played a prominent role in those, to the surprise of modern readers accustomed to the plain, unambiguous language now associated with science.[136] Instead, by reading seventeenth-century rhetorical and poetic treatises in parallel with contemporary natural-philosophical and mathematical works, I reconstruct the role of conceits within the early modern culture of ingenuity, considering them as rhetorical and poetic devices modeled on a view of nature that has become as alien to us as the craze for conceits itself. On occasion, one of the scholars I discuss might trace a clear, explicit link between natural and human ingenuity. Most of the time, however, the parallels between nature and mind fall through the cracks of today's disciplinary boundaries, but are ready to emerge as soon as we read through rhetorical, poetic, and natural-philosophical or mathematical works as if they were—and they indeed were—part of the same world.

Comparing the invention of conceits to the search for truth characteristic of the sciences, Gracián argued that conceits are the result of a rootless, nomadic form of ingenuity—one that, "like a shooting star, lacks a permanent home."[137] Conceits felt everywhere at home in seventeenth-century culture, including in lecture halls, poems, works of astronomy and mathematics, and casual conversation, and yet they lacked a specific and apposite abode. In order to chart the place of conceits within the early modern culture of ingenuity and the role they played in providing humans with a new and appropriate way to engage with nature, a dialogue is needed between fields that now, several centuries later, have fallen on either side of the divide between humanistic and scientific inquiry.[138]

PART I

Comets

CHAPTER ONE

Poetry's Comets

On Novelty and Artifice

In 1618, three comets paraded their unkempt beards and elongated tails before a community of makeshift astronomers and astrologers. On a mission to Isfahan, King Philip III's ambassador García de Silva y Figueroa glimpsed both the second and the third, which also were observed by Mughal emperor Jahangir on his way to Agra. In the earliest observation with the telescope in Asia, Jesuit missionary and mathematician Václav Pantaleon Kirwitzer (1588–1626) observed them from Goa.[1] No one remained indifferent to the events, which scholars from western Europe to China would invoke in retrospect as the harbingers of a climatic, social, and political crisis of comparably global reach.[2] The flagship school of the Society of Jesus, the Collegio Romano, appointed four speakers in different disciplines to address the escalating anxiety and curiosity. Only the lecture by Orazio Grassi (1583–1654), a young professor of mathematics, appeared in print. The initial audience, limited as it had been to the scholars and notables there in attendance, expanded virtually without constraint in a move that, even though the lecture was published somewhat anonymously under the name of "One of the Fathers of [the Society]," could make as much as destroy a career.[3] A few months later, Grassi would recall the day in the winter of 1619 when he took the lectern: overall, it had been a positive experience. Only Galileo found fault, he remembered, "and quite harshly at that."[4]

Unusual activity had taken place in the heavens since August, when, as Grassi later remembered, a comet was spotted licking Ursa Major's "hind feet."[5] When a second, then a third comet became visible, the crowds gathered on elevated spots "with no thought for sleep and no fear of the cold north wind."[6] Others in Rome recorded a similar agitation. "The novelties in the sky in the previous days force out of bed even the sleepiest and laziest in town," scholar and poet Virginio Cesarini (1595–1624) told a sick and bedridden Galileo on December 1.[7] The next morning, Cesarini discussed the news with Jesuit mathematicians. These may have included Grassi.

The focus of Grassi's lecture *De tribus cometis* (*On the Three Comets*, 1619) was the third and last comet, known today as C/1618–W1. Observations from Jesuit institutions in Rome, Parma, Antwerp, and Innsbruck suggested that the third comet had remained visible between late November and January of the following year, surpassing its predecessors in duration, size, and other aspects more difficult to express in quantitative terms: "As long as it lived," Grassi stated with fanfare, it "drew everyone's eyes to its cultivated nature and ways of life." Grassi likened the comet to a courtier of refined manners and behavior who, as a result, had become a target of gossip.[8] As he claimed with a note of humor, Grassi was there to deliver a funeral oration for the comet as one would for a member of the nobility.[9] The mathematician then donned the hat of the orator and mourned the deceased: "I ought not to depart in the least from the advice of the masters of rhetoric," he declared. Grassi would address the comet's date of birth, place of origin, and lineage, following with "its illustrious life," and finishing with "its death."[10] In the continuation, the comet emerged as a character. Its birth, no later than November 29, had occurred at the point of the ecliptic where, mere weeks earlier, the sun had met Mercury in the constellation of Scorpio. "It was proper," Grassi pronounced, "that a very elegant and splendid feast be prepared for such guests and, as well, that a very bright torch be kindled."[11]

To call a comet "a torch" was hardly remarkable. Early modern rhetoricians knew that the boundaries between literal and figurative language were porous, and the language of astronomy, for one, was teeming with tropes. The constellations formed a menagerie of real and mythological beasts, the planets were named after Greco-Roman gods, and even generic terms such as "planet" and "comet" were rooted in metaphors, referring to wandering and long-haired objects, respectively.[12] In addition to that, Grassi's contemporaries were used to linguistic novelty gaining fast-tracked citizenship. In its infancy, a trope seemed bold and far-fetched; soon, no one would recall other ways of expressing the reality in question.[13]

Scholars and poets had a name for the device seen in Grassi's example, which built on a web of relations facilitated by tropes in order to create a new, more intricate reality. They called it a conceit. In *Arte de ingenio* (1642), the Jesuit Baltasar Gracián explained that calling a comet "a torch" merely on account of its fiery nature qualified as a trope. A trope indicated a similarity in, for instance, form, matter, or quality, but it became a conceit only when it entered into contact with "a special circumstance," a matrix that lent additional significance to the relation posited by the metaphor.[14] For instance, Grassi remarked that, shortly before the comet's birth or first appearance in "the house" or constellation of Scorpio, the sun had briefly met Mercury there—metonymy made gods and planets interchangeable—and they had enjoyed dinner by the light of the comet. The banquet rendered the comet's transformation into a torch meaningful, turning a trope into a conceit.

Though Grassi was there to testify as a mathematician, he knew enough physiognomy to deduce the comet's parents from its features. "Its face and color," he observed wryly, represented Mercury's "eyes, hands, and countenance."[15] Then, however, as if he had had enough of the conceit, Grassi called on the astrologers to finish the work and, "as good midwives, receive the fetus."[16] Grassi's lecture went on to reject Aristotle's view that comets were hot, dry exhalations ignited by ether as they rose to the upper regions of the atmosphere.[17] He demonstrated that the third comet of 1618 had been a celestial, superlunar body, the same conclusion reached by Danish astronomer Tycho Brahe (1546–1601) about the comet that was seen in 1577.[18] Months later, Galileo's friend Mario Guiducci delivered a lecture in Florence in response to Grassi, marking the start of a long and protracted controversy that lasted until 1626, reaching its peak with the publication of Galileo's *Il saggiatore* (1623), today considered a landmark in the history of the Scientific Revolution. Grassi's *De tribus cometis*, conversely, is rarely discussed by historians of science except for its role as the conflict's indisputable *casus belli*.[19] One of the few aspects of Grassi's lecture to have received attention is its style, so full of conceits that one scholar did not hesitate to call it "artificial."[20] Mario Biagioli has remarked that Grassi's conceits shaped a new, courtly audience for mathematics modeled on patterns of expression and communication that were in vogue in Rome's academies.[21] Grassi used conceits to fashion himself as a refined and playful speaker, adept at arousing wonder in anyone capable of keeping pace with his elaborate and intricate figurative arguments. William Eamon has argued that philosophical and courtly discourse differed in that the latter "did not aim to result in necessary logical conclusions, but to present paradoxes and riddles, just as nature continually surprised the observer with its wonders."[22]

The present chapter builds on the remarks by Biagioli and Eamon, with their emphasis on scholarly self-fashioning and the suggestion that rhetoric looks to nature as a source of inspiration for ways to impress and sway the audience. Going one step further, I argue that conceits—which enabled Grassi's self-fashioning at the level of style—and comets—which played a similar role at the level of subject matter—were closely knit in the seventeenth-century imagination. Conceits were "poetry's comets," as indicated in a letter that poet Claudio Achillini (1574–1640) sent to Count Ridolfo Campeggi in 1609.[23] He complimented Campeggi because the poem he had sent him was free from conceits, gleaming as it did with "legitimate lights" and "natural stars."[24] Comets, in turn, reminded Achillini of the intemperate and audacious tropes of poets à la mode. They were "adoptive stars," "illegitimate lights," and fake gems adorning the dome of heaven.[25] Achillini based the analogy between conceits and comets on the two aspects that he believed they had in common. To a salient but suspicious ability to strike the mind with novelty, he added a departure away from nature described in terms of parenthood (conceits, like comets, were similar to illegitimate or adopted children) and truthfulness (they were a product of counterfeit).[26]

Achillini's analogy between conceits and comets resists easy interpretation, as it rests on assumptions about nature, art, and the way they relate to one another that are no longer current. The attribution of novelty to both comets and conceits is only mildly surprising. The tantalizing and perilous horizons that the term "novelty" would have opened to seventeenth-century readers may be forever lost on us, yet picturing how both comets and conceits strike the mind as new and unfamiliar remains feasible.[27] Just as comets were particularly adept at rendering people curious and restless, a point that Grassi was eager to emphasize, so conceits likewise triggered a host of altered and inquisitive states of mind, as Zakiyah Hanafi has argued of the monstrous and destabilizing power of metaphor in early modern rhetorical and poetic scholarship.[28] Conversely, the contrast between comets and "natural stars" is likely to arouse perplexity with its suggestion that comets were nonnatural creations (of nature!).

Conceits and comets were tightly interconnected in Grassi's lecture, printed in a booklet that also contained a folded leaf with engravings of the trajectories of the three comets (Figure 1.1). Two poems, presumably by Grassi, reflected on the skill of the illustrator, who had transformed the comet from a sign of disaster into a beautiful, even propitious phenomenon.[29] Grassi's lecture also aimed to ease any concerns that the comet might have raised.[30] With a conclusion grounded on mathematics and reliable measurements gathered in Jesuit obser-

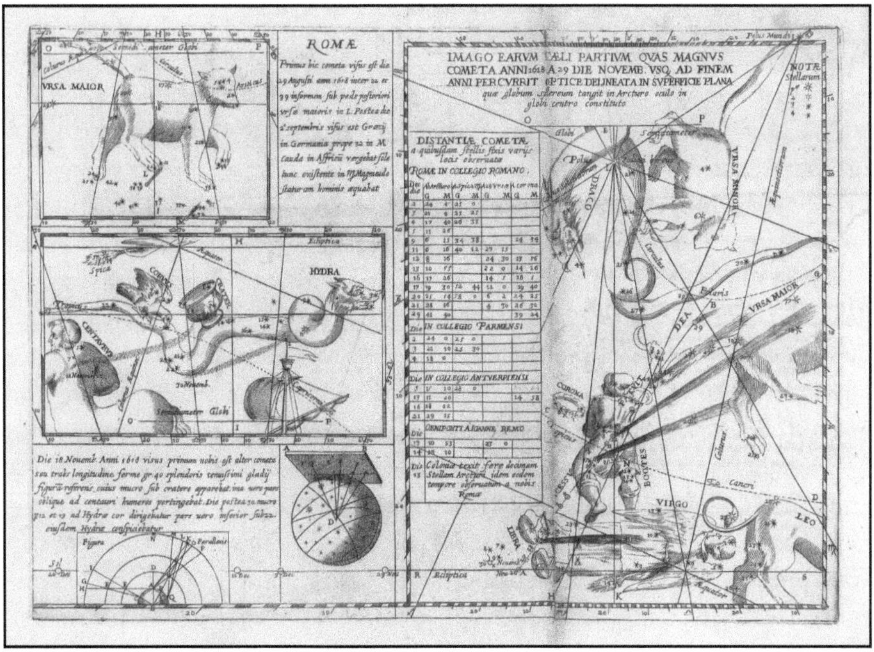

Figure 1.1. Engraving with the trajectories of the three comets of 1618. Orazio Grassi, *De tribus cometis anni M.DC.XVIII. Disputatio astronomica publice habita in Collegio Romano Societatis Iesu ab vno ex patribus eiusdem Societatis* (Rome: Giacomo Mascardi, 1619). The Huntington Library, San Marino, California.

vatories in a variety of locations, Grassi appeared as the herald of a new age in which comets, restored to the realm of nature, would no longer be perceived as omens. Yet even as Grassi sought to free comets from the fearsome and awe-inspiring connotations associated with them, he was also eager to reflect on why they were so powerful at arousing curiosity and excitement.

In what follows, I demonstrate that Grassi's observations about the mind's reaction to comets and similar instances of nature's less usual phenomena closely paralleled the ways in which contemporary writings by Maciej K. Sarbiewski and Matteo Peregrini described the logical and psychological operations set in motion by conceits. I show that comets and conceits appeared as comparable instances of novelty because, as Achillini enigmatically advanced, both were perceived as products of artifice. Conceits were part of a trend that paired natural and human creativity while simultaneously eroding the divide between the natural and the artificial.[31]

Novelty and Spice

A peculiar exercise of introspection greeted the audience of Grassi's *De tribus cometis*. It almost took on the air of a reprimand when the orator claimed that the mind "is so desirous of novelty that at times it grows weary of the long continuance of even that which is good, and desires to improve upon the situation by the alternation of some evil. . . . We have wished—oh unjust desires!—that some portents of this sort should arise which, at least by their novelty, would delight and gratify our eyes."[32] Grassi insinuated, tongue in cheek, that his audience had less-than-virtuous reasons to be in attendance that day: a taste for novelty at all costs brought them there, not tempered by a regard for the good or bad tidings that accompanied it.

Pamphlets for sale in the fall of 1618 cultivated feelings of fear—and, more rarely, hope—by portraying the three comets as harbingers of bad news. Yet as Grassi suggested, wonder at comets was first and foremost due to their novelty. Irrupting against the backdrop of a sky whose past and future had been painstakingly charted in tables and almanacs, comets—and even more so new stars—came nearest to novelty among the experiences of the astronomer and the astrologer. Then, it was only a matter of days or weeks before they vanished from a place where they never truly belonged.

The notion that comets embodied novelty in its purest form was a legacy of Grassi's source, the seventh book of *Naturales Quaestiones* (*Natural Questions*, ca. 63 CE) by natural and moral philosopher Lucius Annaeus Seneca (ca. 4–65 CE). Seneca decried the power of novelty over the intellectual and emotional makeup of humans, deploring the way it determined their attitudes toward astronomy and learning more generally: "We are so constructed that daily occurrences, although they are worthy of admiration, pass us by; on the other hand, the spectacle of even minor events affords pleasure if they appear unusual. Accordingly, the usual collection of stars which distinguishes the immense firmament does not attract people's attention, but when something is changed from the ordinary the eyes of all are on the sky."[33] Seneca was adamant that novelty should not determine what matters or not in the study of nature, yet humans find it hard to resist a condition so deeply rooted in their senses and minds. He lamented that we barely pay any attention to phenomena "as long as the usual order is preserved," focusing on unpredictable and fiery comets instead of the triumph of order over chaos represented by regular, stable planetary orbits and periods. Conversely, "if something is disturbed or flashes out beyond the customary, we look at it, ask about it, point it out—so natural," he concluded, "it is to marvel more at the new than at the great."[34] Plutarch (ca. 46 CE–post-119 CE) echoed Seneca's sentiment. Even though he considered astronomy to be a corrective for vile and ill-advised curiosity, Plutarch was also concerned that even in a lofty and suitable pursuit, the mind kept straying in search of exceptional and rare events.[35]

In 1639, Matteo Peregrini cited Seneca's complaint in *Delle acutezze* to illustrate the notion that "our nature is the friend of novelty."[36] However, the allusion came in a discussion not of comets but of "poetry's comets"—as Achillini had called conceits three decades earlier. Seneca's reflection proved versatile enough to apply to astronomy and rhetoric alike, a symptom of widespread cross-fertilization between disciplines as well as a generalized interest, even concern, with novelty per se. The year after he published Peregrini's treatise on conceits, printer Giovanni Maria Farroni became one of the individuals involved in the dissemination of Italy's earliest printed newspaper in Genoa.[37] Novelty—which was, along with artifice, one of the two features underpinning Achillini's analogy between comets and conceits—struck seventeenth-century scholars and cultural agents as not only intellectually and aesthetically suggestive but also eminently marketable.

Novelty also left a mark on intellectual habits more generally. Brendan Dooley has argued that the development of the early modern news market mattered at least as much as religious reform and the revival of ancient Pyrrhonism with regard to the rise of skeptical attitudes about the possibility of attaining certain knowledge.[38] As suggested by contemporary satires of the newsmonger, the circulation of both manuscript and printed *avvisi*, or newsletters, along with the proliferation of news-exchanging sites such as Madrid's *mentideros*, Rome's *banchi*, and London's "Old Saint Paul's," epitomized the period's avid hunger of novelty despite—or rather because of—its ethical and epistemological indifference, its fraught relationship with the good and the true.[39] Grassi's lecture facilitated the encounter between "poetry's comets" and comets of nature. The orator capitalized on the fascination around the three comets of 1618 through a host of conceits that replicated the effect of novelty at the level of style.

To be sure, not even the playful and capacious frame of the preamble, with its tolerance of almost any form of humor and self-criticism for the sake of *captatio benevolentiae*, was enough to obscure the stakes of openly trafficking in novelty. Grassi relished the ambivalence of a notion that reeked of political and religious turmoil. Since starting as a student at the Collegio Romano in 1603, Grassi must have listened to the condemnations of untutored individuality delivered at the outset of every new academic year. Impatient youths, faculty, and authorities listened to variants on the same story: instead of the limited but well-defined skills of every other animal, humans were gifted with a protean, chameleon-like nature that brought them near to God but could also devolve into a curse if left without adequate guidance.

Drawing inspiration from Giovanni Pico della Mirandola's (1463–1494) *Oratio de hominis dignitate* (*Oration on the Dignity of Man*, first printed in 1496), Jesuit professors declared from the lectern their outright fascination with ingenuity, "the greatest of spectacles," and the power that had transformed a wild and inhabitable planet into a paradise.[40] But ingenuity might wreak havoc if left to wander at large. Set in a career for which he lacked aptitude, a student risked succumbing to frustration. Despairing from certain and useful knowledge, he might end up pursuing novelty per se, his vanity tickled with the satisfaction of bringing forth a new creation or idea to the world, even if—as with the omens spelled out by comets—it was perhaps a bad or evil one. History was rife with examples, ranging from Catiline's conspiracy to the Lutheran and Calvinist heresies.[41] Likewise, by the time of Grassi's lecture, the Jesuit superior general Muzio Vitelleschi (r. 1615–1645) had been immersed for at least two years in a campaign, soon to have continent-wide and transatlantic ramifications, to eradicate conceits from the style of the order's preachers. It was yet another attempt to curtail the pursuit of novelty.[42]

Grassi's perspective was more realistic. To start with, he considered Seneca's disdain for the taste of novelty impractical. Grassi not only displayed little concern about novelty's indifference to good and bad but also was quick to recognize that by reading Seneca's passage against the grain one could clear the path for a new way of approaching both natural philosophy and rhetoric. Instead of eradicating or, at least, restraining the inclination to novelty that is inherent in humans, one should rather make the most of it.

Grassi's unconvincing denunciation of the taste for novelty—"oh unjust desires!"—paid lip service to the indictment of curiosity formulated by theologian Augustine of Hippo (354–430). Augustine viewed curiosity as a risk to salvation, a form of weakness that caused the soul to trade the love of God for the acquaintance of vile and earthly matters. It was even worse than that: although the intellect is superior to the senses, at least the latter have an appetite only for objects that are genuinely pleasant. Conversely, the intellect pursues every piece of new learning, no matter how unworthy, contemptible, or even dangerous, as was the case with comets.[43]

Yet by the time of Grassi's lecture, Augustine's indictment of curiosity had been subjected to reassessment. Early modern natural philosophers had recast curiosity as a virtue that enabled humans to make progress in the study of nature—and, through nature, of God himself.[44] Of those who were resolved to exploit the intellect's penchant for novelty, Juan Eusebio Nieremberg deserves particular attention. A professor of natural philosophy and history at Madrid's Reales Estudios, Nieremberg was Grassi's contemporary and, like him, a member of the Society of Jesus.[45] Nieremberg agreed with Augustine that curiosity was a disorder that affects our intellectual and emotional makeup, but he insisted that it was a useful and innocent one.[46] Subscribing to the principle that the taste for novelty is natural to humans, Nieremberg took a pragmatic approach: instead of seeking to bring change to how the intellect works, why not feed it what it craves? Qualifying Aristotle's claim that humans are naturally fond of learning, Nieremberg suggested that some objects are more capable of awakening curiosity than others.[47] For instance, "a vulgar truth" lying right at hand is often not enough to stir the intellect. If the aim of educators and natural philosophers is to ostensibly prevent students and readers from flying into fiction's novelty-filled domains, then there seems to be a case for bringing novelty into the pursuit of truths about nature.[48] That, at least, was Nieremberg's course of action. In 1629, he told his audience that where Madrid prided itself as a court made up of diverse individuals and commodities originating from every corner of the world, the inquiry into nature was like "a court of learning" in which the intellect never tires of finding novelty, "hunting" for knowledge without incurring the fatigue that invariably follows exercise.[49] Chapter titles in Nieremberg's *Historia naturae, maxime peregrinae* (*A History of Nature, Especially of the Rare or Exotic*, 1635) featured the term "new" and cognates such as "extraordinary" and "wondrous" at a similar rate as those in Miguel de Cervantes's *Don Quixote* (1605–1615).[50]

Nieremberg asserted that God had populated the universe with rare and new wonders merely to stimulate humans. He even created comets and a variety of what Nieremberg, using a similar term as Achillini, called "other illegitimate stars."[51] With a logic that verged on circularity, Nieremberg suggested that God thus circumvented the limitations that he himself had placed on humans, which made it difficult for them to become interested in regular and daily phenomena. Nieremberg's argument was in line with the belief, widespread among Jesuit educators, that each individual should pursue intellectual and artistic endeavors in accordance with his or her nature, only he added a new and crucial twist: the notion of one's nature referred no longer to what made each student different from everyone else but rather to what made humans similar as a whole. It was as though the focus on everyday phenomena forced the intellect to a task for which it was not well prepared. Paradoxically, the interest in novelty and rarity, criticized since antiquity for its cheapening of learning to freak shows and juggling, provided a safer path, and one that was certainly more pleasant, to the knowledge of nature and its creator, God.[52]

Compared to Nieremberg's detailed exposition, Grassi's remarks about novelty were terse. Yet his use of irony and humor rightly insinuated that whatever "unjust desires" he and his audience may have entertained concerning the appearance of one, two, and even three comets in 1618 made them, as humans, part of the same intellectual and emotional community.[53] The same "unjust desires" also rendered his audience capable of understanding and finding pleasure in the devices that, ten years earlier, Achillini had called "poetry's comets," thus reinforcing the bond between the orator and those who listened to him.

However, the claim that both comets and conceits capitalize on the mind's reaction to rarity and novelty accounts for only one of the two aspects in Achillini's analogy. Achillini suggested that both comets and conceits, as creations of, respectively, natural and human ingenuity, were equally "illegitimate" and "nonnatural." Likewise, Grassi may have tried to imply that comets entailed a certain degree of artifice within the ecology of the universe when he chose to compare their sudden, unpredictable irruption in the sky with the use of dissonance in music, astronomy's sister in mixed mathematics.[54] Just as humans long for novelty, Grassi maintained—even when it could be for the worse—so in music we similarly dislike intervals that "are accompanied by a constant and unwavering consonance of voices." Music, he claimed, "is much more pleasing if it has been composed of dissonant and consonant voices and if by reason of conflicting tones the concord and force of the voices have been weakened and, as it were, broken."[55] Lamenting that the discoveries that followed the invention of the telescope in 1609 seemed to preclude further developments, Grassi likened a sky entirely devoid of novelties to the tedium of uninterrupted consonance in music.[56]

Starting in the late sixteenth century, dissonance became increasingly prominent in musical theory and practice, a fact that renders the claim that dissonance awakens curiosity and stimulates inquiry particularly meaningful.[57] Grassi's advocacy of dissonance was hardly exceptional in 1619. In *Harmonices mundi*, published during the same year, astronomer Johannes Kepler elucidated the structure of the universe as a set of harmonical—that is, mathematical—

relationships, in which the speed of each planet corresponded to a specific and unique musical interval.[58] According to Kepler, universal harmony was made up of both consonance and dissonance. Composer and music scholar Giovanni Maria Artusi (ca. 1540–1613), labeled by Kepler as the greatest authority on harmony, defined consonance as a combination of sounds that are heard "gently" (*soavemente*) and "uniformly," passing as one for the sense of hearing.[59] Conversely, dissonance designated intervals that are received "harshly" (*aspramente*) because the sounds in them "could not be united or mixed with one another."[60] As Kepler well knew, Artusi had ruled out certain forms of dissonance as eminently unpleasant; others, however, were disagreeable in themselves but could be successfully accommodated in the composition by means of artifice. Foreshadowing the analogy with music in Grassi's lecture, Artusi claimed that by arranging dissonance and consonance together the composer was able to provide a degree of pleasure that remained out of reach of the use of consonance alone.[61] Artusi recognized that such a claim was nothing short of perplexing: why should harmony increase by incorporating what—in principle at least—looks like its opposite? He invoked two different but interconnected arguments in response. First, dissonance adds variety to the composition.[62] Second, "since nature has created nothing in vain, art accommodates it [i.e., dissonance] so that the sense of hearing may feel infinite delight."[63] The tag "Nature does nothing in vain" (*Natura nihil agit frustra*) was natural philosophy's perennial trump card, readily available when one had to justify any oddity or paradox. Underlying it was a view of nature as "a wise craftsman or artist, who, as Aristotle says in many passages of his studies on animals, proceeds in a rational way, [and] wastes nothing," as Pierre Hadot explains.[64] By reminding us that nature does nothing in vain, Artusi surmised that there was also a reason for dissonance, no matter how absurd it might seem from the narrow and limited perspective of humans that pleasant intervals necessitate the presence of unpleasant ones in order to cause greater delight. Could it be that conceits, being "poetry's comets," also entail dissonance? And could it be that conceits, as was the case with nature's comets when compared to "natural" stars and planets, also reach the intellect in ways that are unattainable for true and logically valid arguments?

Early modern scholars explored a variety of explanations for nature's tolerance of, and even demand for, monstrous and displeasing phenomena, which, like dissonance in music and conceits in language, could pose a threat to universal harmony. It is ironic that Augustine—who, by proscribing curiosity, had cast suspicion on the study of monstrous and portentous creatures in the first place—provided natural philosophers with a response. Augustine was convinced that whatever seems to occur contrary to nature occurs merely "contrary to what is known of nature," a point that he illustrated with insects.[65] Loathed as useless if not entirely harmful creatures without a role in God's creation, insects are similar to the tools that hang on the wall of an artisan's workshop, meaningful and necessary only for one who is familiar with the art. In the case of nature, only God knows the art and its secrets.[66] Still, it is one thing to agree with Augustine that nothing is ever without purpose or place in nature. It is another thing to claim, as seventeenth-century scholars were wont to do, that the beauty of nature was nowhere more present than on its margins and peripheries, namely in the odd, anomalous phenomena that seem to escape reason. Around 1617, humanist and critic Francisco Fernández de Córdoba (ca. 1565–1626) suggested that monsters were the product of nature's attempt to appear varied and hence beautiful.[67] Fernández de Córdoba, who used the analogy with monsters to justify the use of conceits, thus connected Artusi's two justifications of dissonance: variety and the notion that nature does nothing in vain. For Fernández de Córdoba, variety was the secret and unfathomable reason behind monsters, which the ignorant took to be nature's error.

For Kepler, dissonance was likewise necessary to universal harmony. Simply put, dissonance was as much a part of nature as consonance. To illustrate how dissonance blends with consonance in the composition of universal harmony, Kepler drew a parallel with flavor, labeling dissonance as "that primary and chief seasoning of tuneful melody."[68] He advocated using consonance "leavened with a very brief interpolation of melodic dissonances," which were "the yeast" that enabled the composer to "bake" a harmony of varied and delightful consonances.[69] According to Kepler, one should use dissonance to awaken and stimulate the palate, "like salt or vinegar in cooking."[70]

Grassi compared the novelty of rare and unusual phenomena in nature to dissonance, and Kepler likened dissonance to seasoning. As a matter of fact, cross-fertilization among the realms of nature, music, and taste was pervasive even if we pay attention only to the scholars already discussed, as Nieremberg justified the quest for new and rare phenomena by calling the wonder they arouse "the flavor of reasoning and the salt of the intellect."[71] One might be tempted to discard these and similar claims by Grassi, Kepler, and Nieremberg as merely metaphorical, yet their consistent and recurrent character are evidence of the ways in which early seventeenth-century individuals conceptualized the intellectual, emotional, and aesthetic response of humans to whatever strikes them as odd in the creations of both nature and art. Nature resorted to comets to interrupt the consonance of its usual course. To do so was indisputably a form of artifice, a trick to foster curiosity in both scholarly and lay observers, comparable to how musicians used dissonance to make listeners long for consonance and how orators used conceits to awaken the mind of the audience.

Crafting Dissonance through Conceits

Comets and conceits were destined to share a fate, whether condemned as instances of deviation or praised for taking the intellect's proclivities and limitations into account, feeding it what it craves: novelty. If Nieremberg described novelty as the salt that stimulates the intellect to natural-philosophical investigations, Jesuit scholar Pietro Sforza Pallavicino argued that novelty rendered conceits (*concetti*) full of taste.[72] Conceits were a form of seasoning; they had been called "salts" (*sales*) since antiquity. In 1639, on a visit to Bologna, writer Pier Francesco Minozzi viewed salt as the link between two seemingly unrelated features that made the city famous both in and outside Italy: its writers, who excelled in the invention of conceits (variously labeled as *acutezze, concetti, and spiriti*), and its cured meats or *salumi*, a term that involved salt in its etymology as well as in the preparation of the product it designated. Bologna, Minozzi was forced to conclude, was "a metropolis of taste"—*sapere*, a term that could mean both "to know" and "to taste."[73] From a different perspective, however, he compared conceits to sugar, while others likened them to vinegar.[74] The noun "seasoning" and the adjective "spicy" encompass the range of meanings contained in the analogy between conceits and flavor, one that portrayed conceits—like dissonance in music and monsters in nature—as exceptions or alternatives to the norm. Significantly, the analogy also implied that conceits, dissonance, and monsters, which in principle should be used sparingly, had nonetheless come to be felt as essential to rendering the norm appreciated and even noticeable by the intellect.[75]

The analogy between dissonance and conceits helps establish a parallel between Grassi's lecture and contemporary rhetorical and poetic debates. The first to describe dissonance as a feature of conceits was the former Jesuit Agostino Mascardi, in a discourse first published in 1622. The prompt was a poem, now lost, about one of the three comets of 1618.[76] Mascardi did not conceal his irritation at the anonymous author's excessive fondness for conceits, which forged "violent," far-fetched associations between disparate objects.[77] Mascardi conceded that a poem may have flaws. Flaws, he admitted, are desirable even in a composition provided that they occur within reasonable limits. "They delight," Mascardi suggested, "just as a carefully crafted dissonance does in a perfect conceit. . . . For, in fact, that which happens unexpectedly and without the mind anticipating it is received with greater wonder."[78] Mascardi, who would one day spearhead the resistance against the abuse of conceits, was well aware of what the audience so appreciated about them, namely, wonder provoked by the dissonance of novelty. It was the same aspect that, in Grassi's view, had rendered the three comets a refreshing and stimulating experience for both astronomers and lay observers.

Mascardi's claim that dissonance was fundamental to conceits offered the first confirmation in a work of rhetorical and poetic criticism that Achillini's coinage "poetry's comets" was not mere flourish and, more importantly, was not idiosyncratic. Dissonance was destined to remain central to a debate on conceits that would resound through the same walls within which Grassi had delivered his lecture. Comets and conceits became the subject of simultaneous, parallel debates that often took place in the same settings and involved the same individuals. Their fates and posterity, however, could not be more different. The dispute over the celestial or atmospheric nature of comets soon yielded to a debate between proponents of conflicting views of the universe, which became part of the history of the so-called Scientific Revolution. The publication of *Il saggiatore* by Galileo remains one of its most remarkable and oft-remembered feats. The fourth installment in a controversy that started with Grassi's lecture in the winter of 1619, *Il saggiatore* appeared in print with the support of the Accademia dei Lincei mere months after Cardinal Maffeo Barberini (1568–1644), a patron of Galileo and the Accademia, ascended to the papacy as Urban VIII in August 1623.[79] In contrast, the history of the debate surrounding conceits was writ small, a more or less insignificant item in a long and heterogeneous list of seventeenth-century oddities. However, period sources reveal that the debate over conceits was similarly able to consume public attention, placing in the minds and on the lips of writers and scholars a set of questions destined to receive varied and often conflicting answers: What, exactly, are conceits? How do they work, and how do they act upon the intellect?

Even as the debate on conceits became increasingly vibrant and intricate, the centrality of dissonance was confirmed. It was key in the analysis of Maciej Kazimierz Sarbiewski, a poet of Polish origins to whom we owe the earliest treatise on conceits on record. Sarbiewski, who lived in Rome between 1622 and 1625, when the controversy between Grassi and Galileo was raging, was a member of the Society of Jesus like the former. With the latter he shared the protection of Urban VIII, himself a famous poet of epigrams, who appointed Sarbiewski to the team tasked with adapting the Breviary into Ciceronian

Latin. Before he arrived in Rome, where he was to complete his studies in theology, Sarbiewski had spent several years teaching rhetoric and poetics at the Jesuit colleges of Kražiai and Pólatsk.[80]

It was in Pólatsk, where Sarbiewski taught poetics between 1618 and 1620, that he first lectured on the art of conceits.[81] We know this because, after Sarbiewski left Rome and returned to Pólatsk in 1626, he taught a different, slightly updated version of the art, which is now preserved in a tract titled *De acuto et arguto*, containing Sarbiewski's findings on the nature of conceits in dialogue with those of contemporary scholars.[82] In *De acuto*, Sarbiewski renounced certain views that he had endorsed in 1618–1619, revealing a theory in constant revision and pervious to exchange with others. Between the first and the second period in Pólatsk stretched years of continent-wide travel and a stay in Rome that played a critical role in the evolution of Sarbiewski's ideas.

In the summer of 1623, Sarbiewski lectured on conceits "daily" (*in quotidiana frequentia*) and "with open doors" (*ianuis apertis*) in Rome, turning conceits into a currency that stimulated a flow of scholarly and personal communication as he tirelessly discussed the topic with Rome's academic elite, following previous exchanges with "most learned men in Poland, Germany, France, and Italy."[83] He listed the individuals, invariably Jesuits, with whom he conferred about conceits (or statements containing *acutum*, as he called them) either face-to-face in Rome or by correspondence. The names included rhetoricians such as Famiano Strada, Nicholas Caussin, and Matthaeus Rader, the editor who, as his critics used to say, had castrated Martial for use in Jesuit schools.[84] Sarbiewski evoked a republic of Jesuit letters defined by collegiality with a list of interlocutors who could certify that he was the one behind the ideas found in *De acuto*. After discovering that the course he had taught in 1619 had been circulated in someone else's name, Sarbiewski had learned the hard way that ownership over ideas was elusive.[85]

At stake in Sarbiewski's *De acuto* was the question of what made conceits attractive. Sarbiewski illustrated his views with Epigram 1.14 by poet Marcus Valerius Martial, whom he elevated to a paradigm of conceit making.[86] In Epigram 1.14, lines 1–4 narrate the story of a hare hopping unscathed through the open jaws of a hungry lion. Sarbiewski focused his analysis on lines 5–6, which showcased the poet's ingenuity by drawing a striking and unexpected conclusion to the narration:[87]

> We have seen the pranks, the sportive gambols of the lions, Caesar
> (this too the arena offers you), when the hare as often

as seized returned from the gentle fang
and ran at large through the open jaws.
How comes it that a greedy lion can spare his captive prey?
Ah, but he is said to be yours [i.e., Domitian's].
Therefore he can.[88]

Sarbiewski explained that, by affirming that the hare stood a chance of survival as it cheerfully hopped through the jaws of "a greedy lion," line 5 strikes the intellect as *dissentaneum*, a failed, ill-assorted coupling of ideas. Just as dissonance designates two sounds that are not received as one by the sense of hearing, so the intellect is unable to perceive the union between two notions that exist in *dissentaneum*: they simply do not appear as a likely pair. Enter line 6, which posits the argument that the predator has taken its mild and merciful character from its master, Emperor Domitian, and so—flattery aside—what thus far appeared to contain *dissentaneum* resolves into agreement or *consentaneum*. In doing so, Epigram 1.14 offered a form of pleasure to the reader that, as Grassi claimed of the succession of uninterrupted consonance, was entirely out of reach to *consentaneum* alone. Sarbiewski's explanation soon became widespread within the scholarly and educational circles of the Society of Jesus, to the point that, in *Orator extemporaneus* (*An Improvising Orator*, 1640), the young Jesuit Michael Radau (1617–1687) elaborated on Sarbiewski's views on *acutum* as if they were entirely matter-of-fact.[89]

A conceit was akin to a riddle, posed and solved as though in a single move as it revealed that what at first sight seemed contradictory was so only in appearance. A conceit, Sarbiewski determined, needs to "contain a reason that is, on the one side, discordant (*dissentaneum*), and on the other side concordant (*consentaneum*); that is (so that I may speak more clearly), that the statement from the one side does not seem pertinent to the thing of which it speaks, whereas from the other side it seems to pertain to it to a high degree."[90] In relation to Mascardi's claim that a conceit delights through "a carefully crafted dissonance," Sarbiewski's substitution of *dissentaneum* for *dissonans* took the concept from the sensory to the intellectual domain.[91] In a conceit, *consentaneum* and *dissentaneum* denoted, respectively, a plausible and a questionable or implausible pairing of terms, the former striking the reader as a suitable explanation of the latter. As theologian Denis Petau (1583–1652) declared in a letter to Sarbiewski, a conceit hits at the mark "in a wondrous manner" not least *because* it has risked straying in extravagant fashion.[92]

Sarbiewski encapsulated a conceit's inner conflict by defining *acutum* as a statement's "concordant discord or discordant concord" (*concors discordia vel discors concordia*).[93] *Concordia discors*, a coinage of the poet Horace (65–8 BCE), conveyed in antithetical and condensed form the claim by philosopher Empedocles (fl. fifth century BCE) that the world results from the clash between Love (φιλότης) and Strife (νεῖκος).[94] The notion of *concordia discors* reinforced the connection of conceits with music and nature.[95] For Nieremberg, *concordia discors* described nature's inner structure as a balance between the forces of sympathy and antipathy that are responsible for interactions among bodies; for Jesuit Daniello Bartoli (1608–1685), the blend of grave and acute sounds in the harp was an instance of *concordia discors* that evoked the coexistence in matter of elements hostile to one another, a view ultimately derived from Plato's account in *Timaeus* of a demiurge who created the world in line with the proportions of music.[96]

Concordia discors was proverbial for a union of seemingly conflicting constituents that derived value from its precarious, nonobvious character. The notion that the perception of *concordia discors* within a whole suggested beauty to the beholder had a history that went back at least as far as Pico Della Mirandola. In the late fifteenth century, Pico argued that what is simple, like God, is incapable of beauty. Beauty represents instead a dynamic relationship that starts with contrast and manifests itself as the resolution of discord into concord. The world's beauty, as early modern scholars tended to reiterate, takes the form of a friendly enmity or *concordia discors*, replicated at small scale in a conceit.[97]

There is evidence in *De acuto* that Sarbiewski's mention of *concordia discors* effectively tied conceits to a dynamic and multilayered approach to nature, one that viewed comets as "new" and "artificial" elements that played a role in universal harmony by introducing a component of desirable dissonance. Sarbiewski's *De acuto* explicitly associated conceits with what Lorraine Daston and Katharine Park have dubbed the preternatural. Daston and Park have argued that the reception of Aristotle's natural philosophy in the medieval and early modern periods entailed a reassessment of how certain phenomena related to nature's creativity. For the contemporaries of Sarbiewski, the fact that a phenomenon was nature-made rather than either human-made or miraculous—that is, a product of God's direct and unmediated intervention—was not enough for it to qualify as "natural." The term "natural" implied, in its narrower sense, that a phenomenon had occurred according to nature's usual course, that which happens "always or for the most part," according to Aristotle's formulation of the scope of *scientia*.[98] As theologian Thomas Aquinas (1225–1274) once explained,

"The order imposed on things by God is based on what usually occurs, in most cases, in things, but not on what is always so."[99] A small but nonnegligible array of phenomena, which Daston and Park labeled as preternatural "from Aquinas' repeated phrase 'praeter naturae ordinem'" ("beyond nature's order"), remained outside the scope of nature in its narrower sense, even though they had come to be without the aid of either divine or human intervention.[100] Briefly put, they were nature's doing, but they were not "natural."

Praeter naturam phenomena were exceptional in their rarity and defiance of nature's usual course, yet the reader who devoured the publications of the likes of Nieremberg could be forgiven for believing that there was nothing more common in nature than monsters. Indeed, behind the rise of preternatural philosophy lay not only the ability of rare and wondrous phenomena to stimulate the intellect but also the disproportionate and unprecedented heuristic value attributed to anomaly. As Francis Bacon repeatedly contended, nature's abilities were nowhere more evident than in its errors and vagaries, when it pushed through any boundaries and limits.[101]

Comets were, in a way, monsters. In 1557, physician and philosopher Girolamo Cardano distinguished between comets and "the natural and usual changes that happen in the sky."[102] Cardano was not the last to call comets unnatural. Kirwitzer, Nieremberg, and countless others followed suit.[103] A book about the comet that was first spotted in December 1652 in Pernambuco (C/1652 Y1) was published with the title *Coeli Anomalon* (*The Sky's Anomaly*), its subtitle reminding the reader that its topic lay *Praeter uniuersi ordinem* or "beyond the order of the universe."[104] In 1618–1619, comets were rare and entirely unpredictable.[105] The majority of Grassi's audience would have considered them preternatural in the sense just mentioned, and Achillini had argued as much ten years earlier by pointing to the difference between "natural stars" and comets.

Sarbiewski took "poetry's comets"—that is, conceits—to be similarly preternatural, even though it is almost by accident that this was revealed to the reader. Near the end of *De acuto*, Sarbiewski renounced certain views still circulating under his name but which he no longer maintained. Specifically, he retracted a classification of conceits according to their greater or lesser degree of acuity that he had proposed in 1619. Sarbiewski had quantified conceits based on how they related to nature in its narrower sense, understood as its usual course: a conceit was sharpest if it was "against nature" (*contra naturam*) and presented the audience with some form of impossibility; conversely, a conceit "according to nature" (*iuxta naturam*) was completely blunt because it failed to pose any challenge to the mind. In addition to *contra* and *iuxta naturam* conceits, Sarbiewski had identified a third class of conceits "beyond nature" (*praeter naturam*), which were absurd only in appearance.

From the vantage point of several years of reflection and continent-wide exchanges, Sarbiewski eventually came to realize that *contra naturam* conceits lacked *consentaneum*. They appeared absurd because they were, in fact, contrary to reason. *Iuxta naturam* conceits, in turn, were devoid of *dissentaneum* and thus failed to create dissonance. Conceits, Sarbiewski eventually realized, were necessarily *praeter naturam*, with "a balanced proportion between *dissentaneum* and *consentaneum*," resting on the tenuous and fragile harmony of contraries encapsulated in the notion of *concordia discors*.[106] Just as comets "seasoned" universal harmony with their unpredictable and disruptive behavior, so conceits too challenged and stimulated the audience by leading them to realize that a statement that at first seems irrational and absurd contains insights of truth if looked at from a different perspective.

Like rumors, however, scholarly theories may live independently of those who first generated them, and Sarbiewski would have been dismayed at the resilience of views he had ceased to embrace. The distinction between *iuxta*, *contra*, and *praeter naturam* conceits that he had been eager to disavow and patient enough to refute reemerged as a solid and widely accepted view in Radau's *Orator extemporaneus*, first published in the year of Sarbiewski's death.[107]

Yet whether in its earlier or its later form, Sarbiewski's discussion was evidence of a link between seventeenth-century writings on conceits and natural philosophy. It both clarified and questioned Achillini's analogy between comets and conceits. While Achillini's indictment rested on the exclusion of comets from nature in its narrower sense, Sarbiewski restored both comets and conceits to a different and more inclusive version of nature, one in which *praeter naturam* phenomena coexisted with, and even had priority over, those that occur *iuxta naturam*. In Sarbiewski's view, conceits stood in relation to logic just as *praeter naturam* phenomena stood in relation to nature's usual course. As Radau claimed while paraphrasing Sarbiewski, *praeter naturam* conceits were "not against nature," but they were "not conventional or natural" either.[108]

Achillini's analogy belonged to a range of numerous and often conflicting assessments of the parallel between natural and human ingenuity. Conceits could be called "poetry's comets" because both conceits and comets were, in a way, inventions that the poet and God, respectively, crafted with consideration of the intellect's characteristics and specific limitations, namely its eagerness for novelty and its pleasure in dissonance. In addition to novelty, however, comets and conceits—as, respectively, *praeter naturam* phenomena and statements—also shared artifice, which is the topic of the last section of this chapter.

Wonder and Artifice

A conceit, Sarbiewski claimed, brings the audience wonder with pleasure. These were two distinct but interconnected reactions that he associated, respectively, with a conceit's *dissentaneum* and *consentaneum*:

> Wonder is born from what comes unexpectedly (*ex inopinato*), that is, when we hear something that we did not expect and of which the cause escaped us, and for that reason we thought it could not occur. For, as Aristotle says, wonder is born of the ignorance of causes. In turn, if this wonder is conjoined with pleasure—as always occurs in a conceit, for in it we wonder with pleasure—it needs to have, at the same time, some appearance of that cause that was concealed from us and that allowed to occur what we did not expect could occur.[109]

Wonder was the emotion aroused by a conceit's *dissentaneum* and a stimulus for natural-philosophical inquiry.[110] Nieremberg, who subtitled one of his works *Tesoro de maravillas* (*Treasure of Wonders*), believed that God solicited the attention of humans through *praeter naturam* phenomena, which provided the intellect with a ladder to reach knowledge of God.[111] Likewise, a conceit's *dissentaneum* was goading: it spurred the intellect to search for causes that could justify what seemed, at first sight, absurd. Yet there was only a partial or incomplete analogy between the role of wonder in natural philosophy and in conceits. As Sarbiewski acknowledged, to contend that the lion in Martial's Epigram 1.14 had spared a hare because it belonged to the emperor was less a cause than "the appearance" of a cause. The truth of a conceit rested on figurative language and based its validity and effectiveness on a way of speaking.[112] However, despite the analogy's relative shortcomings, there was little doubt that the dissonance created by the inventors of conceits shared with nature's wonders the ability to agitate the intellect into a state of restlessness.

Traditional assessments of wonder had struck a less positive tone than Nieremberg's, owing to its centuries-long proverbial association with ignorance. Daston and Park explain that for Albertus Magnus (d. 1280), "the master of philosophy felt no wonder as he taught, since he already knew the causes—which is what qualified him to teach. Similarly, the student felt no wonder as he was taught, since his teacher's exposition followed the order of certainty, from cause to effect."[113] The rise of preternatural philosophy was accompanied by the embrace of wonder, as natural philosophers took on increasingly rare and strange phenomena and asked *why* and *how* they occurred.

Wonder wove different disciplines together, from the study of nature to rhetoric and poetics, yet despite its transversal and far-reaching scope, it never lost its dubious character. The dedicatory sonnets to a volume published in 1597 celebrated that its author, the Italian physician and philosopher Giovanni Talentoni (1542–1620), had vindicated wonder from its moniker of "the child of ignorance."[114] Talentoni's *Discorso in forma di lezzioni . . . sopra la maraviglia* (*A Discourse in the Form of Lectures . . . on Wonder*) was probably the only early modern work to be published with wonder as its main and only topic. In it, Talentoni recalled wonder's ancient and noble pedigree: both Plato and Aristotle had called it the origin of philosophy.[115] He explained that the emotion we call wonder takes place whenever we are made aware of some object but remain ignorant of its cause. Wonder differs from doubt, because it is compounded by the eagerness to learn that which we do not yet know; it is a tyrant that throws one into a strong and intense examination of the object that awakened it in the first place.[116] One of Talentoni's key contributions was the insight that wonder acted as a threshold between the mind and what lies outside it, a view usually ascribed to René Descartes (1596–1650), who four decades later in *Les passions de l'âme* (*The Passions of the Soul*, 1646) called wonder "the first of the passions."[117] For Talentoni (and later Descartes), wonder functioned as a sort of zero-degree emotion, a primordial and indefinite reaction to the object's sheer novelty and its resistance to comprehension.[118] This is why Talentoni rejected the claim by Aquinas that wonder was a kind of fear, arguing that, without even cursory knowledge of the object, it was impossible to experience love, hate, desire, rejection, or any emotion involving either a positive or a negative attitude toward it.[119]

The intellect's sudden turn to a conceit's *dissentaneum* makes sense in light of Talentoni's claim that wonder takes the soul as its prisoner and forces it to focus on the object of inquiry, only to regain freedom after finding a cause for that which appeared mysterious or unknown.[120] In a conceit, the equivalent occurs when the intellect starts to realize that what at first seemed impossible could be exceedingly likely just by looking at the issue from a different angle. Then, it all strikes the mind as *consentaneum*. For instance, it would be absurd to even suggest that a hungry lion would spare a hare hopping through its open jaws. But, wait: did you say it is the emperor's? How could it then do otherwise? As the intellect considers the situation in a different light, Sarbiewski wrote, it "starts to feel pleasure, while in a way it favors both sides, and becomes dragged into both, and in turn it becomes united in, or assents to the wondrous union of both, and even that which it once thought was discordant now perceives as concordant, or vice versa."[121]

For even the briefest moment, the intellect realizes that the conceit's *dissentaneum* and *consentaneum* are merely two sides of the same reality. To explain this, Sarbiewski resorted to a comparison with mathematics (Figure 1.2). In a triangle, two intersecting lines originating in the base meet "sharply" in the union represented by the angle opposite the base. A conceit embodies a different but comparable form of sharpness or *acumen*, in which the base of the triangle represents the matter or topic of which one is speaking. Two lines of thought concerning the same matter—one concordant and the other discordant with reason—seem to briefly meet in what Sarbiewski called "the union or affinity of the discordant and the concordant lines that grow from the same matter," represented by the vertex.[122] The union, representing a conceit's *concordia discors*, affords pleasure, as the intellect "dilates" in the contemplation of a multiplicity of objects that nevertheless converge in—precarious and unstable—unity.[123]

In music, harmony also causes dilation in the soul. According to Jesuit professor Athanasius Kircher (1602–1680), the union of dissonance and consonance brings pleasure because, as the heart dilates with it, the spirits expand throughout the body. In early modern medicine, the spirits (*spiritus*) were "a subtle vapor or exhalation produced from blood and disseminated throughout the body by the arteries and nerves."[124] They were the soul's instrument, its means of communication with the body. With their inward or outward movements, the spirits were tightly connected with emotions, considered by early modern physicians and natural philosophers as "primarily physical, embodied phenomena."[125] The spirits leave the heart and expand outward in the presence of favorable objects, heating the limbs with moderate joy and bringing a rosy or reddish color to the skin. In the presence of less desirable or even dangerous stimuli, the spirits retreat into the heart, leaving cold limbs and a pale skin. While a pale skin could be associated with wonder, Talentoni maintained that the coincidence between wonder and fear was only apparent. Wonder's liminal position as the first and primordial emotion—one that occurs before the subject casts a verdict on whether the object being considered is good or bad—meant that it could not easily fit into a physiology of emotions based on the opposition between inward- and outward-moving spirits. Wonder stirs the spirits without pushing them in either direction: "In pain and fear," Talentoni claimed, the spirits "flee at once toward the heart, and do not dare to go out, whereas in wonder they do not actually flee, but remain suspended, ambiguous and impatient." The spirits of the individual who wonders remain alive, teeming with desire to comprehend the object.[126]

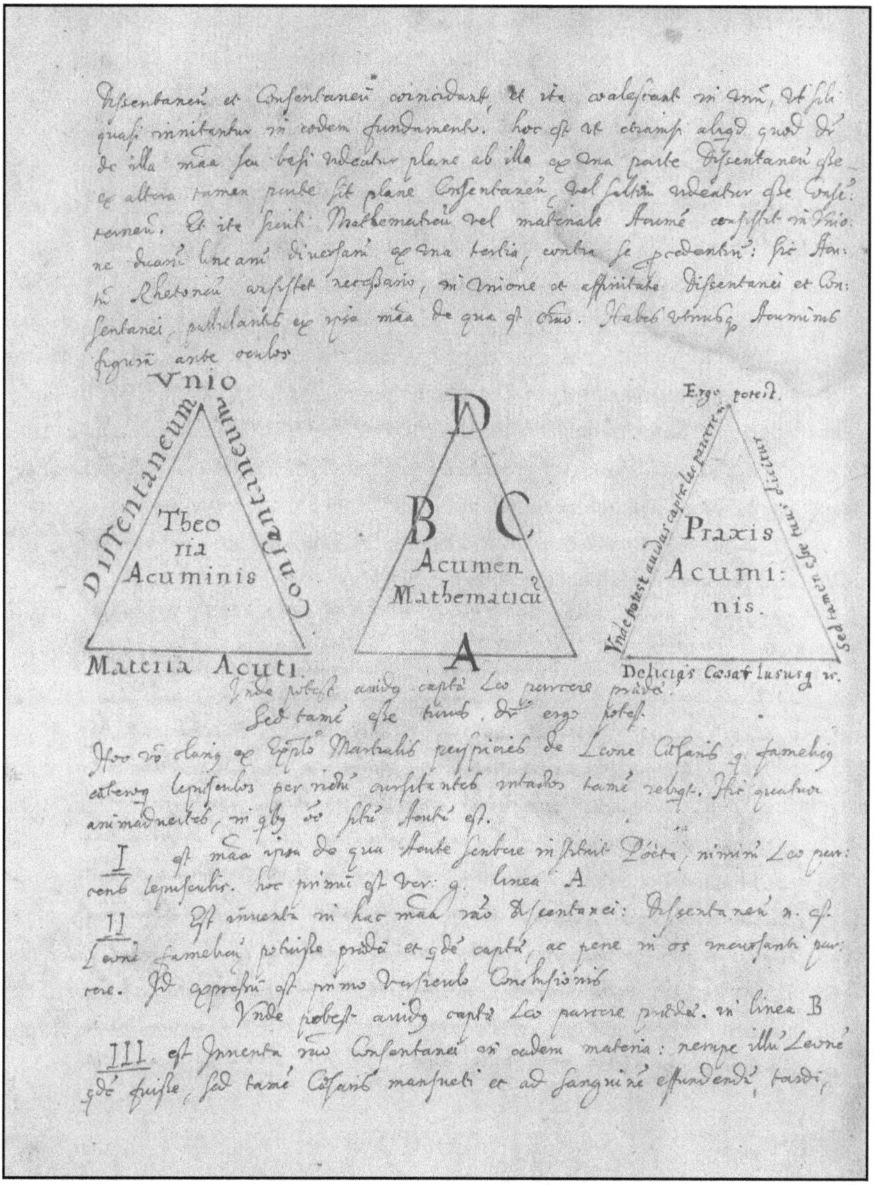

Figure 1.2. The analogy between rhetorical and mathematical acuteness or sharpness. Maciej Kazimierz Sarbiewski, *Praecepta Poetica*, Czartoryski Library, Ms. 1446, 280. Laboratory Stock National Museum in Krakow.

Sarbiewski's description of how a conceit affects the individual, caught between wonder at *dissentaneum* and pleasure at its resolution in *consentaneum*, entailed potential therapeutic or healing consequences for the audience or the reader. In 1623, Bacon argued that wonder or "admiration and light contemplation are very good to prolong life, keeping the spirits busied in delightful matters, and in a peaceable quiet gentle temper: so that all philosophers, and observers of the wonders of nature (as Democritus, Plato, Parmenides, Apollonius) were long-liv'd."[127] Bacon believed that repeated wonder, of the kind experienced by natural philosophers and orators, was a reliable source of good health, in stark contrast to other disciplines that required sustained contemplation of the object of study before the mind could grasp it, providing few occasions for the spirits to dilate.[128] There was nothing healthy to be found in occupations that condemned the individual to perpetual dissonance or *dissentaneum*, without offering any dose of *consentaneum* most of the time. Conceits, conversely, seemed designed to provide wonderment and pleasure in quick succession, offering the riddle and its solution almost at once.

It was almost in passing that Sarbiewski made a remark that had the potential to destabilize the analogy between wonder in a conceit and its counterpart in natural-philosophical investigation. If a conceit were truly about discovering the cause of what seems new and unexpected—that is, finding *consentaneum* for what at first appeared *dissentaneum*—then, Sarbiewski aptly declared, it would be a unique or onetime experience. However, it is possible to read a conceit a thousand times and still enjoy its sharpness, even after we have been made entirely aware of what renders it *consentaneum*. To overcome the obstacle, Sarbiewski explained that a conceit's *dissentaneum* and *consentaneum* were only so in relation to one another, so that "when one is known, the other by its nature seeks to hide and, as a consequence, it is possible not to know it."[129]

Mercedes Blanco was the first to interpret Sarbiewski's enigmatic explanation through analogy with visual representations. She suggested that just as in ambiguous drawings we can see now a rabbit, now a duck, but never both simultaneously, so in a conceit's union of *consentaneum* and *dissentaneum* "the attention we pay to one necessarily conceals the other."[130] In a similar vein, Jon R. Snyder compared Sarbiewski's passage to the paintings by Giuseppe Arcimboldo (1527–1593).[131] When the eye sees the face of Emperor Rudolf II in Arcimboldo's *Vertumnus* (ca. 1591)—made up of vegetables, fruits, and flowers—it can see either the parts or the whole as it switches attention between the two, but it cannot see both at the same time (Figure 1.3). In 1646, Kircher clarified

Figure 1.3. Giuseppe Arcimboldo, *Rudolf II as Vertumnus* (ca. 1591). Oil on panel, 70 × 58 cm. Item Number 11615_SKO. Photograph: Jens Mohr. Skoklosters slott/SHM (PDM).

Figure 1.4. "Campus antropomorphus." Athanasius Kircher, *Athanasii Kircheri . . . Ars magna lucis et vmbrae . . .* (Rome: Ludovico Grignani, 1646), plate opposite p. 807. Library Special Collections, Charles E. Young Research Library, UCLA.

that natural creativity and human creativity were equally able of creating bodies that revealed something different and unexpected when looked at from a certain perspective. Thus, the promontory of Scylla, in the straits of Messina, displays "a head with eyes, a nose [and] a beard."[132] Likewise, Kircher laid down the instructions to craft a depiction in which one can see now a landscape, now a man lying down, but never both at the same time (Figure 1.4).[133] Similar to Arcimboldo's and Kircher's visual examples, in a conceit, which is based on figurative language, the intellect is aware of a variety of ways in which a term can be understood, but "focuses" on just one of them at a time. It is only for a brief but critical moment, in what Sarbiewski dubbed the union of a conceit's *dissentaneum* and *consentaneum*, that the intellect can contemplate the two sides "as one" and experience pleasure.

Early modern writings associated the ability to repeatedly arouse wonder with products that the intellect understood as works of art, regardless of whether they were nature- or human-made. Consider the example of pyrotechnics. As Simon Werrett explains, early modern sources distinguished between two reactions to fireworks. The fear of the crowds, victims of wonder followed by panic, stood in stark contrast to the calm and detached amusement enjoyed by learned spectators. The latter, understanding to a certain extent the specifics of the engineer's art, experienced a form of wonder that, as also happened with conceits, could be renewed without end.[134] Likewise, in 1560, poet Bernardino Partenio (d. 1588) contrasted those who feel wonder at the use of tropes with the learned, who are familiar with the rhetorical and poetic methods that produce them. Instead of wonder stemming from perplexity, the latter experience a greater and virtually endless pleasure, which is reserved for one who knows the secrets of the art.[135] Philosopher Jacopo Mazzoni (1548–1598), a friend of Galileo, claimed in 1587 that the crowds feel wonder at the rare and unlikely events devised by the poet. This is different for the learned, who, privy to rhetorical and poetic wisdom, relish the poet's ability to deceive those who lack education.[136] The crowds feel wonder because they are ignorant of rhetoric and poetics, which are "the cause" of the plot, whereas the learned know rhetoric and poetics from the start. For them, knowledge is not the reason why wonder comes to a halt but a requisite for enjoying a type of wonder that both is and is not the same that the crowds experience. As with Sarbiewski's conceits, the learned feel wonder in a process that is virtually without end because knowledge, the only ingredient able to quench it, is the fuel that kindles it in the first place.

In 1557, Aristotelian natural philosopher Julius Caesar Scaliger (1484–1558) became the first to identify a variation of wonder that required knowledge rather than ignorance of the cause. The context was a discussion of subtlety in reaction to Cardano's treatise on the topic. In 1550, Cardano defined subtlety as the characteristic by which some objects are difficult to perceive through the senses or to comprehend through the intellect.[137] Convinced that nature is always one and the same, Scaliger took exception to Cardano's views, which implied that nature had been selectively subtler in certain creations, while neglecting others. He believed that if humans were more likely to find subtlety in the myriad shapes and colors that distinguish seashells and flowers than in other instances of nature's ingenuity, it was due to their limited sensory and intellectual capabilities, which project onto nature notions of difficulty, worth, and beauty that originate in the consideration of human-made objects.[138]

Scaliger's belief that subtlety belongs exclusively to the domain of art had implications for its relationship with wonder. For instance, it was a mark of subtlety if anything human-made aroused wonder in a person who understood its cause. This was the case with fireworks and poetic plots experienced by well-educated individuals, but also with conceits, which Scaliger described as products of *argutia*, or subtlety in speech.[139] Scaliger claimed that where subtlety is at stake, wonder is no longer aimed at the cause of the object, represented by, for instance, the arts of engineering, poetics, and rhetoric. Instead, it targets "the union between the cause and the effect," or the art and the object it has been able to produce. As a marker of difficulty and worth, subtlety prompts the individual to acknowledge that the object remains wondrous despite one's knowledge of the art. For that to happen, one needs to discern or intuit some form of value in it that lies beyond the application of a set of precepts or rules. Instead of a stimulus to investigate the cause, in similar cases wonder is experienced as "the judgment of a certain subtle realization."[140] Scaliger's formulation was cryptic even for Talentoni; it seemed to suggest that the object's subtlety caused the intellect to enjoy its ability to notice the relationship between cause and effect and, in doing so, to feel subtle in turn. Scholars from Aquinas to Cardano had remarked that the intellect's ability to cope with difficulty (*subjective subtlety*) acted as the counterpart to the object's difficulty (*objective subtlety*). Simply put, it takes subtle individuals to understand subtle objects.[141] Scaliger insinuated that humans erroneously ascribe subtlety to nature because any object or phenomenon—whether nature- or human-made—that resists perception or comprehension prompts the belief that there must be something else to the object than the art that lies behind it. The mind thus experiences repeated and virtually endless wonder at the artificer's ingenuity, be the artificer God, the poet who invents a conceit, or the engineer who has crafted a spectacle of fireworks.

Sarbiewski discovered that conceits aroused a form of wonder characteristic of the reaction to the subtlety of human-made, artificial objects. Nature's creations, however, could elicit the same emotion as long as the intellect considered them "artificial." Did Scaliger's criticism of Cardano not pertain to the intellect's proclivity to project onto nature categories, such as subtlety, originally intended for nonnatural objects? When Talentoni took up Scaliger's insight forty years later, he called wonder that requires knowledge of the cause "wonder of the second type," wonder "of the first type" being the one in which the mind seeks and desires knowledge. Talentoni narrated that the painter Nicostratus, when asked why he stared for so long at Zeuxis's portrait of Helen, replied, "You wouldn't have asked me if you had my eyes." Talentoni clarified that by "my eyes," Nicostratus meant knowledge of the art of painting, the cause of the object that made him wonder. In line with Scaliger, Talentoni contended that Nicostratus wondered at the union of cause and effect, pondering how the art of painting could have resulted in the portrait.[142] Not surprisingly, Talentoni's examples of wonder of the second type included both nature- and human-made objects. For instance, he recalled that the physician Galen (129–216 CE) had wondered at the formative virtue that generated human limbs from the embryo, even though Galen professed to understand how it happened.

The shift from what to how it is made—implicit in the consideration of objects as artificial regardless of their nature- or human-made character—soon found its way into seventeenth-century natural-philosophical works. Nieremberg, for one, compared nature's artifice to that of the statue of the goddess Athena sculpted by Phidias in the fifth century BCE: "Anyone who saw separately the pieces of which Phidias's statue of Athena was made would give them little consideration. However, once put together and attached to one another, they struck wonder in everyone, especially so if one paid attention to the art with which they came together in the shield of the goddess, in which the face of the artificer was portrayed."[143] A tradition held that Phidias's face stood right in the center of the shield, and a mechanism made sure that removing the sculptor's face would cause every piece in the shield to come apart.[144] Nieremberg argued that nature's creations also related to one another in such a way that at its union one could see "a face of God as if enameled, the likeness of its artificer." It was like reading "God made me" (*Deus me fecit*) at the spot in which nature's *concors discordia* became manifest, where different parts joined into a whole.[145] Nieremberg thus insinuated that artifice implied authorship, signified by the allusion to God adding a signature to nature as if he were a painter. Put simply, noticing the union or fitting composition of different parts in a whole led the intellect to realize that it is the work of some artificer or author rather the product of chance.[146] By the same logic, one might assume that the reverse was also true: the intellect took any phenomenon that failed to elicit the idea of some artificer or author behind it as occurring naturally.

Nieremberg's anecdote linked nature's artifice to seventeenth-century writings about conceits in distinct and profound ways. First, it recalled Sarbiewski's claim that the union of a conceit's two parts, its *dissentaneum* and *consentaneum*, elicits wonder at the link between the two that artifice makes possible. Second, the reference to Phidias's face sculpted on the shield sheds light on a perplexing and enigmatic statement advanced some years earlier by the historian and Collegio Romano professor Famiano Strada, one of Sarbiewski's interlocutors in Rome and his colleague in the pope's Breviary task force. In a discussion of tropes found in a course of rhetoric he taught in approximately 1606–1607, Strada wrote the following: "When we see the work of some excellent artificer and experience wonder at it, then, if possible, we always look with pleasure at the face of the individual, for to see the cause from which the work originated seems to satisfy to some extent the feeling of wonder that we have experienced. For, since the orator's ingenuity is also the metaphor's cause, it follows that the mind of the audience feels attracted to it and takes pleasure in the sight of it."[147] The notion that artworks and tropes alike carry the intellect away with a desire to see the author's "face" connects the invention of conceits with the anecdote about Phidias. Read together with Strada's digression, Nieremberg's claim seems to imply that natural philosophers and the curious in general, as they notice the union that is nature's *concordia discors*, make a leap toward considering nature as art—that is, as God's artwork.

However, Nieremberg did more than merely subscribe to the trope of God as artist, or assert some bland and empty equivalence between nature and art. Aligning with the program he advocated of basing the investigation of nature on the intellect's limitations, Nieremberg suggested that nature is never more appealing (at least to humans) than when it passes as art. As he put it, although in principle art receives praise for following in the footsteps of nature, "all of a sudden, the tables are turned, and the most admirable thing in nature seems to be the one that imitates art. I mean its artifice and design."[148] In *Oculta filosofía* (*Occult Philosophy*, 1633), Nieremberg took "nature's artifice" as object of study, by which he understood "its ingenuity, scheme, architecture, order, [and] correspondences; in sum, its art."[149] Nieremberg substantiated the claim that it made sense to speak of nature's artifice with analogies of clockwork mechanisms, of the kind that pervaded countless contemporary works of mechanical and Aristotelian natural philosophy.[150] However, unlike Descartes, for whom the notion that nature was similar to a machine implied that it should follow universal laws, Nieremberg granted latitude to nature. For instance, he claimed that some aspects of nature were particularly "artificial" or "artful," including a set of heterogeneous effects characterized by their relative rarity and their reliance on occult sympathies and antipathies that connected things to one another.[151] For Nieremberg, artifice was greater where the links among the parts in the whole of nature appeared less obvious to humans. He observed approvingly the mind's habit of tampering with the divide between art and nature, which served as a counterpart to its inability to appreciate any object devoid of novelty. Artifice and novelty, the two elements that underpinned Achillini's analogy between conceits and comets, were cornerstones of Nieremberg's natural philosophy because both catered to the intellect's limitations and ways of knowing.

Nieremberg, Kircher, and a range of early modern natural philosophers combined novelty and artifice in a category that de facto linked products of art with nature's rarities and wonders, illustrated by the analogy between conceits and comets.[152] The artificial and the preternatural not only became favored topics of scholarly research for much of the seventeenth century but also shared space in early modern cabinets of curiosities. By juxtaposing human- and nature-made objects and by restricting the latter to a great extent to the preternatural, cabinets underlined the porous character of the divide between *artificialia* and *naturalia*.

The preternatural became a bridge between natural and human ingenuity. Given that *praeter naturam* objects were perceived as products of nature's "artifice," it is easy to understand why they were likely to invite alteration by humans, as if (natural) art called for (human) art to cooperate in a complex, multilayered creative process.[153] In addition, the perception of artifice in nature's creations triggered the emotion that Talentoni called wonder "of the second type," which led the intellect to admire nature's ingenuity. Nieremberg's pairing of artifice and ingenuity as attributes of nature's *praeter naturam* creations might cast doubt on the traditional, centuries-long opposition between *ingenium* and *ars*—two terms that referred, respectively, to innate and acquired talents—or to nature and nurture.[154] However, a similar conflation of nature's artifice and nature's ingenuity emerges from the correspondence of Kircher, for whom corals and fossils were evidence of nature's artifice, while the earth's interior appeared as a system of interconnected workshops or factories (*ergasteria*).[155] Take a letter, dated 1660, discussing a lizard enclosed in a lump of Baltic amber that Duke August the Younger of Braunschweig-Lüneburg had sent to Rome for display in Kircher's Collegio Romano museum (Figure 1.5). Thanking August for the gift, Kircher claimed that the object displayed "the diligence of nature, which has showed a great deal of ingenuity in the arrangement of its artifice."[156] Nature seemed to have sculpted a living creature inside the stone. Its artifice led the intellect to presume that the object was the product of God's creative and imaginative powers.

Recent scholarship tends to emphasize the continuity between natural and human ingenuity. The engagement of artists and artisans with matter often appeared to early modern individuals as a form of partnership between nonhuman and human transformative agencies; interestingly, we also find nature donning the vest of a dutiful artist or artisan, using effort and diligence (*industria*) to distill a product worthy of being considered as "art."[157] This was a far cry from the ideal of effortlessness and naturalness famously advocated by Baldassare Castiglione (1478–1529) in the early sixteenth century; now, nature saw its creations become increasingly art-like, commodified as objects of intellectual and aesthetic appreciation.[158]

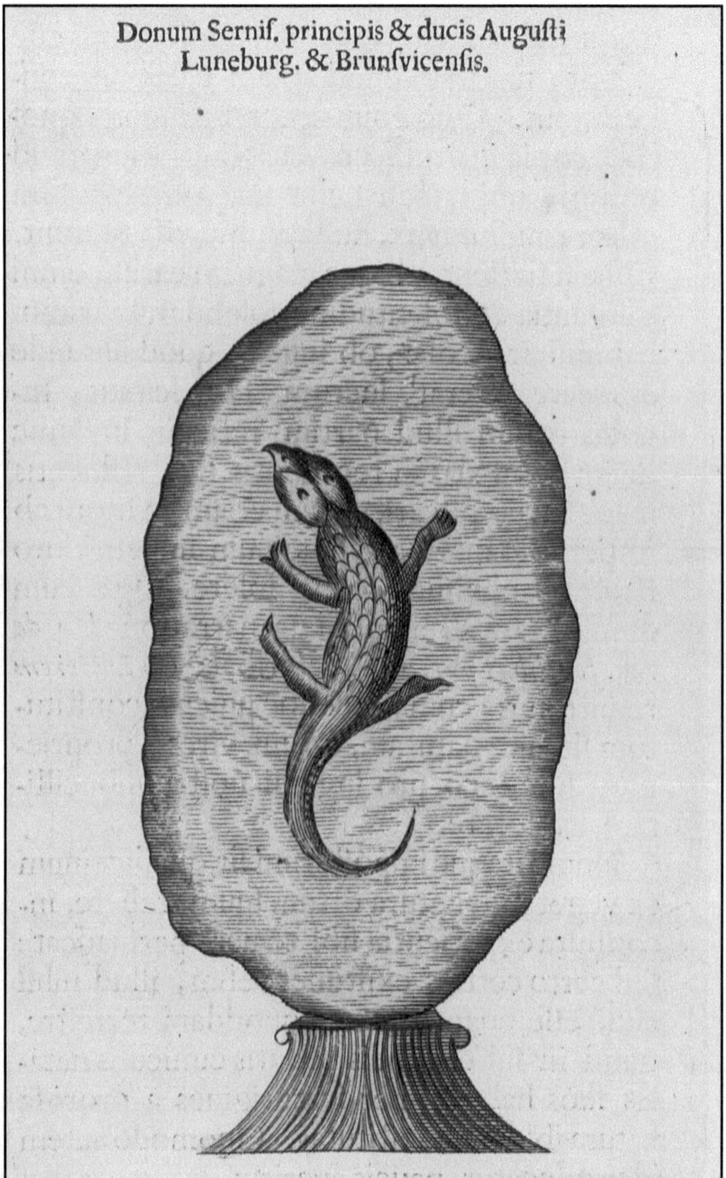

Figure 1.5. Lizard in amber that Athanasius Kircher received from August the Younger, Duke of Braunschweig-Lüneburg. Athanasius Kircher, *Athanasii Kircheri e Soc. Iesu Mundi Subterranei Tomus II.us In V libros digestus . . .* (Amsterdam: Johannes Janssonius van Waesberge and Elizaeus Weyerstraten, 1665), 2:76. Library Special Collections, Charles E. Young Research Library, UCLA.

The preternatural and the artificial also overlapped with each other in writings on conceits. Conceits were identified by Sarbiewski as being by definition *praeter naturam* but were referred to as "artificial" a few years later in Peregrini's *Delle acutezze*. The notions of artifice and ingenuity, which served Kircher in describing nature's involvement with the lizard enclosed in amber, were central to Peregrini's discussion of the pleasure afforded by conceits.[159] *Praeter naturam* beings and phenomena appear "artificial" in the eyes of humans, who construe them in opposition to what is "natural" and what occurs according to nature in its narrower sense. By way of analogy, Peregrini argued that conceits appear artificial to humans in comparison to the language of logic.[160] He concluded that in a syllogism, "it is not possible to form anything rare. In the absence of artifice, there is nothing valuable to be expected other than a good and clear syllogistic connection, such as might greatly satisfy one's intellect but not his or her ingenuity (*l'ingegno*). For that to happen, it is necessary that the middle term, or the end terms, or all of them, be artificial, and, as a rhetorician would say, figurative."[161]

Peregrini equated logic with nature in its narrower sense and conceits with the broad and hybrid sense of nature that had emerged at the intersection between the preternatural and the artificial. The terms of a syllogism, which represents logical, truth-oriented argumentation, exist in what Peregrini revealingly called "their natural condition." Consequently, they fit with each other in "a perfect union" that is nonetheless incapable of waking wonder due to its stable and static structure in which the whole, and the role of its parts within it, make sense from beginning to end. A syllogism resembles phenomena that occur "according to nature" in its narrower sense, including the repeated and predictable movements and positions of planets and stars, which Achillini had labeled "natural lights" for a reason. Conversely, the union of a conceit, which Sarbiewski had compared with the brief and elusive intersection between two sides of a triangle destined to depart again in different directions, was neither static nor stable.[162] In Peregrini's formulation, the terms of a conceit exist in a union that is not "perfect" like the one in a syllogism but is instead "very rare and conspicuous."[163] This is because they are "figurative" or, by another name, "artificial" or "artful" (*artificiosi*) and cause the audience to switch attention from the conceit to its author, who has resorted to ingenuity as he or she crafted the conceit's "artifice." A syllogism conditions the audience to care about only "the truth that it demonstrates," whereas a conceit prompts them to consider how it has been made. In it, the pleasure "does not derive from learning or seeing many things in just a few words, even if this happens as well, but mainly from the particular virtue of the poet's ingenuity, which shines marvelously in it."[164] Like Kircher's lizard in amber, a conceit delights the audience because its artifice acts as a hint of the author's ingenuity. It mattered little to Peregrini's contemporaries whether the author was human or not. After all, were comets not nature's conceits, as Emanuele Tesauro claimed in 1654?[165] The fact that they were not "natural" in its narrower sense may have bothered the likes of Achillini, but it opened new and promising dimensions.

Peregrini's claim that the audience derived pleasure from a conceit because they noticed the author's ingenuity was nothing short of revolutionary. It was a criticism of Aristotle's assertion that smart sayings (ἀστεῖα, a notion that Peregrini took to be the nearest predecessor of conceits in ancient rhetorical and poetic theory) delighted the audience because they used tropes as catalysts for effective and fast learning.[166] Peregrini countered that, in a conceit, figurative language did more than simply accelerate learning; it also rendered the whole artificial: the author forms or shapes a product entirely of his or her own making, rather than merely discovering something that already exists, which is what the intellect does when it arrives at some truth expressed in a syllogism.[167] Peregrini's distinction between making and discovering struck at the core of the ambivalence between the two meanings of the term *inuentio*, which in Latin means both "discovery" and "invention"; by claiming that the author's ingenuity or imagination "shapes" or "forms" something new that can be attributed to his or her own making, he also implied that the inventor of conceits resembled less the philosopher who discovers truths about nature through contemplation than the artisan and the engineer, who use knowledge about nature to create products that are different from nature but nonetheless exploit its properties and laws.[168]

The parallel between conceits and works of art and engineering might explain the concern with property and theft that surrounded these rhetorical and poetic devices. Pamela O. Long has argued that the notion of "intangible property involving inventions and craft processes" had emerged in late medieval guilds, and early modern painters and sculptors increasingly underscored "a view of original authorship as involving the creation of unique products resulting from the ingenuity of the creator or inventor."[169] For centuries, different notions of property and theft had been in use in connection with poetry—even Martial toyed with them to comic effect, lambasting those who dared to "steal" his lines and conceits—but they rarely reached the degree of concern that transpired from Alessandro Piccolomini's digression, in 1572, concerning who deserves to be called a metaphor's "first inventor," or the chapters dedicated by Jesuit professor Pietro Sforza Pallavicino in 1646 to summarize the strategies that enable a speaker to take inspiration from preexisting conceits without stealing them.[170]

Seventeenth-century writings on conceits made explicit the link between property and artifice. In 1652, best-selling writer Giovanni Battista Manzini asserted that statements that do not resort to figurative language and, as a consequence, lack artifice are akin to "common capital" and "money from the public treasury."[171] What is natural is also public, meaning that no one has the right to claim to have invented it, let alone to own it. In today's language, one could say that statements deprived of artifice are simply "not trademarkable."

Another example brings the point home. In a passage borrowed from Peregrini, Tesauro juxtaposed a theorem by Euclid with one of Martials epigrams; he argued that a theorem was true but could not be called Euclid's property in the same sense that Martial's epigram belonged to Martial. Tesauro concluded that only through the tricks of metaphor, with its component of untruth, does it become possible to make an argument "ours."[172] In conceits, figurative language served to shape rather than to reflect reality. Even as both Peregrini and Gracián were adamant that conceits were the result of the intellect's deep and successful "penetration" of a subject, it was by expressing discovery through the linkages of figurative language that they became recognizable as new creations.[173] Figurative language was a tool of imagination that compared and combined facets of reality with one another in ways that escaped logic because it was based less on what the intellect distinguishes as essential than on material similarities and accidents.[174] By rearranging reality through imagination, the inventor of conceits knew reality by transforming it into something new, in a way that placed him or her closer to the artisan who transformed matter through knowledge of its properties than to the philosopher in the act of contemplation.

In the invention of conceits, humans used artifice to become authors: artifice entitled them to ownership in a way that the discovery of truth did not. In doing so, they imitated nature, which Nieremberg saw as using artifice to create novelty for humans—or, at least, what humans would take as novelty—so that God might be recognized as author. If Nieremberg's assertion that "the most admirable thing in nature seems to be the one that imitates art" remained perplexing, writings on conceits suggested some possible keys to understand it. For instance, Peregrini claimed that artifice favored the encounter between the readers' and the author's ingenuity. By noticing the author's ingenuity in the artifice of a conceit, the readers found beauty in the act of recognizing a peer.[175] Just as the eyes find the greatest pleasure in the sight of another's beautiful eyes, "ingenuity [i.e., imagination], which is the mind's eye," Peregrini claimed, "cannot find a ray of beauty so full of flavor to its taste as the one that shines from someone else's ingenuity."[176] For Peregrini, a conceit was, first and foremost, a site at which the readers could see themselves in the mirror as they encountered another ingenious or imaginative self in action.

Manzini took a different path, asking, how is it that we naturally prefer that which is nonnatural? Drawing from Plutarch's *Quaestiones convivales* (*Table-Talk*), Manzini contended that conceits affect the intellect more successfully than true and logically valid statements because humans are "curious" and the artificial is "more sympathetic and agreeable" to them than the natural.[177] Manzini echoed Seneca's claim that nature has made humans lovers of novelty, suggesting that nature has given us a proclivity to prefer the artificial, demonstrating for the umpteenth time that novelty and artifice were two sides of the same reality and confirming that the divide between nature and art mattered less than the one separating nature in its narrower sense from a composite and heterogeneous group comprising nature-made *praeter naturam* beings and phenomena, as well as human-made objects. In fact, Manzini argued that it is natural for humans to prefer that which does not align with nature in its narrower sense. He had read in Plutarch that when a child was presented with two pieces of bread of the same dough, one shaped as a loaf and the other as a dove, he picked the latter at once.[178] The tropes in a conceit, one can surmise, are analogous to the act of shaping bread in unconventional ways.[179]

Ten years earlier, in *Arte de ingenio*, Gracián had turned to a fable to ground the inclination of humans for artifice in historical and even anthropological terms. In a primeval or Edenic moment, Truth was the intellect's legitimate wife and all it needed for pleasure. Accused by Untruth of being uncouth and unpleasant, Truth had to look elsewhere for support until she found counsel in Ingenuity. "Light is harmful to the sight when it strikes it directly," Ingenuity told Truth. "You should behave politically," she advised.[180] At once, "Truth opened her eyes. She decided to use artifice. From then onwards she resorted to inventions; she finds her way through meanders, wins through stratagems, paints as far that which is nearby, represents in a subject what she wants to condemn in another, aims at one target to hit another, sows confusion in people's passions, dissimulates her affects, and by ingenious circumlocutions she always hits the mark."[181]

In 1618, nature resorted to comets for the same reason that orators and poets used conceits: both comets and conceits appealed to postlapsarian humans, who are sympathetic to artifice because they have fallen from a state of innocence in which the intellect was happily married to truth.[182] Similar to Grassi and Nieremberg, who had embraced the ability of novelty and artifice to cater to our sensory and intellectual limitations, Gracián did not see any need to feel regretful. Instead, he embraced conceits in a pragmatic, proud, and unapologetic acceptance of humanity as it is. Like Peregrini, Gracián believed that conceits struck the intellect as beautiful. He was convinced that, in order to accurately understand the taste for conceits, it was important to fashion a theory that took into account the realities of a flawed and imperfect sensory and intellectual apparatus characterized by its loss of taste for truth.[183]

Flemish antiquarian and historian Justus Lipsius (1547–1606), a name familiar to Gracián, wrote in 1589 that political and social life occurred not in the ideal and abstract conditions of Plato's *Republic* but rather "in Romulus's dregs" (*in Romuli faece*), referring to the use of violence and deceit by Rome's first king as described by historian Livy (59 BCE–17 CE) and later commented by Niccolò Machiavelli (1469–1527) in *Discorsi sopra la prima deca di Tito Livio* (*Discourses on Livy*, 1531).[184] Ingenuity's advice in Gracián's *Arte de ingenio* that Truth should behave *politically* echoed a pragmatic and disenchanted view of society that also permeated Gracián's *El héroe* (1637) and *El discreto* (1646). In 1647, Gracián opened *Oráculo manual* by ironically celebrating that humans finally stood at the peak of a transformative process characterized by mounting opacity and deviousness.[185] Gracián's writings on life at court denounced the use of direct and transparent communication as unjustifiably imprudent. The wise can be direct with one another: they have forsaken personal interest and seek only what is good and true. In the world of civil and political interactions, however, one should act and speak with artifice. Lipsius was well-known as the editor of Cornelius Tacitus (ca. 56–ca. 120), a historian that Gracián revered because he was able to lay bare the hearts of rulers and courtiers.[186] Gracián's approach to conceits was eminently pragmatic: he took the minds and the senses of fallen men and women, rather than ideal ones that exist only in myth, as the subject of his investigation into cognition and aesthetics. Gracián sided with the artificial and devious appeal of conceits rather than with logical, truth-oriented language because it suited the realistic and disenchanted take on humanity that he adopted in the works he wrote about life at court. For him, conceits embodied a rhetorical and poetic counterpart to the world that Tacitus described for the realm of politics.

Conclusion

The analysis of natural-philosophical works alongside writings on rhetoric and poetics confirms Achillini's insight that, for seventeenth-century individuals, natural and human ingenuity were entangled with one another, for better or for worse. Achillini's label "poetry's comets" was more than just a condemnation of conceits—which, it is worth mentioning, Achillini also cultivated with both enthusiasm and success. The coinage was also a testament to the belief that all of nature was not equally "natural," a view ratified ten years later when Grassi told his audience that humans are built in such a way that they find nature's digressions from its usual course irresistible. Comets were a prime example of that, and conceits were the rhetorical and poetic counterpart of comets, materializing as they did the efforts of speakers and writers to fashion themselves after nature's example. At stake was the right of humans to exploit, not resist, the intellect's innate fascination for things unfamiliar and new. Why restrict one's appreciation and imitation of nature to its narrower sense, that is, its usual course? António Lopes da Veiga (ca. 1586–ca. 1656), a skeptical and embittered detractor of conceits based in Madrid, adopted a conservative take on the issue in 1641 when he inveighed against contemporary poets who increasingly preferred what was possible over what was likely, abandoning the path of what "usually" occurs in favor of what is rare and far-fetched, albeit "within nature's power." Only what is likely is sure to strike every reader as "contingent rather than monstrous," Lopes claimed.[187]

There was a part of nature's behavior that was possible but not usual. It created dissonance in the intellect in a way that, as Sarbiewski argued of conceits, aroused wonder at the novelty and eventually admiration for the author's ingenuity and the artifice that transpired from his or her creation. As Nieremberg maintained, when nature behaved artificially or artfully it was more deserving of wonder, prompting onlookers to reflect on its role (or that of God) as artificer and author. If Nieremberg, just blocks away from Lopes, put out a number of publications that focused on extraordinary, nearly incredible phenomena, why ask that poets restrict themselves to nature in its usual course? A reasonable but insufficient answer would be that Nieremberg is likely to have subscribed to Lopes's complaint. After all, in the previous year, 1641, he had warned preachers against seeking applause through a display of ingenuity based on conceits. Unlike God, they did not have sufficient reason to call attention to their inventiveness and talent.[188]

Not everyone agreed, though. As he delivered a sermon in 1626, when he was still a member of the Society of Jesus, Tesauro admitted to feeling like a traveler who, "after a voyage full of perils, bringing the spoils of some ultramarine monster, unveils it before a stunned gathering of people and obtains muted applause from astonished lips."[189] Three decades later, he would call monsters "nature's conceits" ("argutezze della natura"), confirming that, in a world shaped by courtly habits and fashions, the speaker who used rhetorical and poetic devices was somehow engaging in a contest with nature, which brings us full circle to Eamon's remark at the start of the present chapter.[190] Grassi probably felt similarly when he paired comets with conceits—that is, poetry's comets—in 1619. Like a menagerie of monsters, conceits enabled the orator to appear as a merchant of curiosities captivating the audience's attention and desire. In Grassi's lecture, conceits and nature—represented by the topic of comets—thus coalesced into a conscious and masterful self-fashioning effort.

Grassi's meditation on the link between scientific inquiry and novelty suggests that he had a keen eye for topics that were easy to market as well as likely to heighten the intellectual and social capital of the speaker and the institution to which he belonged. Grassi's previous publication, printed under the name of a student in 1617, was a lecture on the rainbow prefaced by a parade of conceits similar to the one that he would use two years later.[191] If there was such a thing as preternatural philosophy, one could argue that Grassi enacted a sort of preternatural rhetoric and poetics to go with it.[192] Comets and rainbows had great potential for building nonspecialized audiences ready to wonder at the novelties of nature and of the orator's language, doubly bewildered at two different but analogous forms of dissonance. In 1589, engineer Bernardino Baldi (1553–1617) recalled that ancient myth considered the rainbow, the topic of Grassi's lecture in 1617, the child of Thaumas, meaning "wonder."[193] Baldi also noted that the ancients called engineers "thaumaturges," or "makers of wonders," and the discipline's foundational work, Pseudo-Aristotle's *Mechanical Problems*, claimed that machines worked effects "παρὰ φύσιν," a phrase that early modern scholars from Niccolò Leonico Tomeo to Henri de Monantheuil consistently translated as *praeter naturam*.[194]

Conceits were just one of a number of ways—along with fireworks, automatons, and countless products of art—in which humans could use ingenuity to rival nature's ability to arouse wonder.[195] Grassi himself was evidence that the same individual could pursue different paths as a maker of wonders. He had been living in Genoa for some time when Peregrini's *Delle acutezze* was disseminated there in 1639, and the news of the publication might have reached him as he worked on a project to design and build a ship that could never be sunk. The project never came to fruition, unlike the church of Saint Ignatius adjacent to the Jesuit Collegio Romano, for which he served as the architect in charge of drawing the plans and of supervising much of the construction.[196] Grassi the orator, like Grassi the engineer and Grassi the architect, was not a stranger to the dream of becoming one of Baldi's makers of wonders.

What are deforming mirrors, Tesauro asked in 1654, but conceits of optics? What are fireworks but "a strepitous metaphor" that imitates lightning, thunder, and comets?[197] Comets also resembled conceits. In them, nature "with mysterious artifice" conveyed its secrets by means of figurative language. The comet's shape, in conjunction with the figure represented by the constellation in which it first appeared, created a game of combinations that nature relished, behaving like a poet "full of ingenuity" with pointed disdain for plain language.[198] Yet despite the fact that both nature and humans resorted to artifice, it also had a dark side. "Artifice," a term often synonymous with trickery, elicited suspicions of lurking deceit. The presence of artifice in conceits cast a dubious light on them. It is revealing that in 1648, in a revised and expanded version of *Arte de ingenio*, Gracián replaced the tale of Truth and Untruth with a story borrowed from Mateo Alemán's picaresque narrative *Guzmán de Alfarache* (1599 and 1604). Alemán (1547–1614) belonged to a circle of Tacitus enthusiasts, passionate students of the historian's deft analysis of political maneuvers and secret intentions, at times as ignoble as those of the rascals in *Guzmán de Alfarache*.[199] Gracián was also fond of the similarities between the high and low forms of artifice enacted by courtiers and rogues, and in *El Criticón* he depicted a world in which artifice and trickery infused the actions of humans, regardless of social status.

According to Hans Blumenberg, for Gracián, the climate of opacity that permeated social and political interactions did not affect nature, which, in *El Criticón*, remained a book that humans could read without fear of deception.[200] Yet in the eyes of most seventeenth-century natural philosophers, nature was anything but transparent. Even Nieremberg had to conclude that even if nature does not aim to deceive, it nonetheless walks meandering paths that are often indirect and less than evident, similar to a ruler who resorts to calculation in order to appear powerful. Nature, Nieremberg argued, went out of its usual course in order to "keep its majesty."[201] Nieremberg's language was highly reminiscent of the literature on reason of state or political expediency that filled seventeenth-century bookshop shelves throughout Europe. He certainly meant to be pious when he asserted that, to gain respect, nature kept the extent of its powers secret, so as to catch humans, who are familiar only with its usual course, unprepared when it suddenly resorted to artifice. Yet nature's methods were surprisingly close to Gracián's advice to a courtier in 1637 that he should not allow anyone to fathom "the extent of his capacity," and they certainly aligned with the advice from Ingenuity that Truth should behave "politically."[202]

Artifice, secrecy, and deceit pervaded the language that seventeenth-century individuals employed to speak about nature as much as the language that served to describe people's speech and actions, revealing a deep and complex entanglement between contemporary understandings of natural and human ingenuity. In chapter 2, I explore the relationship between conceits and optical devices, perceived by Grassi's contemporaries as two instruments that were similarly fitting to reveal what lay beyond reality's artifice, deceit, and secrecy. The invention of new technologies of vision, such as the telescope that enabled Grassi to observe the third and last comet of 1618, gave rise to a trope that portrayed the study of nature's artifice as akin to a form of espionage, the furtive peering into secrets to which the observer, perhaps, had no right. In the debates that followed, conceits, pictured as lenses of the mind instead of the eye, also played a role.

CHAPTER TWO

Gossip Made of Glass

On Artifice and Deceit

When poet and diplomat Bartolomé Leonardo de Argensola (1562–1631) pleaded "Look at me with mercy, Phyllis, and let the comet burn / that stares at us with its pale light," it was easier said than done.[1] Even for courtiers playing at shepherds, the fall of 1618 was hardly a time for love. As Europe plunged into a destructive, thirty-year-long war, the sky suddenly came to life with three comets that appeared in rapid succession. Comets were proverbial harbingers of misfortune, and by the start of 1619, the printing workshop of Pedro Craesbeeck in Lisbon issued a cascade of works that addressed the novelties in the sky. Pedro Mexía predicted ominous events.[2] Antonio de Nájera announced calamities among "the barbarians of Africa and the heretics from the north" and prompted Catholic rulers to annihilate the enemy "at low cost."[3] Following a different approach, Manuel Bocarro Francês analyzed the measurements of the second comet's trajectory and reached the conclusion that Aristotle had been wrong to claim that comets were formed in the atmosphere. They were heavenly bodies, as Tycho Brahe had argued in 1577.[4] Mathematician Orazio Grassi also ratified Tycho's position as he lectured on the third and last comet at the Jesuit Collegio Romano in the winter of 1619.[5]

The reaction to the first comet had been modest when it made its appearance in August 1618. "It blazed for the blind," Grassi recalled.[6] Speaking to his audience, he feigned surprise: Was it too dim and small? Or perhaps astronomers had lost interest in a sky recently depleted of its secrets? "No part of the sky escapes our sight anymore, and the beauty of the moon is not as great as it used to be," Grassi recited. "We were able to peek at the dances of Venus and Mercury, and who is there who does not blush at the sight of the sun, who at times shows himself ugly? We laid bare the snares of Mars as it approaches the earth, and exposed the minions of Jupiter and Saturn, who had tucked themselves away in vain until now. Comets alone," he pithily concluded, "remained out of reach for these lynx eyes."[7] With its proverbially sharp and penetrating sight, the lynx had become a metaphor to name the telescope, the device that enabled Galileo Galilei to perform the discoveries he later publicized in *Sidereus Nuncius (A Sidereal Message)* in March 1610. Those in Grassi's learned and partly aristocratic audience were certainly aware of these and similar recent breakthroughs in the discipline. They also knew that astronomy was not out of business and were probably amused by the claim that the invention of the telescope meant that its practitioners were, as Grassi put it, "no longer . . . bleary-eyed from continual watching of the stars," contemptuous of a sky now torn open, and thus failing to pose new challenges to them.[8]

The audience recognized, and some perhaps even appreciated, Grassi's lavish use of rhetorical conceits—statements that, as Matteo Peregrini would explain twenty years later, resorted to tropes to strike wonder and stir admiration for the author's ingenuity.[9] By using tropes, conceits linked different facets of reality through material and circumstantial similitudes, forsaking the rules of logic and thus creating mere appearances of truth. Yet appearances are at times as meaningful as reality, if not more, and by imagining the telescope exposing the secret lives and intentions of celestial bodies, Grassi struck a note with the audience as he drew on connections that were socially and culturally relevant to early seventeenth-century Europeans. Grassi's passage reworked the discoveries recently made by Galileo into a tale of counterfeit beauty, peppered with episodes of thwarted political or tribal violence. To start with, the telescope revealed embarrassing truths about the face of the moon, which the etchings in *Sidereus Nuncius* had transformed from a smooth, polished surface into one "covered all over with huge bumps, deep holes, and chasms."[10] The moon, Grassi cleverly suggested, had been using distance to conceal the imperfections that now it could no longer deny. The sun was similarly at a loss now that its face was known to be covered by a number of blemishes. When Christoph Scheiner, a claimant of the honor of first discovering them, went to extraordinary lengths to demonstrate that these were actually satellites in its orbit, he managed to persuade part of the community of astronomers, but the sun's reputation had been damaged forever.

It was bad enough to disguise oneself. Yet the actions of Venus and Mercury, caught orbiting the sun rather than the earth, were more transgressive, surpassed only by a pair of discoveries that Grassi, strategically, left for the end, and which he saw fit to describe in military and political terms.[11] Alluding to a recent consensus that Mars's orbit was elliptical rather than circular, he spoke of the planet's "snares" in reference to the moment in which Mars—who was, let us not forget it, the god of war—threateningly approached the earth at remarkably close quarters.[12] Last on Grassi's list was the discovery of four moons orbiting Jupiter, which Galileo had dedicated to Cosimo II Medici in 1610. In a celestial court animated by gallantry, dance, and both military and political scheming, it was hardly surprising to catch the king of the gods and planets, Jupiter, accompanied by what the orator depicted as a band of minions ambushed in a thicket of interstellar bushes. Grassi dramatically portrayed the discovery of the satellites as if a raid on Jupiter's environs had found the band of accomplices ready to commit some crime, luckily exposed by the acuity of the telescope. The same applied to Saturn, which appeared to have two attendants hanging around, in reference to two bodies that, as Christiaan Huygens concluded in 1655, were actually rings.[13] Grassi crafted the telescope as a device that pried into the lives of the planets, conveniently imagined as a host of courtiers and rulers.

As indicated in chapter 1, conceits were self-fashioning tools, capable of striking surprise by subverting earnest and logical reasoning through the use of one or more tropes. Conceits followed in the footsteps of nature, which was likewise fond every now and then of forsaking its usual course and bringing about phenomena—such as comets—that were indubitably nature's offspring but remained outside the purview of what humans, constrained by their intellectual and emotional limitations and habits, recognized as nature's usual course, its more familiar and predictable behavior.

Yet there was a dark side to the analogy between natural and human ingenuity. For instance, when poet Claudio Achillini claimed in 1609 that conceits were poetry's comets, the coinage implied not only that both conceits and comets were similarly apt to arouse wonder because of their novelty but also that they were both tainted by suspicion of their artful or artificial character.[14] Humans treasured artifice, but they loathed it as well, due to the belief that no one goes out of his or her usual course unless for a reason. As Aristotle once claimed, people distrust artifice because they fear it is meant to deceive them.[15]

In what follows, I explore the role of deception in a provocative and widespread analogy that thrived in the decades after the invention of the telescope, which maintained that conceits and optical devices used different but comparable forms of artifice to prosthetically supplement one's "natural" visual and mental abilities. Through analysis of writings on natural philosophy, optics, and conceits—especially Baltasar Gracián's *Arte de ingenio* (1642) and Emanuele Tesauro's *Il cannocchiale aristotelico* (1654)—in parallel with works of fiction, I demonstrate that conceits often appeared to Grassi's contemporaries as telescope- and microscope-like devices that benefited from increased acuity or sharpness—intellectual rather than visual—to penetrate a reality permeated by concealment and secrecy. Put simply, if optical devices allowed humans to see beyond the reach of the eye's possibilities, and tropes were rhetorical and poetic counterparts of the magnification provided by the lens, then it followed that conceits enabled the intellect to reach where the syllogisms of logic could not.

In the first section, I argue that the claim that conceits attained otherwise inaccessible recesses of reality resonates with contemporary intellectual and ethical concerns that optical devices breached the boundaries of perception that God had set for a reason. In the second section, I focus on the claim that lenses, operating in the guise of conceits, displayed objects in a more distinct and accurate way, while also showing them to be "different" and "new." I claim that artifice short-circuited the equation between sight and understanding that had been crucial since Plato.[16] With a certain circularity, scholars justified using artifice—either conceits or optical devices—as a reaction to the belief that the objects they sought to see or understand more fully were themselves products of artifice. Put simply, it was imperative to see beyond appearances in a world in which everywhere everything—human or not—seemed intent on passing unnoticed, or appearing to be something it was not. The use of indirectness to undo the object's attempts at eluding the beholder, however, as I show, complicated the distinction, central to Peregrini's explanation of what made syllogisms different from conceits, between discovery of what was already there and invention of something new. What does it mean for conceits to penetrate reality and, in doing so, to create something new? In the third section, I claim that the possibility of countering artifice with artifice became, for both Gracián and Tesauro, a reason to celebrate rather than lament human sensory and intellectual limitations. Instead of bewailing the loss of Paradise, and the fall from a state of nature in which humans supposedly enjoyed direct access to nature and one another's hearts, Gracián and Tesauro espoused conceits as invitations to transform the challenge posed by a secretive and elusive reality into a source of pleasure. The omniscience of God and the ability of angels to communicate directly and without language never appeared less alluring than in the age of artifice and subtlety.

A Breach of Secrets

Grassi's list of discoveries probably took its cue, even if only indirectly, from Galileo's *Sidereus Nuncius*. The title, which one might translate as *A Sidereal Message* or *A Sidereal Newsletter*, connected Galileo's publication to the world of *avvisi*, papers that carried intelligence and gossip from place to place. When the ambassador in Venice to King James I of England hailed the book as "the strangest piece of news," he merely acknowledged the blend of familiarity and surprise that Galileo's book could cause to contemporary readers who found, coded in a format usually dedicated to secrets of state, diplomacy, and war, a series of previously unknown facts about the secret lives of planets and stars.[17]

Galileo's title conjured a link between natural philosophy and politics as regimes in which information sharing and suppression are acts of power. None of this was lost on Grassi, as the string of conceits depicting the sky as a court clearly proves. Enthusiasm and concern surrounded *avvisi* in equal measure. The constitution *Contra scribentes*, signed in 1572 by Pope Pius V, decried the dangerous and unpredictable effects of news circulation on political and social stability, comparing it to libel and *lèse-majesté*. "Government officials," it has been argued, "sought to control what they regarded as a dangerous breach of secrecy and threat to reputation."[18] Likewise, with the publication of *Sidereus Nuncius*, the deniers, sensing a fatal blow to long-held beliefs about the reality of celestial bodies, condemned Galileo's news as slanderous and harmful to God's name as the creator of the sky.[19]

The astronomers at the Jesuit Collegio Romano deferred a verdict on the novelties until they could see for themselves with the aid of a reliable telescope.[20] After they did so, on May 18, 1611, with Galileo as a guest, Flemish astronomer Odo van Maelcote (1572–1615) delivered a lecture in which he listed and updated the discoveries of *Sidereus Nuncius*.[21] Yet when another professor speaking at the same institution celebrated Galileo before the end of 1611, the newsletter from the sky, by then a relatively familiar and well-codified genre in European academic circles, looked markedly different. From the lectern, the anonymous orator spoke of a moon sculpted with angular forms and irregular textures, a crowd of watchful stars where a mist of light used to be, Venus imitating the moon as she displayed growing and waning light, and Saturn transformed into a threefold star. Last, "the planets" dedicated to Cosimo II Medici accompanied Jupiter.[22] The anonymous orator named each of Galileo's discoveries by means of conceits, effectively entertaining listeners who no longer took them as news. He capitalized on a thirst for novelty that straddled the consumption of news, natural-philosophical discoveries, as well as wonder-arousing rhetoric. Grassi, who as a student might have been in the audience in 1611, would follow the same model in the winter of 1619.[23]

Ofer Gal and Raz Chen-Morris have argued that early modern representations of astronomers and other individuals performing scientific or artistic tasks requiring instruments often showed them absent-minded, as if they were mulling on the implications of instrumentality and its mediation between subject and object.[24] Grassi's list also activated associations that were merely latent in Galileo's fashioning of *Sidereus Nuncius* as news, with conceits transferring the attention of the audience from the discoveries to the artifice of the telescope that had made them possible in the first place. Portraying the moon, the sun, and the planets as courtiers was not just comical within a culture keen on conflating natural philosophy with political and social forms. Grassi's conceits simultaneously voiced and dismissed with humor concerns over the effects of artifice on the observation of nature. The telescope peeked through the boundaries that until recently had blocked humans from access to nature's secrets. It resembled a spy in charge of unmasking the dissimulation that enables courtiers and rulers to create social and political capital—only now it was about undoing the dissimulation of nature. As a metaphor, optical devices channeled social and political anxiety.[25]

Others had raised alarm, either jokingly or in earnest, about the telescope's transgression. Answering to Scheiner's *Tres epistolae de maculis solaribus* (*Three Letters on Sunspots*, January of 1612), a treatise *De maculis in sole animadversis* (*On the Spots Observed on the Sun*) published anonymously in Leiden mere months later and now attributed to Dutch astronomer Willebrord Snellius (Willebrord Snel van Royen, 1580–1626) prefigured Grassi's concerns and conceits. Snellius suggested that, armed with a telescope, humans had lost any reverence for what lay beyond the limits of eyesight: "With little fear, we inspected the throne of Jupiter," he wrote. "We explored in great detail the full face of the moon with all its stains, and we counted almost every star . . . just as one would count coins."[26]

The telescope enabled direct observation of a range of phenomena that for millennia had remained the province of hypothesis. Snellius wrote only three decades after Michael Maestlin (1550–1631) declared that astronomy would remain doomed to conjecture unless it enjoyed direct access to the realities it studied. Maestlin likely had in mind the trip to the sky of philosopher Menippus imagined by satiric writer Lucian of Samosata in the second century CE.[27] In Lucian's *Icaromenippus*, the moon is barely able to dissimulate her irritation as

Menippus's colleagues incessantly craft baseless speculations about her, and begs to be left in peace.[28] The invention of the telescope carried arrogance and curiosity even further. The way Snellius described it, its artifice reeked of both sedition—with the assault to Jupiter's throne—and greed, as it transformed the stars into coins. Now it even dared to target the sun.[29]

Snellius was ironic, even playful, and the awareness that discovery implied a performance of power did not make him anxious. In *Tiphys Batavus* (1624), a treatise on navigation that gestured in its title to the pilot of the ship *Argo*, he opened with a panegyric of Phoenician ingenuity. The city of Tyre had first supplied Mediterranean ports and cities with a surfeit of commodities. One day, hubris seized them. "Overcoming geographical restrictions and slipping away through the Red Sea, [Tyre] dared to survey the Asian and African coasts and try the outward seas, with nature almost resisting and unwilling."[30] Resorting to a conceit, Snellius construed the absence of a passage from the Mediterranean to the Red Sea as a sign of nature's reluctance to allow Tyre to extend its reach. Landmasses posed an obstacle to navigation as distance did to astronomy, and in a feat of arrogance, artifice overcame what in either case seemed impossible.

Snellius's ethical and theological misgivings were only a pretense. They allowed him to stress more effectively the benefits of Phoenician hubris, which included the invention of the alphabet and astonishing developments in arithmetic, navigation, and geography. A republic of merchants and navigators needed to look no further for a model. Dutch authorities should want nothing else than to support a treatise that was ultimately aimed at fostering exploration and trade in the context of imperial expansion.[31] In 1611, friar Tommaso Campanella (1568–1639) similarly called Galileo "a new Tiphys" comparable to Christopher Columbus and Amerigo Vespucci. Astronomy and navigation built on a similar entanglement of knowledge and violence articulated as a transgression of nature's boundaries.[32] Snellius suggested as much as he devised conceits that portrayed both the telescope and navigation as forms of artifice enabling humans to go beyond the state of nature—restricted to the naked eye, in the case of the telescope, and to autochthony and travel by land or near the coast in that of navigation—and to do so despite nature's opposition. The mention of astronomers counting stars as if they were coins reinforced the link between the two disciplines, since it was greed that, as moralists and poets often repeated, made humans risk danger at sea.[33]

Before *Sidereus Nuncius* transformed the telescope into the loud and provocative hero of a revolution in astronomy, the notion of a device that might enable humans to see from afar populated countless rumors and legends. Consciously or not, the conceits by Snellius and Grassi invoked desires and anxieties from a time when the destiny of the telescope was far from clear even to Galileo, who, as late as the summer of 1609, demonstrated it from the top of Saint Mark's bell tower with the aim of persuading the senate of Venice of its use as a tool of surveillance. In fact, Galileo's display was just a sliver of a history, studied in detail by Eileen Reeves, that goes back to ancient and medieval legends and purported intelligence regarding a device, variously described as a mirror, a lens, or a combination of the two, which could be used to spot enemies from a distance and spy on others.[34]

Grassi's conceits, however, differed from those of Snellius in the distribution of agency between the telescope and its object, celestial bodies. In Grassi's conceits, what Snellius depicted as arrogance in the observer appeared partially justified on the grounds that celestial bodies also shared in the guilt. Humans merely used artifice to undo nature's artifice, its use of simulation and dissimulation to appear otherwise than it actually is. Rather than a scene of violence and exploitation, Grassi portrayed a chapter of subject-object interaction marked by sophisticated competition. The telescope appeared as a form of artifice apt at fighting nature's artifice, which period sources were keen on describing as a labyrinth.[35]

It was appropriate that Grassi used conceits to comment on the telescope's artifice. Like the telescope, conceits were instruments that resorted to artifice ostensibly to undo nature's artifice. Owing to a trite but productive analogy between sight and intellect harkening back to Greek philosophy and materialized in the metaphor of "the eye's mind," to see and to understand were often treated as interchangeable terms. Conceits served as counterparts of optical devices in the mind of Grassi's contemporaries because conceits applied artifice, in the form of tropes, to intellectual reasoning, while optical devices transformed sight by means of mathematical, specifically optical, artifice. Thanks to artifice, the telescope was able to penetrate otherwise inaccessible aspects of reality but in a way that rendered it suspect. Likewise, artifice damaged the chances that conceits would be accepted as valid competitors of logically valid, nonartificial intellectual penetration, embodied by the earnest, truth-seeking pursuits of philosophers and theologians, who also fashioned themselves through analogies with sight. For instance, in 1598, preacher Francisco Terrones del Caño (1551–1613) told a congregation mourning King Philip II of Spain that the acuity of sight associated with the eagle symbolized wisdom. Terrones would have balked at the possibility, materialized a few decades later in Gracián's *Arte de ingenio*, that the eagle could be included in a bestiary of sharp-sighted animals signifying the invention of conceits, which, for Gracián, had replaced wisdom as "the intellect's most beautiful offspring."[36]

Although the nomenclature of conceits included references to their sharp or acute character—*acutum* in Latin, *acutezza* in Italian, *agudeza* in Castilian—what exactly was sharp about them remained a matter of debate. In 1626, Jesuit poet and rhetorician Maciej K. Sarbiewski imagined a conceit to be a pointed and piercing statement. Just as wedges, swords, and arrows are said to be "sharp" because they have two sides that converge into a sharp or acute angle, so "a material similitude" made it acceptable to claim that a conceit was likewise acute. A conceit "pricked" the mind in a fresh and delightful way by manifesting the agreement of two sides of the same reality that otherwise seemed to contradict one another.[37] It was like a thorn capable of awakening a dormant mind, as suggested by one of Sarbiewski's interlocutors in Rome.[38] There was also a second and more controversial interpretation of the issue that would gain currency in the wake of Matteo Peregrini's *Delle acutezze*. Peregrini located a conceit's sharpness or acuity in the author's intellect at the moment of invention. He claimed that a conceit "pricked" the audience's intellect because it proved the author's "ingenuity to have penetrated [the topic] greatly and successfully, a condition of sharp/acute objects."[39] Sharp tools are suitable to carve sharp objects; likewise, conceits are sharp only if the mind that invents them is also sharp.

In a world in which philosophers and inventors of conceits competed for the title of "sharp-minded," the former rued the idea that conceits, the toys of orators and poets, could serve as evidence of intellectual penetration. For instance, in 1644, Jesuit professor Baltasar López praised the "sharp," "lynx-eyed" intellects who, investigating the truth of a subject, firmly "enter into what is deepest and innermost about it, becoming immersed in the marrow of things."[40] López warned the audience at Mexico City's College of Saint Peter and Saint Paul that he was not referring to conceits (*argutiae*), which are sharp only the way a needle is sharp. Unlike a sword, they tickle the skin but penetrate nothing.[41]

The insistence on the lynx and its proverbial ability to see through solids, with Grassi referring to the telescope as "these lynx eyes" and López praising the philosopher's "sharp," "lynx-eyed" intellects, is evidence of a paradigm that understood both visual and intellectual apprehension as forms of penetration, with its attendant distinction between out- and inside. In Latin and in European vernaculars, penetration expressed comprehension, on the assumption that what can be seen from outside is, by definition, incomplete or untrue.[42] To see inside is to grasp fully. Thus, in 1619, Francesco Ingoli balanced the criticism of what he believed were Aristotle's erroneous views about comets with a hint of respect by calling him the one who "saw from within the hidden and innermost aspects of nature."[43] One needs sharpness or acuity in order to fully understand a complex and intricate reality. One needs to be subtle too. "Subtlety" denoted the quality of a delicate and thin body that insinuates itself "inside," as if sliding through pores and interstices.[44] "Acuity" and "subtlety" were often used in reference to logical or mathematical abilities, enriching but also complicating their application to the invention of conceits.[45]

As avatars of the lynx and its perspicacity or ability to "see through," the telescope and the microscope inherited a history that predated them by thousands of years, fraught with layers of meaning and often contradictory expectations and concerns.[46] First adopted as a personal and intellectual emblem by natural philosopher Giambattista Della Porta (1535–1615), the lynx later became the namesake of the Accademia dei Lincei (Academy of the Lynx-Eyed), founded by Federico Cesi, Francesco Stelluti, Anastasio De Filiis, and Johannes van Heeck in Rome in 1603, which in 1610 inducted Della Porta as one of its members.[47] In 1630, Stelluti (1577–1653) clarified at length the connection between the lynx and the academy's program for the investigation of nature (Figure 2.1). According to him, Cesi chose the lynx

> so that it might be a lasting stimulus and goad to keep in mind the acuity of sight—not of the body's eyes, but those of the mind, which is indispensable for the contemplation of nature that we profess. This is even more so because that contemplation is all about penetrating the interiority of things in order to learn about their causes and the operations of nature, which works internally—just as, with a beautiful simile, it is said that the gaze of the lynx sees not only the outside of things, but also that which hides inside them.[48]

The lynx, which Pliny the Elder (23/24–79 CE) associated with the acuity of sight, conjured the notion of nature as working "inside," aloof from prying eyes.[49]

With the exception of Pliny, however, the ancients signified perspicacity by referring to the ship *Argo*; just as its pilot Tiphys was Snellius's emblem for the exploration of uncharted territory, so its lookout Lynceus recalled visual acuity. It was either the influence of Pliny or the lack of familiarity with the figure of Lynceus—and perhaps a popular, unsubstantiated etymological connection between "Lynceus" and "lynx"—that eventually caused the animal to displace the character. Thus, by the thirteenth century Albertus Magnus could state

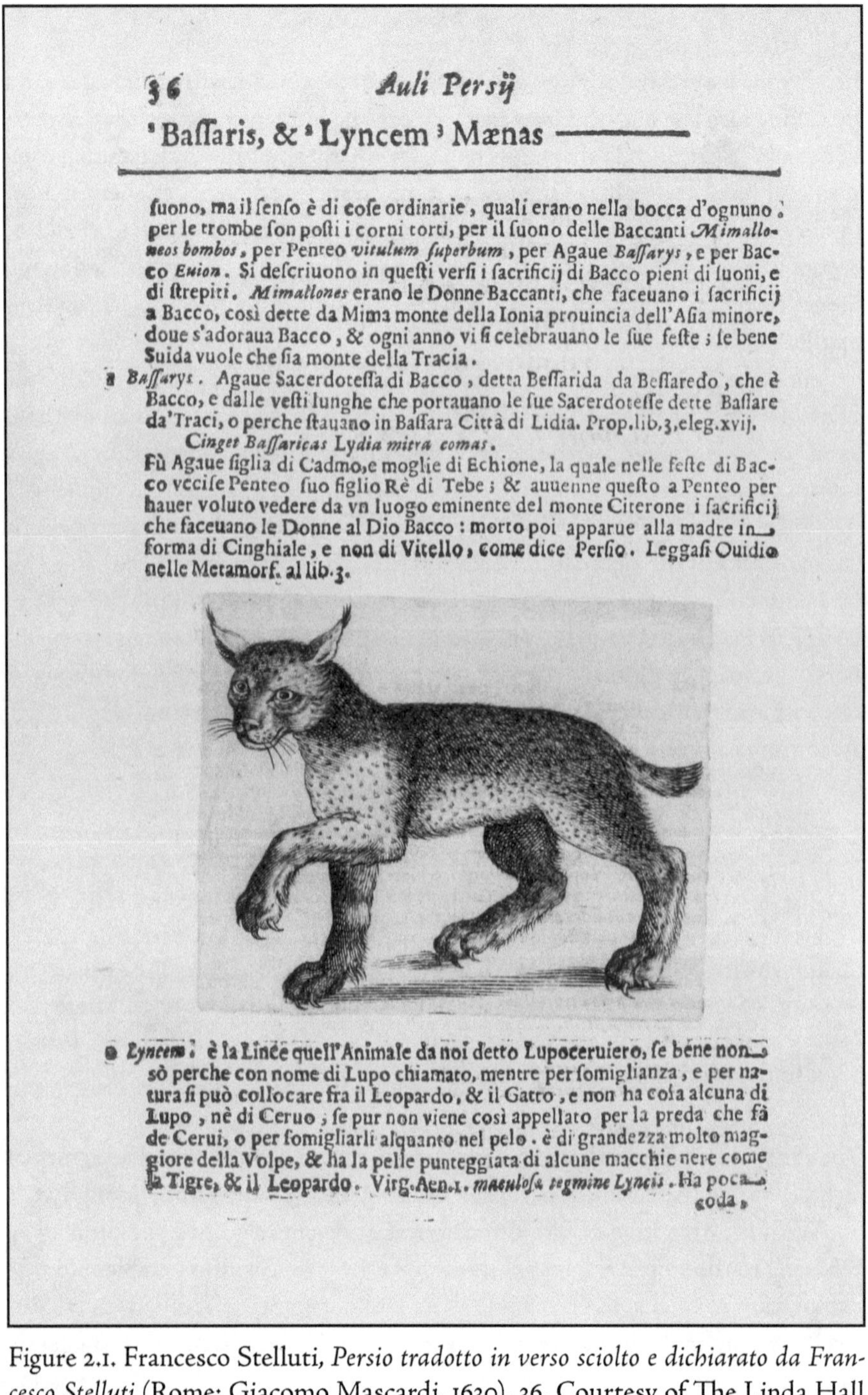

36 *Auli Persij*

[1] Baſſaris, & [2] Lyncem [3] Mænas ———

ſuono, ma il ſenſo è di coſe ordinarie, quali erano nella bocca d'ognuno; per le trombe ſon poſti i corni torti, per il ſuono delle Baccanti *Mimallo-neos bombos*, per Penteo *vitulum ſuperbum*, per Agaue *Baſſarys*, e per Bacco *Euion*. Si deſcriuono in queſti verſi i ſacrificij di Bacco pieni di ſuoni, e di ſtrepiti. *Mimallones* erano le Donne Baccanti, che faceuano i ſacrificij a Bacco, così dette da Mima monte della Ionia prouincia dell'Aſia minore, doue s'adoraua Bacco, & ogni anno vi ſi celebrauano le ſue feſte; ſe bene Suida vuole che ſia monte della Tracia.

[8] *Baſſarys*. Agaue Sacerdoteſſa di Bacco, detta Beſſarida da Beſſaredo, che è Bacco, e dalle veſti lunghe che portauano le ſue Sacerdoteſſe dette Baſſare da' Traci, o perche ſtauano in Baſſara Città di Lidia. Prop.lib.3.eleg.xvij.

Cinget Baſſaricas Lydia mitra comas.

Fù Agaue figlia di Cadmo, e moglie di Echione, la quale nelle feſte di Bacco vcciſe Penteo ſuo figlio Rè di Tebe; & auuenne queſto a Penteo per hauer voluto vedere da vn luogo eminente del monte Citerone i ſacrificij che faceuano le Donne al Dio Bacco: morto poi apparue alla madre in forma di Cinghiale, e non di Vitello, come dice Perſio. Leggaſi Ouidio nelle Metamorf. al lib.3.

[9] *Lyncem*. è la Lince quell'Animale da noi detto Lupoceruiero, ſe bene non sò perche con nome di Lupo chiamato, mentre per ſomiglianza, e per natura ſi può collocare fra il Leopardo, & il Gatto, e non ha coſa alcuna di Lupo, nè di Ceruo; ſe pur non viene così appellato per la preda che fà de Cerui, o per ſomigliarli alquanto nel pelo. è di grandezza molto maggiore della Volpe, & ha la pelle punteggiata di alcune macchie nere come la Tigre, & il Leopardo. Virg.Aen.1. *maculoſæ tegmine Lyncis*. Ha poca coda,

Figure 2.1. Francesco Stelluti, *Persio tradotto in verso sciolto e dichiarato da Francesco Stelluti* (Rome: Giacomo Mascardi, 1630), 36. Courtesy of The Linda Hall Library of Science, Engineering, and Technology.

with confidence that the lynx could see through solids, a quality the ancients mostly associated with Lynceus.[50] By the early seventeenth century, Lynceus and the lynx were interchangeable emblems of visual and mental acuity, and Snellius and Grassi used both indistinctly to name the telescope.[51] Stelluti also leaped from one to the other: "His gaze," he wrote of Lynceus, "penetrated the thickness of walls. . . . From Sicily, he counted the ships as they left from the harbor in Carthage."[52] He compared Galileo favorably to both Lynceus and the lynx in a poem published in 1623: "What nature conceals from everyone else in its bosom," Stelluti waxed lyrical, "it reveals openly" to Galileo. Compared to him, the lynx would be like a mole, "a blind Argos, like Lynceus without eyes."[53]

The reference to Lynceus connected natural philosophy not only with surveillance and military surveying but also with mining. Regarding the proverb "More perspicacious than Lynceus" (*Lynceo perspicacior*), Erasmus of Rotterdam (1466–1536) remarked that nonmythical explanations for the character's ability to see beneath the ground included the belief that Lynceus "was the first to find mines of metals."[54] The link with mining suggested that intellectual penetration was comparable to the invasion of nature's otherwise untouched "inside." The associations that surrounded the myth of Lynceus thus floated the possibility that perspicacity was, of necessity, violent and resonated with concerns that Snellius and Grassi also projected, tongue in cheek, onto the telescope's artifice.[55]

Ecofeminist scholars in the wake of Carolyn Merchant's *The Death of Nature* (1980) have underscored the component of violence that often characterized early modern approaches to nature, from Francis Bacon to the so-called new science, which portrayed experiment as torture and knowledge as penetration of a body imagined as female. These and similar attitudes, often shared by early modern Aristotelian natural philosophy, are in line with what Pierre Hadot has called "the Promethean attitude," which considers knowledge as a source of power over nature.[56] Indeed, while Snellius and Grassi approached the transgression of boundaries attendant on the act of knowledge with irony and humor, others seemed to relish the connotations of overbearing power over the object. Thus, Juan Eusebio Nieremberg found Lynceus emblematic of natural-philosophical research, not less because he promised access to otherwise forbidden and secret spaces. Although Nieremberg believed all the abilities attributed to Lynceus—from seeing through tree trunks and rocks to reaching above the moon and finding people concealed anywhere—to be mythical, they all symbolized the access to the secrets of nature of which he dreamed.[57] In a language that exemplifies the linkage, criticized by Merchant, between "the subjugation of nature as female" and "the scientific method as power over nature," Nieremberg wrote that

> if nature is generally soothing and pleasant to contemplate even at first sight and by considering it only from the bark . . . it becomes much more enjoyable and delightful when we penetrate its secrets and get access to the bottom of its mysteries. . . . We will violate its most shielded retreat and reach to that which is arduous, difficult, and inaccessible about it—its greatest sacrament, which lies in the sympathy and antipathy, as the Greeks call them, that is, a secret conformity and aversion that either appears or exists in things, which triggers effects that arouse wonder due to the extraordinary and anomalous character that they offer to the sight, and their invisible and occult causes.[58]

It must be noted, however, that what Nieremberg expected to find upon closer inspection of nature differed from anything that the microscope would show to a scientist today. For Nieremberg, a body's interior allowed natural philosophers to "see" its relationship with everything around it rather than anything intrinsic. When the eyes of Lynceus pierced through the bark, he saw a web of relationships expressing "a secret conformity and aversion" between things that was invisible from outside: that was nature's artifice.

The invention of conceits resembled natural-philosophical penetration in the way Nieremberg modeled it after Lynceus. Consider Gracián, for instance, who resorted to Epigram 1.47 by the poet Marcus Valerius Martial to illustrate what he called "conceits of correspondence and proportion" (*conceptos de correspondencia y proporción*). In these, he wrote, the author "pays attention to the correspondence of the terms with one another. It is the kind of acuity that we find in a certain epigram by Martial that has deserved more rounds of applause than it has syllables. He compared Diaulus, once a doctor and now an undertaker, and said: *Diaulus was a doctor and is now an undertaker. What he does as an undertaker he did too as a doctor.*"[59] Martial's perspicacity enabled him to see inside the notions of "doctor" and "undertaker." Though Diaulus's new and old professions were superficially unrelated, a look beyond the crust revealed that "doctor" meant less a healer than a practitioner who, either by incompetence or malice, made sure that people ended up beneath the ground. Since that was also the purpose of the undertaker, Martial wryly concluded that Diaulus had changed profession only in name.

According to Gracián, as orators and poets penetrate the subject on which they speak and write, the latter appears to the intellect's eye as a center surrounded by causes, effects, attributes, contingencies, and circumstances, a group of categories that he called "the entities" of the subject and which was reminiscent of the topics used by rhetorical handbooks to organize the invention of arguments.[60] The intellect, he claimed, casts "lines of subtlety" that cross the subject and the crown of the entities that lies around it in search of a match, as when Martial realized that "doctor" and "undertaker" were professions that shared the same purpose or final cause.[61] The intellect, Gracián concluded, "puts [the entities] face to face with the subject, and then with one another. Having found that they have a certain conformity or proportion, it expresses it with subtlety."[62]

The verb *carear,* which I have translated as "to put face to face," carried the same amalgam of fascination and disquiet that the conceits by Snellius and Grassi had associated with the hubris of the telescope. In judicial language, it meant "to bring one person face to face with another," as judges used to do in trials in order to extract the truth out of discrepant opinions.[63] A few decades earlier, Bacon also resorted to judicial language when he referred to the examination of nature "according to the practice in civil causes . . . upon interrogatories."[64] For Bacon and a range of natural philosophers throughout the seventeenth century, there was a close analogy between torture and experimenting with nature. In a context deeply influenced by alchemy, experiment was often synonymous with the application of fire, often dubbed a form of torture.[65] In Gracián's view, the invention of conceits, understood as trial, likewise affirmed the intellect's power over its object in a way that was never entirely peaceful.

Yet Martial's act of intellectual penetration, as Gracián knew only too well, relied on figurative language. Burying people was the purpose or "final cause" of the undertaker; for the physician, however, doing so was merely a contingent and, hopefully, accidental result. Even if one were to maliciously presume that the physician killed on purpose, he would nonetheless not be burying his patients like the undertaker. The same leap in the argument that made it invalid for the logician was of the essence in a conceit, which provided a reader disposed to fiction with a new perspective on reality, made up of connections between different aspects of it that had previously passed unnoticed. Should the term "doctor" imply one who actually buried people, the observation would have failed as a conceit. Departure from the rules of sound argumentation, and a linkage that implied bending reality to a certain extent, were marks of what Gracián called "acuity of artifice," the talent characteristic of the inventor of conceits, which was different from the philosopher's "acuity of perspicacity."[66] The use of artifice in the form of tropes, as Peregrini had claimed in 1639, transformed a conceit's act of intellectual penetration into one of making rather than merely knowing.[67] Intellectual penetration involving artifice, of the kind occurring in a conceit, should differ from truth-oriented and earnest reasoning just as looking through the telescope differed from looking with the naked eye. A conceit, like the telescope, offered a way to get closer to reality that rendered reality hardly recognizable and new.

To understand how penetration and artifice become entangled in the invention of conceits, one should turn to Tesauro's *Il cannocchiale aristotelico* (1654). At first sight, Tesauro might seem to associate penetration and artifice with two different intellectual virtues operating more or less in isolation from one another. The first, perspicacity, "penetrates the farthest and minutest circumstances of any subject . . . these lie all crammed and concealed in any subject," hidden within "just like metals in the mines."[68] Foregrounding the role of extraction associated with Lynceus as the inventor of mining, Tesauro asserted that one could not devise conceits (*argutezze*) without first penetrating the subject, just as it was impossible to build a cannon without first digging for the metals needed in its construction.[69] Accordingly, he laid down a method to excavate deeper and find new materials in seemingly exhausted subjects, just as early modern machines brought abandoned mines back to life in hope of exploiting previously untapped ore deposits.[70]

In addition to perspicacity, there was "versability" (*versabilità*), which—in a formulation reminiscent of Gracián—placed the circumstances excavated through perspicacity "face to face with one another and with the subject." Tesauro's depiction of versability portrayed the intellect basking in a playful and not entirely trustworthy mode of argumentation, in which tropes passed one thing off for another, augmenting and diminishing objects at will in a way he compared to how the juggler's cup-and-ball routines use quickness and dexterity to trick the audience.[71]

However, perspicacity and versability, referring to the intellect's ability to, respectively, penetrate and connect, were two aspects of a single act rather than consecutive steps. The instrument that enabled the intellect to "see" inside the subject, metaphor—by which Tesauro meant tropes generally speaking—allowed the intellect to discover circumstances in the subject only insofar as they could be placed in relationship with one another. Metaphor enhanced perspicacity and versability at the same time.[72] There are figures of speech that "stop at the surface of the term," Tesauro averred, but metaphor "penetrates with reflection and investigates the most abstruse notions [in the subject] in order to connect them."[73]

A conceit's reliance on tropes rendered it artificial. Tropes, Gracián maintained, served as instruments to the intellect as it searched for connections.[74] Like optical devices, tropes served as prosthetics that altered the way humans interacted with the world. With the telescope, they could see beyond the reach of the naked eye, which "naturally" embodied visual—and, often, intellectual—ability. Similarly, the inventor of conceits used tropes, instruments of imagination, to attain what he or she was unable to reach "naturally," that is, through syllogism.

The description of tropes as instruments for enhanced understanding of reality is seemingly at odds with Peregrini's assertion that conceits shaped something new, as opposed to syllogisms that merely discovered what was already there. The contradiction, however, is only apparent, a consequence of the gap that separates what seventeenth-century scholars understood by artifice from what the term suggests to us today. To better understand how the analogy between intellect and sight might shed light on seventeenth-century notions of artifice, in the next section I focus on pieces of fiction that linked specific conceits to specific optical devices. Their eccentric but highly perceptive imaginings reveal that invention and discovery had become increasingly entangled in the eyes of early modern individuals. I show that conceits, as instances of intellectual instead of visual artifice, played a critical role in the logic of sight-as-transformation that became associated with the telescope and the microscope. Yet on a deeper level, as I also demonstrate, the logic of sight-as-transformation found validation in the belief in a secretive and elusive nature, which justified artifice as a deceitful but necessary means to understand and interact with reality.

Truth as Transformation

"The astrologer, however lynx-eyed he might be, cannot peer into the mind's inner thoughts," Lothario Sarsi remarked bitterly in the fall of 1619, "not even with his telescope."[75] It was Sarsi's response to a recent accusation by the consul of the Accademia Fiorentina, Mario Guiducci, that Grassi's lecture on the three comets tacitly endorsed Tycho Brahe's system of the universe, with its compromise between heliocentric and traditional geocentric models, instead of that of Nicolaus Copernicus.[76] The debate was a masquerade: Guiducci acted as a speaker for Galileo, and the man who signed as Sarsi was none other than Grassi himself. As Sarsi disingenuously noted through the allusion to mind reading, the way Grassi's lecture calculated the position of the third and last comet of 1618 was the only hint that might have led Galileo to conjecture that he was adopting Tycho's system, a subject on which Grassi had remained silent.

The notion of the telescope as a spy privy to inner thoughts and intentions haunted Grassi's contemporaries. It left a mark, for instance, in a poem titled "Adulatio perniciosa" ("Pernicious Adulation"), which Cardinal Maffeo Barberini, the future Pope Urban VIII, sent to Galileo in the summer of 1620. Barberini portrayed the sky as a court while he celebrated Galileo's discovery of "Jupiter's attendants," "those of Saturn his father," and "the black stains" defacing the sun.[77] A praise of the astronomer who had torn open the interior and secret rooms (*penetralia*) of planetary lives, "Adulatio perniciosa" is even more a fable about flattery. It depicts, close to the king, a crowd of attendants leading the way, as "courtiers follow seeking office."[78] Deceit lies in ambush around the corner, and adulation taints the ruler's happiness, similar to spots on the sun.

A story that became increasingly popular in the late sixteenth century recalled that Momus, the emblem of fault finding in Greek myth, had criticized the god Vulcan for not adding a window to man's chest to ensure that his words matched his thoughts.[79] With every conversation and exchange tarnished by suspicion, the voice in Barberini's poem wished that there were some equivalent to the telescope's perspicacity in the realm of political and social interaction. He asked, who is Lynceus (or, perhaps, lynx-eyed) to such a degree that he could see into people's hearts?[80]

The notion of a device applying optics to mind reading was not entirely new. Both Grassi and Barberini drew on the precedent of Traiano Boccalini's (1556–1613) mordant, best-selling satire *De' ragguagli di Parnaso* (*News from Parnassus*, 1612). In a work entirely set in Apollo's court on Mount Parnassus, the episode in question narrated the arrest of historian C. Cornelius Tacitus on the charge that "out of the seditious matter of his *Annals* and *Histories*, he crafted spectacles of a certain kind with the most pernicious effects for princes; for placed on the nose of the common people, they turned their sight so subtle that they could see into the inner and hiddenmost thoughts of others, reaching even into their guts."[81]

The accusation against Tacitus drew on the opinion of Justus Lipsius—both were characters in Boccalini's satire—, who in 1574 commended the author of the *Annals* and *Histories* over Livy because he cared less for the events than for the motives leading to the events. Tacitus, according to a formulation destined to enjoy great success, showed his readers into "the halls of the princes, their inner lives, counsels, orders, and facts."[82] Turning praise into blame, the princes in Boccalini's *News from Parnassus* complained that the spectacles or glasses made out of the *Annals* and *Histories* revealed their intentions and thoughts to their subjects, whom it was of utmost importance to keep in the dark. Dissimulation was "the artifice" of ruling, and it was only fair that Tacitus should pay for ruining civil and political life.[83]

To some, Tacitus appeared as a threat, a catalyst of political unrest who had laid bare the intentions of those in power. That was also the view of the anonymous orator who, in a lecture he delivered at the Jesuit Collegio Romano in the early seventeenth century, decried the proliferation of learning about the court (*aulica doctrina*), feigned prudence (*prudentia adumbrata*), and criticized the newfangled science of preserving and augmenting the state (*scientia firmandi augendique status*). Those were all facets of reason of state, a doctrine that took shape in the wake of sixteenth-century political thinkers and historians such as Niccolò Machiavelli and Francesco Guicciardini, and which the anonymous orator condemned as not only a distraction from the advancement of true and legitimate learning but also the cause of the seditious and heretical tendencies that permeated early modern European social and political life.[84]

The art of "preserving and augmenting the state," as the princes in Boccalini's satire were keen to underscore, entailed a lack of transparency that was as "necessary" for the sake of expediency as it was ethically dubious.[85] It would be fatal for the ruler's reputation if the subjects came to know about his or her intentions, yet that was precisely the effect of Tacitus's glasses, which cleaned the dust with which the princes blinded people's eyes.[86] The problem, as the anonymous orator was keen to underscore, was that thanks to the success of Tacitus, reason of state was no longer restricted to princes and counselors. It had become "a common good," a fever that moved freely through markets and squares, such that, both in private meetings and in public discourses, individuals high and low spent time discussing "secrets of state, as they call them" (*arcana, ut aiunt, imperii*).[87] One would say that the age of discovery and invention had devolved into a parody of itself: after the compass, the printing press, and the exploration of unknown oceans and lands, all that was left was a population entirely dedicated to the crafts of manipulation and deceit, satisfied with peeking into other people's minds.

While Boccalini imagined that optics could serve the purposes of mind reading, Gracián suggested that the act of unveiling people's intentions was, in fact, a conceit. Soon, reason of state, and especially the part of it concerned with the arts of simulation and dissimulation, became a strange and unexpected bridge between optics and conceit making. In *El Criticón* (*The Critic*, 1651–1657), Gracián's allegory about life, the two protagonists arrive in Madrid and enter a bookshop to purchase a guide to life at court. Inside, they encounter a character who scoffs at old and outdated sixteenth-century manuals of good manners. All of them would be of little help to the travelers, he points out, because they are silent on the topic that has become fundamental to survival at court: the art of seeing into people's hearts while preventing others from doing the same to us. The present, he notes in a tribute to Boccalini, is not "an age of weapons" but one "of glasses."[88] Yet taking a step ahead of Boccalini, the character casts doubt on Momus's remark about the window in the chest. In a world populated by increasingly lynx-eyed—and "lynx-minded"—populations, there was little need for the window in the chest, let alone Boccalini's glasses.[89] Due in part to the craze for Tacitus, whose *Annals* and *Histories*—if we are to believe Gracián's translator into French—inspired countless aphorisms of Gracián's *Oráculo manual* (1647), seventeenth-century individuals had become impervious to manipulation thanks to a mastery of indirectness and opacity without precedent in previous periods of history.[90] When threatened by lynx-eyed intellects, individuals high and low had learned to conceal their intentions and thoughts, retreating like the cuttlefish that hides behind a curtain of ink in the presence of danger.[91]

For Gracián, it was a mark of prudence to be unremittingly suspicious and never take anything at face value.[92] Gracián advanced a kind of "experimental science" based on the notions of *reparo*, which designated the ability to perceive and stop a blow, in the face of a reality that is always intent on deceiving us—and *concepto*, a term denoting intellectual grasp of a situation.[93] "The wise," he wrote, "gets a full grasp (*haze concepto*) of everything, but especially where depth makes him alert. And he considers that, at times, there is more than he thinks, and arrives by reflection where the first apprehension could not reach."[94] That *concepto* was likewise the term for a rhetorical and poetic conceit in *Arte de ingenio* signals the continuity, characteristic of Gracián's approach, that tied political and social behavior to rhetorical and poetic ingenuity.[95] To see inside people's hearts—what Boccalini pretended, Grassi denied, and Barberini wished that optical devices could do—was, for Gracián, "to make a conceit."

Arte de ingenio designated as "crisis" (*crisi*, from the Greek word κρίνω, meaning to pass one's judgment on a subject) a conceit that penetrated the motive behind some action or statement that had remained without explanation, or which had been imputed to intentions that the author now challenged and proved to be false.[96] For early modern Aristotelians, knowledge was knowledge of the cause, and the cause of people's statements and actions was the motive leading to them.[97] Crisis rectified incomplete or erroneous knowledge. According to Gracián, Martial was one of the two masters of crisis. He discovered that Fabulla always surrounded herself with old or ugly women in order to appear young and beautiful by way of contrast.[98] Then there was Tacitus. Reflecting on why Emperor Augustus had elected Tiberius to succeed him in the throne, Tacitus argued that Augustus did not think Tiberius was well qualified for the position, yet his character was unpleasant enough to make sure that Augustus would be favorably compared to the new prince. Tacitus, Gracián wrote, "was never satisfied with the vulgar and simple narration of history. He did not stop at the bark of events, but proceeded until the innermost rooms of intention."[99] Crisis was a form of competition between two forms of ingenuity: the agent's, who acted by concealing his or her heart, and the writer's, who, prompted by superficial anomalies in actions or statements, triumphed over the agent by revealing that which was meant to remain concealed, more pleasant when the challenge was also greater.[100] "Just as acting with artifice and reflection is evidence of superior ingenuity," Gracián pithily remarked, "to reveal and expose that artifice attests to a form of ingenuity that is twice subtle over."[101]

According to a number of seventeenth-century testimonies, the telescope did to nature what Tacitus had done to Tiberius. For Tesauro, Galileo's device outsmarted God's efforts at concealing celestial bodies using distance. Tesauro corroborated the continuity among conceits, optical devices, and mind reading by means of a strange but revealing anecdote. Following the invention of the telescope during the Dutch Revolt, the stadtholder Maurice of Nassau introduced the device before the senate as a tool to "learn from a distance about the deceits of the Spaniards."[102] In a new avatar of Boccalini's spectacles, Nassau hinted with humor at the telescope's ability to detect dissimulation and learn whether the enemy intended to respect the truce they were about to sign.

The invention of the telescope gave rise to speculations about the existence of optical devices that, similar to Gracián's conceits of crisis, could see inside people's hearts. In turn, natural philosophers increasingly envisioned access to nature as a form of espionage of intentions and thoughts in which a personified and elusive nature struggled to keep its (or her) secrets out of the reach of humans.[103] Thus, the assertion advanced by Lipsius that Tacitus let his readers into "the halls of the princes, their inner lives" informed Gracián's language as he argued that crisis provided access to "the innermost rooms of intention." Yet Lipsius also shaped a range of statements about the study of nature, including Bacon's hope that natural philosophy might find its way "into [nature's] inner chambers" and Nieremberg's promise to deliver the secrets concealed in nature's "most protected retreat."[104]

The craze for Tacitus, a subject that was once part of Bacon's instruction to the future King James I, normalized the belief that the study of nature, much like conceits of crisis, entailed the encounter between two forms of artifice: one concealed a secret, and the other uncovered it, justifying Bacon's claim that nature liked to play hide-and-seek with humans.[105] To a society living under the spell of Tacitism, nature appeared as a series of increasingly withdrawn spaces marked off by differentials of power and information, and permeated by references to social and political dynamics. Religion, however, also played a part. For instance, works addressed to courtiers regularly alluded to the king's "mysteries" to designate that which he refrained from sharing with others.[106] Historians also spoke of "the sacrament" of the archives that preserved government papers, and that was also Nieremberg's term for nature's inner realm, where the natural philosopher aspired to find its artifice, its secret, consisting of sympathies and antipathies.[107] The transfer of "mysteries" from religion to politics, as well as to natural philosophy, obeyed the same logic that placed *arcana Dei* ("God's secrets") side by side with *arcana status* ("secrets of state") in seventeenth-century writings.[108]

Conceits of crisis transformed statements and actions by attributing motives to them that were not only unheard of but, at times, outright unthinkable. Conceits of crisis did not resort to metaphor, and they consequently operated metamorphoses that were less spectacular and colorful than Grassi and Barberini's portrayal of the sky as a court, but they were nonetheless shocking. As implied by Gracián's assertion that conceits of crisis approach the topic being discussed "in clever panegyric or ingenious crisis, that is, in praise or blame," it follows that they see inside people's actions and statements in the act of transforming them into something either better or worse than they seem to be from outside.[109] As was the case with Tesauro's description of metaphor, in conceits of crisis, penetration and transformation represented two sides of a single act, inextricable from one another.

While admitting that "every great wit is ambidextrous," able to "gloss" any statements and actions for the better or the worse, Gracián avowed that malicious interpretations were naturally more pleasant and struck the audience as a mark of the author's education.[110] "Anyone with a superior taste likes it because it hurts," he concluded wryly.[111] Yet the same inclination to find fault that Gracián found so seductive about Tacitus failed to impress Jesuit professor Famiano Strada. In a lecture first published in 1617, Strada accused the historian of consistently interpreting statements and actions for the worse.[112] Where Gracián saw a deeper layer of causality and meaning, Strada recognized a "bad" or "false" reason of state. Strada believed that even though self-interest and deceit were universal among humans, that was not enough reason to justify perennial suspicion and to normalize the belief that appearances are deceptive by necessity. Similar to the princes in Boccalini's satire, Strada claimed that Tacitus eroded trust (*fides*), the thread that wove the fabric of society. Without it, individuals turn "against each other, full of suspicion, fearing the perfidy of the rest."[113]

In Strada's opinion, making suspicion the option by default was both politically and socially corrosive. Doing so in the study of nature was, as Snellius suggested, hubris against God. Indeed, early assessments of telescopic and microscopic magnification often suggested that optical devices normalized suspicion toward the surface of nature. Enthusiasm for the telescope coexisted with actual or feigned anxiety toward the discoveries that it enabled: actual, as in Scheiner's case, who dedicated a life of research to optics and yet felt compelled to find alternative explanations for the appearance of sunspots, a notion that he found insulting; feigned and fairly lighthearted in the case of Francisco de Quevedo (1580–1645). Around 1636, Quevedo imagined a group of Araucanians in Chile who rejected the telescope presented to them by Dutch navigators. The reply of the natives sums up the malice associated with optics and artifice more generally: "An instrument that discovers stains on the sun and finds out lies in the moon, and reveals what the sky has concealed must be a rebellious one. It is gossip made of glass, and the heavens cannot possibly like it."[114] In 1657, astronomer Giovanni Battista Hodierna likewise spoke of a century "made of glass" in which mortals attack the heavens "armed with lenses, capturing stars as they conquer."[115] One decade earlier, Francesco di Lucà had depicted the effects of Hodierna's research in disquieting albeit humorous terms. "Nature herself," he claimed, "is stupefied, and shows / Sadness and grief because she has born a child / Who reveals and exhibits each and every of her secrets."[116] The problem was not only with the telescope, though. A lecture delivered in Valencia at the end of the seventeenth century criticized the microscope for making that even seemingly clean and white milk appear full of minute worms. The device, the speaker was forced to conclude, was "the true emblem of the slanderers," who like to find fault everywhere.[117]

Conceits informed seventeenth-century understandings of the operation of optical devices, a striking fact considering that we are contrasting a discursive and linguistic device with one that is the product of optical, that is, mathematical, principles. Optical devices and conceits were consistently portrayed as comparable forms of artifice that humans could use as antidotes to nature's artifice, its secretive and elusive character. Consider Argensola, who condemned Tacitus's exploration of hearts as a form of anatomy. Argensola's analogy cast a negative, even ominous outlook on conceits of crisis, while advancing at the same time a link between conceits and medicine.[118] In *De arte* (*On the Art of Medicine*, a work of the school of Hippocrates from the fifth century BCE) we read that without direct access to the causes of illness, which nature has blocked from view within the chest and the abdomen, the physician needs to look for external, superficial signs (*semeia*) to conjecture what could be occurring inside the body.[119] Like the physician, Argensola's analogy seemed to suggest, Tacitus identified clues and derived from them interpretations that, for the skeptical, were the product of mere whim and lacked any certainty.[120]

Interestingly, conjecture from signs also underpinned the mechanism of Boccalini's spectacles. In a way, one could say that they were conceits disguised as optical devices. Indeed, both enthusiasts and detractors of Tacitus described the divination of motives as a process of semiotic rather than mathematical, specifically optical, character. The question of whether signs should serve as a lead into inner truths permeated a range of sciences and arts. For instance, it became a matter of discussion in connection with water prospectors and dowsers, who embodied a model of acuity strongly associated with the lynx. Stelluti had read about dowsers—*zahoríes* in Spanish—in a work on magic published in 1599–1600 by the Jesuit Martín Del Río (1551–1608). "We should call them lynceans," Del Río argued, a term that referred ambiguously to both Lynceus and the lynx. People claimed that *zahoríes* "see things that are concealed in the depth of the earth's bowels, such as water courses and treasures of metals, as well as the corpses placed beneath gravestones."[121] While some attributed the power of *zahoríes* to melancholy, Stelluti was skeptical that they could see through the earth at all.[122] In a similar vein, mathematician Ginés Rocamora y Torrano (1550–1612) suggested in 1599 that *zahoríes* merely conjectured from signs that anyone could perceive with the senses, including "the colors, humidity, and grass that grows on the surface; just like those who seek mines."[123] Nieremberg likewise saw there a semiotic model, based on the search for "some grasses, brambles, canes, and other green trees" as well as gnats and "subtle vapors." Like Rocamora, Nieremberg pointed out the similarities between the practice of *zahoríes* and the art of finding mines, of which there were "natural signs, and certain grasses that signify them."[124] The manual on metalwork and mining by Vannoccio Biringuccio (ca. 1480–ca. 1539) had taught prospectors "the practice of signs offered to us by nature's kindness" as early as 1540.[125]

For Gracián, conceits of crisis were the work of *zahoríes* of hearts, who also read signs. Take King Henry IV of France, a master of "the anatomy of spirits, natures, and inclinations."[126] Individuals like him did not need "even a crack to enter into the hiddenmost interior."[127] What Gracián called crisis was central to a culture that conceived of penetration—imagined as conjecture based on signs—as the artifice through which the intellect countered reality's elusive and deceptive behavior. Crisis was a clash between two forms of artifice, one apt at ciphering reality and the other at decoding it. In terms that strikingly resemble Tesauro's description of metaphor, Gracián explained that *zahoríes* of hearts "see things from inside. They do not stop at the vulgar surface and are not satisfied with exteriority."[128] Crisis implied in-depth understanding as the intellect "saw" how cause and effect, or intention and action, corresponded with one another in a way that appearances sought to belie. For Gracián, however, crisis was not about knowing the true, verifiable motive that had led to actions or statements but rather about suggesting one that would appear striking and plausible considering the circumstances that the intellect had penetrated. Crisis denoted the ability to generalize and abstract, and, as such, it lay close to the invention of metaphors. In fact, Gracián suggested that one should not refrain from "twisting, and even fabricating" intentions and thoughts as long as they enable the author to cast thought-provoking perspectives on statements or actions. "At times," he acknowledged, the intellect "is not satisfied with discovering artifice in intentions but goes as far as to fabricate it."[129]

To the extent that seventeenth-century individuals pictured optical devices performing conceits of crisis, they imagined them operating according to semiotic rather than mathematical models, as Rocamora had intimated about *zahoríes*. Yet this was also true of a range of seventeenth-century approaches to the relationship between sight and intellect. Paolo Galluzzi has pointed out that despite the links between Galileo and the Accademia dei Lincei, a gap separated Galileo's mathematical and quantitative understanding of the book of nature from Cesi's views, influenced by Giambattista Della Porta and the belief that nature was made up of resemblances.[130] Seventeenth-century scholars often acted as if optical devices, similar to the eyes of *zahoríes*, did not see reality but merely traced semiotic and analogical links that enabled them to conjecture what reality is like. It seems safe to argue that scholars, and even more so writers of fiction, felt inclined to understand the form of artifice at work in telescopes and microscopes by analogy with a different, older example of artifice: that of figurative language.

The impact of semiotics on seventeenth-century conceptualizations of optical devices becomes even more obvious in what Gracián called conceits of "correspondence and proportion." Gracián illustrated correspondence by means of Martial's above-mentioned Epigram 1.47, which claimed that Diaulus the undertaker still had the same job as when he practiced medicine, namely, burying people. Martial's conceit became associated with optical devices in a short story titled *Los anteojos de mejor vista* (*The Spectacles for Improved Sight*), published around 1625 by Rodrigo Fernández de Ribera (1579–1631).[131] When the narrator of Ribera's piece reaches the top of *La Giralda*, a former minaret doubling as the bell tower of Seville's cathedral, he looks down on the square as friars, ministers of justice, bailiffs, and merchants conduct business as usual, while a doctor on a mule marches through the gates of the city. As he looks again a while later, however, the sight has become hardly recognizable. Instead of judges and lawyers, birds of prey now fill the square. Scavengers have replaced the friars who stood among them, chasing peaceful doves that only a minute before looked like people conducting business. The bailiffs have turned into line fishermen, and even the pedant who had climbed the tower with the narrator appears now like a large and ridiculous hen outfitted with a sword and a mustache. Riding a mule, the executioner appears to be where the doctor had been just a moment earlier. Could they be the same person?[132]

They certainly are. In the intervening time, old man *Desengaño* (Disabuse) has handed the narrator a pair of spectacles ground by Experience using the glass of Truth, which enable anyone to see things "as they truly are."[133] Ribera's spectacles are optical devices that disabuse the narrator by replicating the mechanism that Gracián saw at work in Martial's conceit of "correspondence and proportion." Even if Ribera replaced the undertaker with the executioner, the spectacles resemble Martial's conceit in suggesting that the art of medicine leads to death rather than health if one considers it "from inside," either through mathematical artifice, in the case of optical devices, or through rhetorical and poetic artifice, in the case of conceits.

As with crisis, the analogy between optical devices and conceits of correspondence was fraught with challenges. As Enrique García Santo-Tomás has explained, Ribera's spectacles replaced error in perception with "a new and bare reality."[134] And yet the sense in which the sight afforded by the spectacles is "new" and "bare" is far from obvious for today's reader. Ribera's spectacles display a world "so strange, new, and prodigious" that the bell tower itself risks falling, shaken by the surprise.[135] They unveil the truth by weaving a new veil for it with the thread of metaphor, transforming the object into something else.

Early accounts of telescopic and microscopic observation raised similar doubts regarding the relationship between discovery (of the truth) and invention (of something new). In 1624, when Johannes Faber received the microscope sent to him by Galileo, he celebrated the astronomer as "a second Creator." With the microscope, Galileo made things appear "that were not known until now to have been created."[136] Likewise, in a poem included with *Il saggiatore*, Faber had compared the discoveries enabled by the telescope with the explorations that had opened up a world previously unknown to Europeans. As with Ribera's short story, novelty turned out to be a product of perspective. One could speak of novelty in connection with objects that were as old as Columbus's "New World," which, as was the case with Galileo's discoveries, was new only for those who had not known it before.[137]

Along with novelty, another feature that tied Ribera's spectacles and Gracián's conceits of correspondence together with early modern accounts of telescopic and microscopic vision was transformation. John Wedderburn, Galileo's student in Padua, recalled that his teacher used to peer through the microscope at the eyes of insects, covered by a thick membrane "perforated with holes like a warrior's visor."[138] It soon became fashionable to report that the microscope turned fleas into elephants, suggesting that its promise to counter nature's secrecy passed through the metamorphosis of one thing into another with which it shared one or more traits.[139] Metaphor penetrated the topic being discussed and linked what it found there to something else. In a similar vein, early users of the device reacted with bewilderment as the object's identity dissolved within a web of relations with a variety of other things. Margaret Cavendish (1623–1673) was among those who expressed skepticism that artifice could be of help in the observation of nature, as when in 1666 she ridiculed the perplexity of scholars who balked as they saw every insect and worm transformed into a monster of seemingly gigantic dimensions when looked at through the lens.[140] To others, however, the lens offered confirmation of long-held beliefs about nature.

It might be somewhat unsettling to see Stelluti's dream of finding nature at work beneath the surface yield to the discovery that each thing is chiefly something else. In 1646, Athanasius Kircher, a scholar with a keen interest in optics, pithily summarized the feeling when he stated that "all things seen by many of us are in truth other than what they seem."[141] As Nieremberg maintained, nature's artifice and subtlety resided in the links that connected its creations to one another. Early modern Aristotelians peered inside nature in search not of matter's minute and invisible components but of "the elusive and secretive workings of nature, her trickery, and the occult interconnectedness" between different objects, as Christoph Lüthy has suggested. According to Lüthy, the microscope was slow to produce discoveries not only due to limitations in the design of the device and glass-grinding techniques but also because there was a disconnect between the realities it enabled users to see and what they expected to find out about nature.[142] Conversely, what they expected to find out led them easily to imagine optical devices in the guise of conceits, which illuminate the similarities between different objects by transforming them. Like Tesauro's metaphor, optical devices turned one object into another based on shared features that became apparent only to those who would undo nature's artifice by using a different but comparable form of artifice in the act of penetrating it.

Transformation gave visual embodiment to interconnectedness. In conceits and optical devices, the resemblance depicted in contemporary works of fiction between conceits (both of "crisis" and of "correspondence") and optical devices offers a window to understanding how seventeenth-century individuals envisioned the relationship between sight and knowledge, and the role of artifice in it. It also reveals that the boundaries between discovery and invention were highly problematic. Like conceits, optical devices used artifice to transform the object as they showed it from the inside, supposedly undoing nature's artifice. In conceits, as was also the case in optical devices like those imagined by Ribera, artifice was synonymous with metaphor, a product of imagination that both enhanced and disrupted the penetration of reality. Yet early modern scholars knew only too well that imagination was not bound by truth. It incessantly sought connections and did not hesitate to make them up if it could not find them. As I argue in the third and last section of this chapter, however, it was imagination's freedom from truth that entitled conceits to lay claim to beauty.

An Ingenious Perspective

Seventeenth-century accounts of telescopic and microscopic observation shared with contemporary works of fiction the view that optical devices provided access to nature's "inner chambers" by transforming appearances into a new and different sight—one that also was, in a way, more accurate. Conceits, as counterparts of optical devices, likewise straddled discovery and invention. In 1642, Gracián claimed that conceits "exist already within the object." The author merely "arrives and lifts the prey." By likening conceit making to hunting, a proxy for natural philosophy, Gracián implied that the inventor did not create conceits but found them ready-made.[143] It was a bold and potentially misleading suggestion, which imperiled Peregrini's distinction between syllogisms—in which the intellect "discovers" a truth that already exists—and imagination's prerogative to shape novelty and beauty by resorting to conceits.

Imagination connected different objects in new ways, using tropes as the instruments that enabled oblique and circuitous connections. By defying the linear and obvious links argued by logic, imagination brought about products that did not preexist the act of invention.[144] Peregrini believed that the presence of tropes alerted listeners to the artifice of conceits, with the result that they stopped paying attention to the truth of the statement and were instead carried away by the odd and far-fetched, yet fitting connection being put forward, eventually rejoicing as they noticed the author's ingenuity behind it. Because they involved tropes, conceits offered not a direct, unmediated view of reality but "an ingenious perspective" on it, as Peregrini claimed in a formulation that officially sanctioned metaphor's engagement with optics. It also confirmed that conceits belonged within the realm of aesthetics rather than truth.[145]

Taking up Peregrini's cue, Tesauro carried the analogy with optical devices a step further. He recalled that Socrates once laid the blame on nature for creating man without a window in the chest to see what he thought "face to face . . . without the interpretation of a lying tongue."[146] It was a variant of the story about Vulcan and Momus invoked by Gracián, yet while Gracián believed the window in the chest to be redundant in a society of lynx-minded individuals, for Tesauro the possibility of removing language from the picture was simply uninviting. The window represented a world of natural—that is, unmediated—understanding and communication that, as he candidly put it, would deprive humans "of the pleasure of all the arts of speech."[147] Let individuals proclaim their fondness of transparency; they would loathe it as soon as they see it at hand. Because humans admire "a clothed, rather than a naked truth," they communicate through artifice, using conceits instead of "the short and natural language of angels."[148] And they are not entirely wrong, Tesauro adds, given that angels and even God play poets and inventors of conceits every now and then. The aspiration to transparency denoted by the window in man's chest strikes Tesauro as perhaps effective, but wholly insipid. Add only a lens to the window, however, and the excitement suddenly comes in: "Even a disgusting and ugly object . . . such as a tarantula, a toad, a serpent, a scorpion, it looks harsh from close-up, yet you see it with pleasure through the lens due to the novelty of the ingenious device that represents it."[149]

As any learned reader would have recognized, Tesauro was borrowing Aristotle's reflection on the pleasure that we experience in portraits—and, supposedly, poetry—as products of imitation. Yet by applying Aristotle's meditation to optical devices, Tesauro also suggested that the latter conveyed to the eyes representations mediated by a form of artifice that was not unlike that of painters and poets.[150] In portraits, the painter's artifice lies between the sitter and the onlooker. As a result, the sitter is distinct from his or her portrait, in contrast with what happens with the direct, unmediated view that the window supposedly affords.[151] Likewise, the lens creates products that resemble but differ from reality, triggering a process of recognition followed by pleasure as the onlookers realize that they know who the sitter of the portrait is and yet that what they see is not the sitter him- or herself. By referring to the lens as "the ingenious device" (*l'ingegnoso ordigno*), Tesauro underscored that the cause of pleasure was the mediation that transformed the object, creating something new.[152]

In conceits, the mediation is metaphor's doing. Expanding the analogy with optics, Tesauro clarified that metaphor is comparable to watching a multiplicity of objects appear in simultaneity through "a perspectival aperture" (*vn' istraforo di perspettiua*) instead of one after the other.[153] Tesauro's reference to "a perspectival aperture" could allude to anamorphosis, a projection that makes playful use of the laws of optics. Anamorphosis characteristically presents the onlookers with seemingly monstrous and nonsensical images that become transformed into perfectly normative and recognizable ones if reflected on a mirror or contemplated from a specific vantage point.[154] Others hold that Tesauro had in mind a sort of *camera obscura*, a room or a box in which light enters through a small hole or lens, projecting onto the opposite wall the image of the objects outside.[155] In either case, Tesauro likened metaphor to a projection in which the mediation of artifice transformed reality by recombining its components instead of providing direct access to what is naturally visible. As Tesauro put it, metaphor transformed reality by showing "one thing inside the other."[156] It was a startling but effective way to envision how conceits created novelty. They caused certain aspects of reality that somehow echoed or resembled one another to morph into one another thanks to the trope's plastic and metamorphic potential, as occurred with the spectacles in Ribera's novella.

Like Peregrini's allusion to "an ingenious perspective," Tesauro's analogies with optics and the art of perspective suggested that conceits provided a rare, unusual perspective on reality that made the whole appear new. In addition, they emphasized that, unlike the arguments of the sophist, conceits did not seek to instill false opinion. After all, the claim to a benign, nonmalicious transformation of reality had come to embody, for Tesauro's contemporaries, the operations of perspective, as summarized by the tag *Citra dolum fallimur* ("We are deceived without malice") inscribed in the frontispiece of Guidobaldo Del Monte's (1545–1607) *Perspectivae libri sex* (*Six Books on Perspective*, 1600; see Figure 2.2).[157] Minim scholar and artist Jean-François Niceron (1613–1646), who authored the treatise on anamorphosis *La perspective curieuse* (*The Curious Perspective*, 1638), went on to paint the tag *Citra dolum fallimur* on a wall. He did so at the convent of Trinità dei Monti in Rome, where he put anamorphosis to use in a fresco created near the end of 1639, just months after the publication of Peregrini's *Delle acutezze*.[158] It was also Niceron who, in 1642, built a device in which a set of different heads painted on a table morphed into a portrait of Grand-Duke Ferdinando II of Tuscany when contemplated through a tube containing a polyhedric lens. Of all the early modern examples of anamorphosis that are extant, Niceron's artifact is perhaps the one that most closely resembles Tesauro's description of metaphor.[159]

In 1651, Gracián also suggested that artifice allowed natural philosophy to "renew" its object of study, nature, in order to render it appealing to the intellect. *El Criticón* opens as shipwreck survivor Critilo meets Andrenio, a man raised by wolves in the depths of a cave. Before the two characters arrive at the court in the episode that I discussed earlier, Critilo learns that the first time that Andrenio had the opportunity to contemplate the sun and the moon, he was already capable of reason. The situation was anomalous, Critilo observed. For the most part, humans become capable of reasoning when they are already too familiar with nature's usual course to even notice regular and cyclical phenomena, which they take for granted. Less fortunate than Andrenio, they need to "philosophize with artifice," "going back" through "reflection" to look at objects as if they were new for them, even though they are not.[160] In a similar vein, Robert Hooke claimed in 1668 that natural philosophers imagine themselves to be foreigners who look at familiar phenomena "as if they were the greatest rarity."[161] For both Gracián and Hooke, the mediation of artifice was key to cast the object of study as new.

Seventeenth-century scholars often welcomed the telescope as a prosthetic capable of restoring humans to a state of nature prior to the fall from Paradise. Alternatively, it raised them to the hierarchies of angels.[162] Yet Tesauro's remarks about "the short and natural language of angels" suggest that, at least for him, the appeal of prelapsarian and angelic ideals was only relative. The artifice of optical devices and that of conceits did not erase the gap between humans and either Adamic or angelic sensory and intellectual abilities. Quite to the contrary, artifice served to underscore the limitations of postlapsarian humans, but in a way that also afforded opportunities for sensory and intellectual pleasure that

GVIDI VBALDI
E MARCHIONIBVS
MONTIS
PERSPECTIVAE
LIBRI SEX.

CITRA DOLVM FALLIMVR.

PISAVRI.
Apud Hieronymum Concordiam,
M. DC.
SVPERIORVM PERMISSV.

Figure 2.2. Frontispiece of Guidobaldo del Monte, *Gvidi Vbaldi e Marchionibvs Montis Perspectivae libri sex* (Pesaro: H. Concordia, 1600). Courtesy of The Linda Hall Library of Science, Engineering, and Technology.

the direct and unmediated communications of angels sorely lacked. It makes sense that Tesauro imagines angels occasionally demoting themselves as they indulge in conceits, forsaking the ideal of nonmediated, nonlinguistic communication that has survived through present-day brain-to-brain transmission dystopias. Even God, Tesauro avowed, might have reasons other than being understood by humans when he invents conceits, like the ones that populate Scripture. In fact, it might be that he simply enjoys doing so. How could anyone resist metaphor, after all?

As a way of conclusion, Tesauro reminded the reader that metaphor in conceits is pleasant not only because it renews reality as it penetrates it but also because it reflexively "teaches . . . that one thing is another," so that "by hinting at a certain feature, allows you to understand another one with your intellect."[163] In addition to conciliating discovery and invention by introducing the awareness that "one thing is another," which the mind receives as "a novelty," metaphor imparted to the audience the responsibility to actively reconstruct the connections outlined in the process of discovery and invention. "With your intellect" was Tesauro's reminder that the audience plays a role in conceits that is both active and cooperative, shedding further light on the claim that the window in the chest would deprive humans of the pleasure of "all the arts of speech." Metaphor artificially penetrates the topic in search of connections between its circumstances, revealing its inner secrets and truths. However, whatever it finds out, it expresses it only in outline. In doing so, according to Tesauro, metaphor encourages the espionage of intentions and thoughts, inviting the audience to, in turn, penetrate the speaker's heart in search of what he or she actually meant and left only half said. If nature likes to play hide-and-seek with humans, it follows that in order to imitate nature they, as inventors of conceits, should adopt similarly secretive and elusive manners, playing hide-and-seek with the audience.

In *Arte de ingenio*, Gracián stated that conceits lift humans to the hierarchies of angels.[164] However effective in terms of praise, the allusion to angels was also misleading. Metaphor was not about achieving direct and unmediated expression. Quite to the contrary, it was the instrument responsible for infusing rhetorical and poetic activity with the cult of indirectness, opacity, and suspicion that defined seventeenth-century interactions at the court. Just as the craze for Tacitus and reason of state literature permeated natural-philosophical language, it also shaped Tesauro's understanding of speaker-audience relationships, based on the use of conceits.

For the most part, writings on conceits were the work of individuals well-versed in the life at court, even if they, as was the case with Gracián, felt or at least affected a distaste for it.[165] Like Peregrini, who made a reputation as a political writer and counselor, Gracián became famous for a career devoted to teaching the tactics of defensive and offensive self-fashioning at court.[166] Gracián imagined Tacitus engaged in a competition with Tiberius. It was a game of hide-and-seek in which the ingenuity of the historian surpassed that of the emperor. Yet conceits placed the reader in a similar position in relation to the author. For instance, we call statements or actions "suspicious" if the motive behind them appears to be different from the one stated by the agent. Likewise, seventeenth-century rhetoricians often called a statement "suspicious" if it seemed to signify more than it said explicitly, a feature that Peregrini called "understood" (*sottointeso*), and Tesauro, *Laconism*.[167] For Gracián, it was a mark of prudence to always suspect that there is more to statements or actions than what appears on the surface. Conceits invite the audience to be similarly suspicious, and always search for what the author has left unsaid. Even Tacitus was, according to Gracián, one "who signifies double of what he says."[168]

The detractors of Tacitus were among the first to connect the historian's habit of being constantly suspicious with a desire to fashion himself as ingenious. When Argensola claimed that early modern imitators of Tacitus wished to appear as "brave, and with a gallant brain," he knew that the ability of historians à la mode to transform statements or actions by attributing far-fetched and unheard-of motives to them acted as a form of prestidigitation on the audience.[169] Yet one can trace the connection between suspicion and metaphor even further by going back to Strada's criticism of Tacitus as a catalyst of mistrust and social and political ruin. According to Strada, Tacitus was prey to a fury that prevented him from taking appearances at face value: he was persuaded that every action or statement, by definition, answered to motives other than they appeared on the surface. Interestingly, Strada's diagnosis of the historian's habit or mental illness coincided with Tesauro's characterization of metaphor: "Mad people," Tesauro wrote, "are better conditioned to create conceit-like metaphors in their fantasy than sane people. In fact, madness is nothing but metaphor, which takes one thing for another."[170] Orators and poets who took one thing for another by means of metaphor did so because they suspected that reality concealed something from them. The audience, suspecting in turn that conceits left a part unsaid, likewise penetrated the author's heart. According to Tesauro, they "read the remainder in the heart of the person who wrote them."[171]

Conceits invited the listener or the reader to look for unstated, merely sketched meanings "within the author" just as courtiers searched for intentions and thoughts in the hearts of rivals. Admittedly, conceits only replicated the dynamics of interpersonal interaction in a playful and largely innocuous fashion. Unlike courtiers, who always made sure to keep others in the dark about their intentions and thoughts, inventors of conceits looked forward to being exposed. They produced conceits that were like "a picture in outline rather than finished, where [the audience's] ingenuity needs to understand what the tongue does not say"—and it was of the utmost importance that they manage to do it if conceits are to succeed.[172] Conceits—and Tesauro, like Gracián, also considered heraldic devices or *imprese* (singular *impresa*) as visual, rather than linguistic, conceits—fostered a bond that doubled as a challenge between two individuals. They gave rise to "the double enjoyment" in which the speaker feels "pleased to give life in someone else's intellect to a noble birth," while the listeners are "happy to steal with their ingenuity what someone else's ingenuity furtively hides."[173] Tesauro summarized the rationale behind the pleasure of reciprocity by maintaining that "to interpret a sharp and ingenious device commands sagacity no less than to compose it," a formulation that closely recalled Gracián's claim about Tacitus and the art of performing crisis of intentions and thoughts: "Just as acting with artifice and reflection is evidence of superior ingenuity, to reveal and expose that artifice attests to a form of ingenuity that is twice subtle over."[174]

Conceits fostered a relationship between speaker and audience, one that reenacted in playful terms the espionage of hearts that Gracián imagined taking place among courtiers. Consequently, just as Tacitus had become the author of conceits of crisis by exposing the ruler's artifice, so the listeners who spied the speaker's meaning could be considered at least coauthors of conceits. In the first century, rhetorician Marcus Fabius Quintilian scorned orators who made a virtue of speaking obscurely and believed that being ingenious implied testing the audience's *ingenium*. "They consider a mark of ingenuity," Quintilian lamented, "the fact that we too need ingenuity to understand them." Quintilian found it especially worrying that the listeners seemed to "delight in their own cleverness when they understand them, and rejoice as if they had not so much heard them as thought of them for themselves."[175] They felt as if they were the authors. Unlike Tesauro, who probably had Quintilian in mind when he explained to the reader that metaphor invites understanding "with your intellect," Quintilian believed that the audience should never engage in a process of divination.[176]

Quintilian's reception psychology directly impacted the views of Tesauro and Peregrini, even if they praised audience cooperation as much as Quintilian loathed the idea. They represented the point of arrival in a process that normalized the belief that conceits should engage the audience in a process of active and creative response to the author's language and reasoning. Peregrini, for one, cited verbatim Quintilian's description of how the listeners become, or believe themselves to be, coauthors of conceits. Pleasure, Peregrini claimed, results from "the audience's awareness of the part they play in understanding . . . the audience must cooperate somehow with their own sagacity, which is in fact a pleasant thing. Witness the fact that many people enjoy hearing, or reading obscure things, because (as Quintilian says), when they understand, they rejoice as if they had not so much heard them as thought of them for themselves. For, after all, as Thomas Aquinas said, those things are pleasant by which we get some knowledge and appreciation of our own value."[177]

Without a hint of criticism, Peregrini described a pleasure based on vanity and self-love. It is because of vanity, the Jesuit José de Ormaza (1617–ca. 1676) wrote in 1648, that we enjoy conceits: "We are pleased not so much by the ingenuity we discover in the author as by the one we imagine we display as we understand; and we enjoy it because in each clause we secretly flatter ourselves."[178] Without crediting any source, Ormaza paraphrased Quintilian's insights about audience cooperation, and compared the listener faced with "suspicious sayings" to one who probes deep waters and fathoms possible treasures everywhere. Following in Gracián's footsteps, Ormaza spoke of mysteries that await discovery, with greater difficulty resulting in greater pleasure. He nonetheless advised orators that they need to help the audience penetrate the part of the meaning that they leave half-said: in conceits, words "need to reveal the things they conceal, and to say the things that they do not say. As with good paintings, which, even though they do not show the back of the body, the shoulders and the sides appear in such a way that they allow us to see even the back that they hide from us."[179] Building on Gracián's language of "crisis," Ormaza foreshadowed Tesauro's radical perspective on audience cooperation as a playful and seemingly innocuous reenactment of the courtier's struggle to penetrate people's hearts.

Coauthorship, however, struck others as a dangerous and arrogant fiction. Just as Boccalini's glasses—and the *Annals* and *Histories* of Tacitus—made secrets of state available to high and low subjects, conceits likewise threatened to democratize authorship and the display of ingenuity by providing audiences with the belief that they, and not the speaker or the writer, were capable of reaching a deeper layer of reality. Some sought to relegate conceits to the troupes of mountebanks that entertained the people in marketplaces and fairs.[180] Others argued that most listeners did not understand conceits at all.[181] Yet evidence seems to suggest otherwise. When Tesauro labeled the kind of "preachable conceits" used in sermons as "popular," as opposed to theological, arguments, he did so to extol their wide appeal and reach.[182] Tesauro maintained that God, speaking like a poet, filled Scripture with riddles. The church fathers then penetrated God's enigmatic and poetic language, and now the preacher pierced through what the church fathers had expounded, bringing God's message to the audience as a sort of "exotic merchandise," which the audience, in turn, penetrated by suspecting that the preacher's language, full of tropes as it was, meant more than it seemed to say. In a chain of active and cooperative reception, Tesauro wrote, it was only fair that everyone should have a share in the applause.[183] In Tesauro's enthusiastic and successful description, conceits embodied the positive and joyful acceptance of a world in which the window in man's chest was not only superfluous but also a misguided and unwelcome denial of what made humans human.

Conclusion

Part I has demonstrated that seventeenth-century writings on conceits contain a striking and persuasive love letter to the sensory and intellectual limitations that rendered postlapsarian humans essentially different, and supposedly inferior, to the ideals embodied by Adam and angels. Conceits were evidence that humans should be grateful for sensory and intellectual conditions that previous authors and scholarly traditions, based on ethical, metaphysical, and theological grounds, had portrayed as shortcomings.

Chapter 1 revealed a pragmatic, even celebratory approach toward the inclination to novelty and artifice that seemed to be connatural to humans. Natural philosophers and writers on conceits alike justified pursuing the focus on rarity and unfamiliarity by advancing a provocative argument, namely that nature itself, by indulging in seemingly impossible or, at least, unusual phenomena, catered to the appetite for novelty and artifice that alone could elicit intellectual and emotional responses from humans. Rather than curb inclinations that were of nature's making, natural philosophers such as Nieremberg deemed it wiser to exploit those for the sake of learning. Conceits, which poet Achillini labeled as "poetry's comets," were rhetorical and poetic counterparts to nature's deviations from its usual course. They represented the intellect's ability to imitate nature when it becomes unpredictable: its subtlety or—as contemporaries often put it—artifice.

With a focus on artifice, chapter 2 has investigated what it meant for conceits to confront the awareness that nature—which they were supposed to imitate—had a taste for secrecy and the habit of deceiving with appearances. Writings on conceits measured up against nature's artifice by advancing two conclusions. First, if nature loves to hide, then it could be that any reality created by conceits was new and different only as a matter of perspective. Conceits were tools of invention as much as of discovery. Just as the microscope did not lie when it transformed fleas into elephants, so conceits in a way "revealed" the startling and unfamiliar truth that they had been able to penetrate, but which they had also created, in a complex and dramatic entanglement between making and knowing. Conceits became emblematic of a logic of sight-as-transformation that normalized the period's fascination with, and anxiety about, artifice. Second, if nature's devious behavior constrained humans to be always suspicious of appearances—especially because they had also learned to be indirect in their relations and negotiations with peers—conceits offered the opportunity to transform the barriers and opacity that made up reality from a problem into a new and promising opportunity. Writings on conceits made the case for a radical and historically unparalleled reassessment of suspicion as a structure underlying political and social reality. While Strada and other detractors of Tacitus deplored the notion of communities permeated by the erosion of trust among individuals, the likes of Tesauro celebrated suspicion as a site for humans to engage in a pleasing and harmless form of competition based on ingenuity, a trait that all of them shared, albeit in different degrees. The opacity that permeated the world and the language available to humans appeared to Tesauro not as a threat with the potential to sever social bonds, but as the ingredient that kept communication open, and society alive.

Part II of the book continues the exploration of conceits as imitations of nature's subtlety. Specifically, it claims that seventeenth-century writings on the topic pictured conceits intent on mimicking nature's ability to work at small scale and to create a variety of countless and minute creatures teeming with life. In doing so, I argue that writings on conceits actively engaged with contemporary natural-philosophical developments, which brought a share of scholarly attention away from celestial bodies to the study of insects and a number of other creatures that had long lived in obscurity.

PART II

Fireflies

CHAPTER THREE

At Small Scale

Multiplicity, Variety, and Play

Clouds of fireflies lit up the fields at dusk as Pier Francesco Minozzi took the road to Bologna. "Passing away so soon after birth," he observed, "they seem to be carrying both their funerals and their torches."[1] In the transient and flickering sparkle of fireflies, the young Minozzi glimpsed life pulsating in response to the vicinity of death.[2] Why give such rare and extraordinary light, he seems to have wondered, to creatures that die with the summer? Or was the presence of light necessary precisely because of their short and precarious existence? It was as though fireflies knew that they would perish sooner rather than later, Minozzi insinuated. By calling their lights "their torches" and suggesting that fireflies celebrated funerals as if they were humans, Minozzi had crafted a conceit. In *Delle acutezze*, published in Genoa in April 1639, Matteo Peregrini argued that a conceit was a device aimed at beauty rather than truth, one that short-circuited logical reasoning and shaped a new reality through figurative language.[3] Fictitious or real, Minozzi's encounter with fireflies took place two months later, in June 1639. Reported in a letter dated August 30, it was published only in 1641 in a cheap, pocket-sized volume titled *Sfogamenti d'ingegno* (*Outbursts of Ingenuity*).[4]

In 1639, fireflies were seldom merely insects. For Minozzi and Peregrini, they were code for conceits. In fact, Minozzi staged the encounter with the insects as a response to what he saw as Peregrini's ambivalence, if not contempt, for the rhetorical and poetic devices.[5] The analogy between fireflies and conceits originated in a family of Italian proverbs expressing the preference for a desirable and genuine object over a less valuable surrogate, identified with fireflies. The documented variants may be reduced to two basic formulations: "to take fireflies for lamps," that is, to take one thing for another as a result of misapprehension, and "to give fireflies for lamps," or pass one thing off for another in a deceitful manner.[6] The analogy with fireflies suggested that conceits were trifles, pale imitations of a solid, logically sound statement. That was true, at least, during the earlier stages of a controversy that took place in writings produced in Genoa, Rome, and Bologna from 1636 to 1639. Then, Minozzi's encounter with the insects outside the gates of Modena turned the tables of the debate. Minozzi confirmed that conceits were indeed like fireflies, only to insist that this was actually a strength that spoke in favor of both. The analysis of Minozzi's intervention in the debate and of its context demonstrates that conceits mattered not only because of all the noise they generated in rhetorical and poetic debates but also because, to seventeenth-century eyes, they were part of a new sensibility toward the study of nature that was in the process of redefining what counted as a topic worthy of study.

In Minozzi's hands, conceits and fireflies became protagonists in a new entanglement between natural and human ingenuity.[7] In chapters 3 and 4, I show that for Minozzi's contemporaries fireflies and conceits resembled each other in three aspects. The first was smallness. It was common for the less enthusiastic to depict conceits as small and insignificant units of rhetorical and poetic composition. Yet smallness had become increasingly fashionable: a reassessment of the place of insects and other minute creatures within the hierarchies of nature was underway, giving rise to new disciplines and groundbreaking attitudes. Minozzi's contemporaries approached conceits through the lens of this new interest in nature's creativity and craftsmanship at small scale. Insects, small in size but abundant in number and varied in form and color, provided a model for the inventor of conceits. As we follow in Minozzi's footsteps from Modena to Bologna and then to the city's creative hubs and learned circles, we find a web of interconnected threads that enable us to imagine a multifaceted and transdisciplinary approach to the period's ecologies of ingenuity. What could seem at first a circumscribed dispute on rhetorical and poetic devices—conceits—becomes a way to gauge the various and conflicting ways in which seventeenth-century individuals understood their intellectual and creative powers vis-à-vis those of nature. It also provides a different but complementary angle on the thin and porous border separating nature and art previously explored in part I.

Smallness, the subject of this chapter, is only the first of three aspects that conceits and fireflies had in common. Their vivacity and aliveness, as well as their weak but mysterious light, will be the subject of chapter 4.

The Beginnings of the Controversy

For seventeenth-century Europeans, comets lived in close quarters with fireflies. When poet Andrew Marvell (1621–1678) dubbed glow worms "country comets," he had in mind a resemblance between the insects and the fiery and ominous protagonists of part I.[8] Like fireflies, comets described capricious and unpredictable trajectories of light and were proverbially short-lived. Minozzi's meditation on the brief and precarious existence of the insects, in fact, resonated with the quest for measurements of the ephemeral and inconstant lives of comets that permeated writings on astronomy.[9] As Galileo and others experienced, it was only too easy to miss the spectacle if one happened to be ill or failed to be in the right place at the right time. Last, both fireflies and comets posed a challenge to nature's usual course, raising questions more often than answers. Conceits were "poetry's comets"—following Claudio Achillini's formulation in 1609—as much as they were "poetry's fireflies."

Conceits, comets, and fireflies had a fleeting but momentous encounter in a discourse by Agostino Mascardi, first published in 1622, that I discussed earlier. Reacting to a poem on one of the three comets that plowed through the sky in 1618, Mascardi criticized the author's excessive use of conceits, which he compared to the light of fireflies: "Conceits used to be like lightning. Pulled away from the wheel of the sun, they came down to kindle the intellect. Now, they are fireflies that in a single pulse of their wings both give birth to their innocent light and bury it; now they are flashes that vanish before they can be seen; now they are sparks that disappear at once, never solid and never mature."[10]

The claim that conceits had degenerated in likeness from sun bolts to trembling fireflies owed considerably to rhetorician and historian Famiano Strada, Mascardi's former professor at the Jesuit Collegio Romano. Strada's oration *De stylo oratorio* (*On Oratorical Style*), first published in 1617, recreated a conversation with scholars Lorenzo Cenami and Niccolò Tucci at Villa Guinigi in Matraia, near Lucca. In a friendly dispute regarding the craze for conceits—and following the conventions of Ciceronian dialogue—each interlocutor embraced a different, contrasting position in a process of shifting alliances. In the first exchange, Strada sides with a modern, enthusiastic Cenami, who disputes Tucci's assertion that conceits (*acumina*) were unknown to the ancient Greeks and Romans. Presented with example after example of Cicero's elegant and masterful conceits, Tucci concedes defeat not without adding a crucial caveat: Cicero's solid and magnificent conceits were fundamentally different from the false and vain counterparts cultivated by seventeenth-century orators and poets.[11] As a reader of Strada's oration, Mascardi was indebted to Tucci's distinction, which Strada, as a character in the dialogue, fully supports by siding with Tucci in the continuation of the debate. When Cenami dares to suggest that conceits need not be solid as long as they manage to briefly charm and deceive the audience, Strada—foreshadowing Mascardi's allusion to "sparks that disappear at once"—dismisses a device destined to crumble almost at once under the weight of deceptiveness.[12]

Strada decried a style based on the juxtaposition of brief clauses, mainly conceits, which sought to strike surprise in the reader at every juncture. It was a "broken," "loose," and "unchained" way of writing that others preferred to call "the laconic style" or "laconism," to bestow on it the pedigree of the neat and incisive brevity that Plutarch ascribed to the ancient Spartans (*Lacedaemones*).[13] Laconism was a controversial alternative to the intricate and long-winded periodic prose promoted by those who, like Strada and Mascardi, styled themselves as Ciceronians.[14] Modeled on the *Annals* and *Histories* of Tacitus and Seneca's writings on moral philosophy, the laconic style enjoyed the favor of the Flemish humanist Justus Lipsius and his disciple Herycius Puteanus (Errijck de Put, 1574–1646), who penned a passionate defense of it at the turn of the seventeenth century.[15]

With its nuanced but resolute criticism of laconism and conceits, Strada's oration provided a blueprint for Mascardi's more direct and vocal intervention in the rhetorical and poetic arenas. In 1622, Mascardi protested disingenuously that he did not condemn conceits altogether, or else he would have to count "all modern writers" as enemies.[16] Any trace of ambivalence, however, disappeared by 1636, when in *Dell'arte historica* (*On the Art of History*) he doubled down on the attack and cautioned historians to stay away from the laconic style. It was then that fireflies finally earned a second appearance at center stage as Mascardi elaborated on a letter from Jerome (ca. 342–420) to his friend and would-be priest Nepotian. Jerome's rejoinder against the use of "sharp and concise statements" and "flowers of maxims" reminded Mascardi of the laconic style in vogue in seventeenth-century Europe. At the same time, Jerome's indictment of a puerile or childish style probably conjured in Mascardi the image of children chasing after fireflies in early summer, too immature to realize they have been fooled by faint and false luminaries.[17] For Mascardi, conceits distracted the mind from the solid, truth-oriented forms that befit a prudent speaker, just as the insects enthralled the young and inexperienced. In a sinister turn to the analogy, Mascardi clarified that laconism was childish in a way that recalled not the vigor of youth but the doddering regression of old age.[18]

Mascardi's analogy between conceits and fireflies had gone relatively unnoticed in 1622.[19] Conversely, his remarks in *Dell'arte historica* unleashed a wave of reactions from both detractors and supporters.[20] Minozzi's digression on fireflies and conceits was indeed a reaction to the controversy initiated by Mascardi, hailed as "the prince of all living writers"—he would die within a year, in 1640. Minozzi granted that historians should stay away from conceits. However, what about writings that fell into the epideictic or demonstrative genre, which offers praise and blame? Why ban conceits from Minozzi's travel narrative and essay, or from saints' lives, a genre based on hyperbole, which could benefit greatly by deviating from logically valid argument and truth in the way that was typical of a conceit?[21] Minozzi remained largely deferential to Mascardi, even if he took time to refute the claim that conceits were evidence of regression.[22] His impatience and even his venom were reserved for Peregrini's *Delle acutezze*, a work that boasted a degree of hesitation and ambivalence he found difficult to stomach. Peregrini placed conceits at the top of the spectrum of beauty and pleasure but advised orators to mostly stay away from them.[23] Without citing *Delle acutezze* or even naming Peregrini, Minozzi signaled his call to arms by quoting verbatim several sentences from the treatise. One of them, it must be noted, was from a passage that reworked Mascardi's analogy between conceits and fireflies.

Like almost everything about *Delle acutezze*, the role of fireflies in it was also ambiguous. The insects appeared in a letter, quoted at length in the preface to the reader, in which Olivetan monk and mathematician Vincenzo Renieri asked for Peregrini's opinion on conceits: "When I am wretched enough to run into that kind of composition, the work of writers who care nothing about the study of the most solid eloquence and focus all their attention on the brilliance of such spirits, I hesitate whether to call them wits full of spirit (*spiritosi*) or rather possessed (*spiritati*). And it seems as if their authors had been reduced to the simplicity of children who, under a sky teeming with beautiful lights at dusk, do not raise their eyes even once, intent only on frantically chasing the flicker of fireflies."[24]

According to Elisabetta Graziosi, part of the ambiguity surrounding *Delle acutezze* dissipates if we interpret the treatise as Peregrini's response to

Mascardi. Despite the appearance of criticism, Peregrini investigated Greek and Roman views on conceits in order to build a theoretical and critical apparatus to support the advocacy of conceits by the members of Genoa's Accademia degli Addormentati, or "of the sleeping ones."[25] If Peregrini's treatise was, in fact, a response to Mascardi and if Peregrini was not as averse to conceits as he pretended to be, then one should take Renieri's seeming endorsement of Mascardi's analogy between conceits and fireflies with a grain of salt. That perhaps Renieri was not condemning conceits altogether—or, at least, that he was not entirely sincere—is consonant with the consensus among scholars that Mascardi, Peregrini, and Minozzi merely advocated for different frequencies in the use of conceits or preferred to confine their use to some genres but not others. If Renieri—and Peregrini with him—was mocking Mascardi's argument, it would follow that, unbeknownst to Minozzi, *Delle acutezze* already laid the foundations for renegotiating the terms of the analogy between conceits and fireflies and rethinking the link between the intellect and nature—that is, between the creation of conceits and that of fireflies.[26] Put simply, Minozzi took Peregrini as a target to knock down, when he should have taken him as a moderate but insightful predecessor.

The contrast between the sun and fireflies provided a plain and effective formulation to Mascardi's complaint that conceits fell short of logical reasoning or truth. Astronomer Renieri simply replaced the sun with the radiance of the stars in a clear night, leaving the rest of the analogy untouched. Renieri would have been reluctant toward the use of seemingly endless series of conceits, like the ones in which Minozzi often indulged. However, conceits also were part of Renieri's style. In 1640, he confessed in a letter to Galileo that a cold was not enough to prevent him from waking up often at night "to make love to the stars, who return my affections quite well."[27] More importantly, a learned reader would have heard in Renieri's passage the echoes of a tacit, unacknowledged source with the potential to subvert its meaning and implications.

A student of the motions of Jupiter's moons, Renieri had reason to prefer a bright and starry sky to the glitter of low-flying fireflies. Yet the silent presence of a third, unnamed scholar in Renieri's reply to Mascardi suggests that he was reluctant to dismiss the insects without due process. Back in 1622, Mascardi had referred to fireflies that "give birth to their innocent light and bury it in a single pulse of their wings." By describing perennially lit creatures that turn dark only by briefly covering their abdomens with their wings, Mascardi revealed he was following the first of two passages on fireflies contained in Pliny the Elder's *Natural History*.[28] Conversely, it was Pliny's second and longer discussion of the topic that played a veiled but potentially critical role in Renieri's response, rebalancing the contrast between stars and fireflies. Pliny recalled that peasants called "such star-like flying creatures" (*stellantes uolatus*) *cicindelae*—in Greek they were called λαμπύριδες. Fireflies were evidence of nature's providence, for in them nature had duplicated the signs that announced the time to reap the barley and sow the millet. Fireflies, as another set of stars descended to the earth, made it unnecessary to look for the Seven Sisters or Pleiades in the sky. It was as if nature was crying out loud, "Farmer, why gaze at the heavens? Why, rustic, search for the stars? . . . Look how I hurl stars at your feet to be your own."[29]

Renieri's complaint echoed Pliny's passage even while inverting its logic: "why look down?" replaced "why look up?" With the fireflies transformed into Pleiades strewn at the farmer's feet, Renieri's full-throated complaint became as ambivalent as Peregrini's subsequent caveats against conceits. It has been observed that "the regularity and predictability of Pliny's heavens seem rather lifeless" in comparison to the author's portrayal of nature's restless activity on the earth and underground.[30] With its inversion of hierarchies and its preference for the world around us, the passage from Pliny haunted Renieri's adoption of the analogy between conceits and fireflies, suggesting that the insects stood a chance against traditionally more revered sources of light—and that, perhaps, the same held true about conceits. The injunction to keep one's eyes down implied that fireflies related to human minds and hearts in ways that remained out of reach for the magnificent but irremediably distant brightness of celestial bodies. By analogy, it also suggested that conceits could touch the audience in a way that the stable and conclusive character of logically valid and true statements rarely, if ever, attained.[31]

Renieri's echoes of Pliny failed to impress Minozzi, who cited—and reversed—each of the arguments that thus far had served to disparage both conceits and fireflies. Minozzi admitted that he was just as eager as the children in Renieri's analogy to be caught in the thrall of fireflies and conceits. He found both attractive precisely for the same reasons that rendered them suspect to others, not least their unusual and ephemeral nature. "They truly are Venuses," Minozzi wrote, "the briefer the sweeter. Lively and sharp statements are ingenious Circes that charm my intellect."[32] He confessed a weakness for the works of Count Carlo Della Lengueglia (ca. 1600–post-1681), who presided over a canon of what one might call "firefly writers." "They entertain me, flatter me, and enchant me," he enthused in reply to Renieri's rebuke of conceits as products of a soul possessed by the devil.[33] Conceits were "spirits," Minozzi argued. This is why they were also called "sparks of life" (*vivezze*).[34]

The insistence on life was part of Minozzi's careful and multifaceted attempt at reversing and repurposing the analogy between conceits and fireflies, while subscribing to a sentiment that Renieri's Plinian subtext evoked only timidly. By preferring fireflies—and conceits—Minozzi aligned with a new sensibility toward nature, and a new scientific and cultural habit that, just as in Pliny's rejoinder to the farmer, looked down and around at a world teeming with restless and wonder-arousing creatures that existed at small scale and, like conceits, deserved attention for different reasons than the sun and the stars.[35]

A Time for Fireflies

Originating in a proverb, the analogy between conceits and fireflies first advanced by Mascardi conjured up a hierarchy that for centuries had informed both scholarly and popular perceptions of nature. The sun and the stars, which Mascardi and Renieri placed in contrast with conceits, were obviously positioned above fireflies in spatial terms. Yet celestial bodies were also granted higher scientific consideration for reasons that were, to a great extent, theological and metaphysical rather than strictly natural-philosophical. The views of Aristotle re-elaborated across textbooks and lecture halls in early modern universities held celestial bodies to be incorruptible substances made of ether, removed from the contingent and constant transformation characteristic of life on earth.[36] Enlisting fireflies in service of a new and dignified status for conceits, therefore, implied questioning the order of nature itself and the value attributed to each of its creations.

Insects occupied the lowest tier among the animals right above zoophytes, which straddled the class of animals and that of plants. Aristotle had claimed that warmer creatures, up on the ladder, gave birth to living beings, while cooler ones laid eggs or even larvae. A more or less fluctuant taxonomy, well established by the time of Albertus Magnus in the thirteenth century, described nature according to a ladder that organized species by degrees of perfection.[37] Insects were small, short-lived, and, by most standards, insignificant; they were also "imperfect" in the sense that they did not comply with the rule of generation of like by like.[38] By the time of Minozzi's encounter with fireflies, it was still widely believed that the propagation of insects did not involve intercourse between two parents. Insects belonged to a lower order in which a bewildering, but fascinating, set of rules replaced the principles of sexual reproduction. Substances gave birth to creatures that were unlike them—mud could engender a toad, and a bull, a bee. A number of species were believed to arise by "spontaneous generation" out of corrupting matter, a belief that developed as a result of difficulties in observation due to the minuscule, almost imperceptible size of the eggs, and misconceptions such as the assumption that butterflies and moths were born from the caterpillar's dead body.[39]

The views on spontaneous generation, however, were far from monolithic, and the jury was still out as to the species that resorted to it as a mode of propagation. Take, for instance, the case of Aristotelian natural philosopher Julius Caesar Scaliger, who in 1557 claimed to have caught a pair of fireflies mating.[40] Even with the evidence in hand that fireflies were not animals "born from putrefaction," but reproduced in a way similar to mammals and birds, Scaliger did not see any reason why he should extrapolate his observation to insects in general. Both the public and the scholarly community were likewise reluctant to abandon the belief in spontaneous generation as a possibility. It would take decades of research, experimentation, and heated controversy to bury forever the fascination with a realm of existence marked by indistinct boundaries and constant change, populated by alternative models for understanding generative and creative processes of both nature and art.[41]

The role of fireflies in Minozzi's unconditional endorsement of conceits is a symptom of deep, ongoing transformations in early modern attitudes toward nature. Among those was the emergence, among scholars and artists, of growing interest in insects, the appreciation of which would shift in the period under consideration, as one scholar put it, "from revulsion to fascination."[42] In Bologna, Minozzi visited the house of naturalist Ulisse Aldrovandi (1522–1605), a new Pliny, "a secretary of nature" who, "in almost colossal volumes in speaking of terrestrial, aerial, and aquatic animals, as well as of insects, has made public to the world a new world."[43] With *De animalibus insectis* (*On Insects*, 1602), a work self-described as the first treatise exclusively dedicated to God's smallest creatures, Aldrovandi played a crucial role in the renewed interest in insects that threatened to bring down preexistent hierarchies. Aldrovandi did so, however, motivated by reasons that were natural-philosophical as much as they were aesthetic, giving rise to a series of significant intersections with Minozzi's advocacy of conceits.

A web of scholarly references populated the margins of *De animalibus insectis*. In particular, those in the first chapter revealed that theologian Augustine of Hippo had served Aldrovandi as a source of much-needed arguments to dignify the discipline we today call entomology. As has been mentioned in chapter 1, Augustine suggested that insects are taken to be a nuisance only because the role they play in nature remains elusive to us; he also believed that small bodies are more apt than larger ones at eliciting wonder at God.[44] Augustine's explanation, transcribed verbatim by Aldrovandi, originated in Pliny's eulogy of nature as a skillful and meticulous artisan. Pliny had argued that, in the case of insects, smallness compounded the challenge of shaping matter. The limbs in the body of a gnat or a moth were diminutive, requiring a great deal of ingenuity (*ingenium*) and subtlety (*subtilitas*) in order to determine the place of the senses in such a small body and to handle and attach each part in such a way that they would fit together and function properly.[45]

As has been discussed, in 1550, Girolamo Cardano defined subtlety as the quality that made a substance difficult to perceive by the senses or understand by the intellect. Seven years later, Julius Caesar Scaliger retorted that nature's subtlety was merely a function of human sensory and intellectual shortcomings. Put simply, humans call subtle any object that they are unable to properly grasp.[46] It was, without a doubt, a perceptive remark, yet Scaliger was swimming against the current, clinging to a superseded distinction between the natural and the artificial entirely out of touch with early modern sensibilities, only too prone to judge nature's creations as artificial. Artists and natural philosophers marveled at small, almost invisible bodies equipped with the same properties and functions as larger ones.[47] In the wake of Pliny and Augustine, Aldrovandi argued that God and nature had placed in insects "far greater reasons to wonder than in the rest of larger animals, as if to teach us that we should measure things less by their dimensions than by the amplitude of their virtue." Just think of how a sizeable rock compares to a minute but precious gem.[48]

In the works of Augustine, Aldrovandi found a second reason to esteem insects, in addition to artifice. Augustine had claimed that a fly was more representative of God's power than the sun that allows for the growth of every creature. The fly, Aldrovandi claimed following in Augustine's footsteps, had senses, whereas the sun was inert. It also had a soul that made its minuscule limbs alive and vibrated its wings as it flew through the sky. The fly should be preferred to the sun, "just as what is intelligible is to be preferred to what is sensible," he concluded.[49] Even by Aristotelian standards, Aldrovandi implied, only humans were placed above insects, which boasted superiority over celestial bodies.

Aldrovandi set the tone for the next generation of scholars. In the years leading up to Minozzi's encounter with fireflies, justification for the interest in insects often entailed appreciating God's artifice or subtlety and the life associated with the presence of a vegetative and a sensitive soul.[50] Jesuit professor Juan

Eusebio Nieremberg elaborated on the same two points in the lecture he delivered in 1629 at Madrid's Reales Estudios. Nieremberg remarked that God's ingenuity was more salient when he worked at small scale. Ancient Greek philosopher Anaxagoras (ca. 500–ca. 428 BCE) claimed that the sight of the sun could fill a philosopher's entire life, yet in a passage with Plinian and Augustinian undertones, Nieremberg retorted, "He needed not look so high. He would have been more satisfied with the sight of any small creature: a bee, a little worm, a gnat, in which he would be better able to see the ingenuity of the artificer."[51]

In addition to being evidence of God's subtlety, for Nieremberg, insects also surpassed celestial bodies because they were living creatures. Life coupled with sense placed insects above the stars and the planets. In a remarkable but hardly innovative display of conceits, Nieremberg reminded the audience that God "did not ask [Adam] to contemplate the stars as they gather for evening events in the salons of the sky, or the dances of the planets, or the moon as it changes its mood; he mustered before his eyes only the animals, and the smallest of birds."[52] According to Genesis, Nieremberg concluded, living and sentient natures were the first to receive God's blessing.[53]

It was not the first time that Augustine had played a role in early modern controversies about nature and its study. Natural philosophers, including Nieremberg, had had to come to terms with Augustine's view that a curious and inquisitive approach to the processes of nature was a manifestation of intellectual wantonness.[54] Unlike in the battle over curiosity, however, Augustine came to the aid of scholars interested in insects. For him, the presence of life coalesced with the relation of inverse proportionality between artifice and size, a notion best encapsulated in a tag from Pliny that would become central to seventeenth-century natural-philosophical thought: "The nature of things is nowhere more present in its entirety than in its smallest creations."[55]

Although insects in general embodied artifice and life, fireflies possessed, in addition to that, a source of light that seemed to defy nature's usual behavior, rendering them attractive in ways that, among celestial bodies, only comets or new stars could even approach. Appropriating the analogy between fireflies and conceits was merely the first step in Minozzi's argument. Despite the comparison's more or less casual origins in a proverb, Minozzi must have imagined both fireflies and conceits to share the three properties outlined so far. Both were small and teeming with life; more importantly, they irradiated a strange and mysterious "light" that commanded attention and curiosity. I devote the rest of chapter 3 to smallness, leaving artifice and life for chapter 4.

Smallness

Minozzi's love letter to the minute, small-scale charm of fireflies corresponded to his taste for conceits and the laconic style.[56] To prove that these were two sides of the same reality, he burst into a string of conceits that applied diminutive suffixes to the insects: "luminous small animals," "little flying stars," or "small flames with soul."[57] Smallness had become a target of affection as well as a category of praise.

The connection between conceits and life at small scale transpired from two orations delivered fifteen years earlier by Collegio Romano professor Giovanni Battista Ferrari, which Minozzi certainly read in one of their several editions and reprints. In fact, Minozzi's rant against critics who "scrape and gnaw" at conceits borrowed a pun on two verbs (*radere, rodere*) from Ferrari's lament over the loss of a work in which Cicero's freedman Tyro allegedly compiled a repertory of conceits or *sales*. The fact that Tyro's collection was no longer extant, Ferrari claimed in reference to the sharp and piercing effect of conceits on the audience, had left the present without evidence of the ancients' "scraping and gnawing" style.[58]

The Polish Jesuit Maciej K. Sarbiewski was one of Ferrari's interlocutors in Rome and recalled him claiming to be "sharp (*acutus*) in his speech."[59] Sarbiewski tied the smallness and the beauty of conceits incidentally when he described them as "that small and most beautiful part of human eloquence."[60] However, it was Ferrari who, either during Sarbiewski's stay in the city or soon after, placed the link between beauty and size at the center of a web of theological and natural-philosophical considerations applied to conceits and laconism. It was then, around the year 1625, that a number of meditations on conceits emerged in Rome, at the same time as Emanuele Tesauro codified in Milan some of the principles that three decades later would crystallize in *Il cannocchiale aristotelico*.

The frontispiece to the 1627 edition of Ferrari's orations provides a point of departure to explore the connection between smallness and conceits (Figure 3.1).[61] Two female figures sit on opposite sides of the page. The one on the left holds the boat of Saint Peter, with a bee from the coat of arms of Maffeo Barberini, who had become Pope Urban VIII in 1623. The one on the right, holding a rose, might represent Ferrari's position as a consultant for the gardens of Palazzo Barberini as a client of Cardinal Francesco Barberini, but it more likely signifies the rose in the coat of arms of Ferrari's former patron, Cardinal Alessandro Orsini, who had passed away in 1626.[62] The inscription *In tenui labor / aculeata placet*, visible at the bottom of the page, reunites the bee and the rose in connection with Ferrari's

Figure 3.1. Frontispiece of Giovanni Battista Ferrari, *Orationes* (Rome: Francesco Corbelletti, 1627). Biblioteca Nazionale Centrale "Vittorio Emanuele II" di Roma.

views on conceits and the laconic style. Taking cue from Vergil's *Georgics* (first century BCE), which referred to apiculture as "a work on a small subject, but of not small glory" (*In tenui labor, at tenuis non gloria*), Ferrari's modified adage could be translated as "a work on a small subject is pleasant when full of stings"—or thorns, following the definition of the adjective *aculeatus* in early modern lexicons: stinging like a bee or a horsefly, or the thorns of a rose.[63] Ferrari advocated a sharp, pointed style, a choice that he described as modest but effective. As he dedicated to Orsini a dictionary of Syriac in 1622, and then the first edition of *Orationes* in 1625, Ferrari jested with the rose in Orsini's coat of arms: his style was like a thicket of thorns that saved the flower of his patron from attacks. The allusion to thorns hinted at the lack of refinement of Ferrari's language but also its pungent and spicy character, which awakened the reader the way a conceit does.[64]

Thorns and stings linked Ferrari's advocacy of the laconic style with conceits to his status as a client, first of the Orsini and now of the Barberini families. Yet a further and presumably deeper layer to the frontispiece connected aesthetics with theology. A child ostensibly representing Christ stands surrounded by grotesques at the center of a ribbon, in which one can read: *Verbum breviatum fecit Dominus*. The motto, translated in the 1582 Catholic Douay-Rheims Bible as "a word abridged shall Our Lord make upon the earth," connected theology and conceits in a way that became clear only to one who, going beyond the frontispiece, decided to open and read Ferrari's volume.[65]

A professor of Hebrew and rhetoric, Ferrari vindicated the smallness of conceits through a peculiar comparison between himself and Jerome, the translator of Scripture with whom he shared a love of Hebrew.[66] Ferrari paired Hebrew and conceits in two orations titled "The Safety of Hebrew Letters" and "Hebrew Style," both printed for the first time in the 1627 edition.[67] To express preference for brevity over the prolix and long-winded periods of Ciceronian eloquence, the orator resorted to a parallel with Jerome's conversion from Ciceronian Latin to the language of Scripture. In a letter that has become famous, Jerome avowed a certain distaste, even rejection, for the uncouth style and language of the prophets, which constrained him to set aside the eloquence he had worked so hard to gain in Latin. Jerome even found Hebrew's sounds strident. In a dream on a certain night, a judge appeared to him and accused him of being a Ciceronian more than a Christian (*Ciceronianus es, non Christianus*).[68] Jerome's eventual conversion to Hebrew brought about in him a new sensibility for the language of Scripture, based on a gradual appreciation of its ambiguity and elegant wordplay, itself a product of the multitude of meanings concealed in each term.[69]

In a shrewd and far-reaching maneuver, Ferrari adopted the position of the later Jerome and ascribed the views of the earlier Jerome to those among his contemporaries who despised Hebrew and refused to exchange Cicero's Latin for a language that they deemed perhaps sacred but utterly barbaric.[70] Between Ferrari's perspective and that of his rivals there was the gap that separated one who has already seen the truth from those who remain in denial. Ferrari likened the brevity and the ambiguity of meaning that Jerome had praised in Hebrew to the virtues that excited the intellect of those with a flair for conceits and laconism. He saw in Hebrew a model of tight and abridged diction, rich in equivocity, and entirely unlike the verbose and circuitous Ciceronian style.[71]

Ferrari's eulogy of laconism and brevity culminated in a theological argument that also clarified the meaning of the frontispiece. In Scripture, Ferrari claimed, God expressed his infinity through brevity, conveying much with little. Using a conceit, Ferrari explained that the brevity of God's language materialized in the incarnation of Christ, his word, in which God's immensity became circumscribed, through "a most capacious contraction," to "the small period of a child's body." It was a formulation in which syntax became corporeal.[72] Ferrari compared God's incarnation in human matter and form, which he called the greatest of miracles, with the ideal of the orator's task: to create a brief and condensed conceit brimming with a latent and unexpected meaning.[73] Christ the child, Ferrari claimed, could persuade the temple's loquacious priest, who stood for early modern Ciceronians in the analogy.[74] As it turns out, the representation of Christ in the frontispiece and the verse in the ribbon placed God's incarnation in the body of a child into close connection with the brevity and ambiguity that Ferrari considered characteristic of Hebrew as a language, and, by analogy, the ability of conceits to condense virtually endless meanings in a brief statement. Ferrari's frontispiece was a remarkable instance of the intricate and transdisciplinary character often assumed by the debate on conceits, which here combined theological lore, natural-philosophical emblematics, classical and biblical philology, rhetorical and poetic theory, and visual language.

Moving forward, Ferrari resorted to Pliny as he remembered that "the nature of things is nowhere more present in its entirety than in its smallest creations."[75] Ferrari found himself at a turning point in his scholarly and professional career. A consultant for the gardens of Palazzo Barberini, perhaps Ferrari had already started work on *De florum cultura* (*On Flower Cultivation*, 1633), a treatise on ornamental agriculture. In it, a plate featuring both whole and half-cut hibiscus seeds represented one of the earliest instances of a printed illustration made with the aid of a microscope, to which Ferrari had access thanks to his connections with the Accademia dei Lincei (Figure 3.2).[76] Smallness, as a characteristic of conceits, thus resonated with Ferrari's increasing interest not only in nature at small scale but also in the ability of art to create miniatures, among which he included "Homer's extremely long *Iliad* writ on a parchment so small that it could be housed in a nutshell" and Dante's *Divine Comedy* "locked into a codex that hardly surpassed the size of a human nail."[77]

Ferrari's praise of miniatures drew on a traditional catalog of perplexing and often improbable objects, mostly borrowed from ancient authors, including Pliny, Solinus's third-century epitome of Pliny, and the list of "idle arts" (ματαιοτεχνίαι) in Quintilian's *Institutiones oratoriae*. It was not without risk to mount a vindication of a controversial topic such as conceits through a comparison with objects that were similarly problematic. Idle arts were, for Quintilian, the useless results of squandered talent, useful only for the exhibition of the artist's ingenuity.[78] In 1596, a few years before Ferrari's arrival at the Jesuit Collegio Romano, Francesco Maria Fattori cautioned students against wasting their talent on useless ventures, such as the chariots that Myrmecides once built out of ivory, so minuscule that they could ostensibly be covered by the wing of a fly.[79] Was it not common knowledge that Aristotle had warned against the perils of scale? For the eye to experience beauty, it needs to distinguish each of the parts in a body and the way they fit together.[80] Miniature artworks made that nearly impossible.

The target of countless satires, miniature artworks were nonetheless ubiquitous and highly appreciated in the early modern cabinets of curiosity fashioned by European aristocrats and scholars. Two carriages "like those of Myrmecides," with a driver, four horses, and ladies inside, such as could be comfortably covered under the wing of a bee, were part of the collection established in Milan by Manfredo Settala (1600–1680), a former student of Tesauro. Settala had shaped them with his hands when he was only a student in Pisa, and they were so appreciated by Cardinal Flavio Chigi that he filled his museum with some of Settala's miniature artworks.[81] Among his creations was also a cherry pit made out of ivory with a hundred skulls on its convex side and a game of chess inside, a camel standing on top of forty armed men fighting each other passing through a needle's eye, and a loom that could be used to weave a spiderweb. There were also "a few other minute works, which also served as evidence of great ingenuity," as Pietro Francesco Scarabelli declared in his catalog of Settala's collection.[82] In 1664, the Jesuit Gaspar Schott (1608–1666) boasted that the moderns were even more diligent than the ancients in both fashioning and recording wonders at small scale.[83] Yet enthusiasm

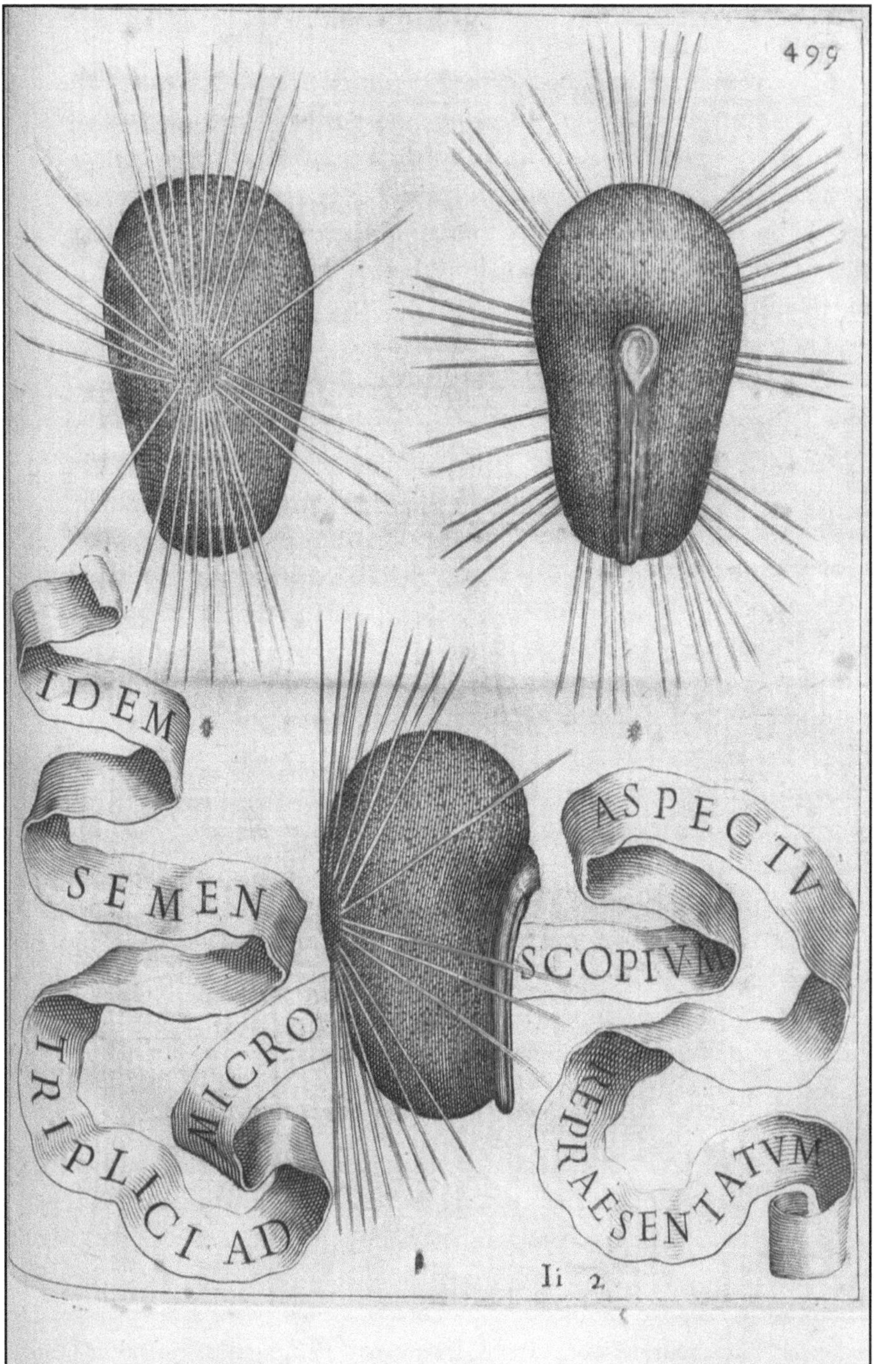

Figure 3.2. Cornelis Bloemart. Engraving of whole and half-cut hibiscus seeds seen at the microscope. In Giovanni Battista Ferrari, *Io. Bapt. Ferrarii . . . De florum cultura libri IV* (Rome: Stefano Paolini, 1633), 499. Courtesy of The Linda Hall Library of Science, Engineering, and Technology.

was far from universal. When painter Vicente Carducho visited the house of collector Juan de Espina in Madrid, he was both admired and perplexed by Espina's assortment of miniature artworks, so much so that he exclaimed, "I do not know what all this can be good for."[84]

On the threshold of visibility, miniature artworks occupied a liminal position in both aesthetic and moral terms. Fattori had warned the audience that in failing to guide one's ingenuity toward a purpose, as happens when the only goal in view is to put one's capabilities to the test, the intellect falls prey to arrogance and looks only for applause: "useless," in brief, was merely one step before "dangerous."[85] The association with miniature artworks placed conceits in risky territory. Even a supporter, writer Giovanni Battista Manzini—who knew Ferrari's orations well—refused to link ingenuity and miniature artworks. The inventor of conceits, Manzini averred, should prefer audacity and grandeur over the subtle and painstaking craft of objects that exist only at a subhuman, inferior scale.[86]

Ferrari praised a brief and sharp style by analogy with a group of objects that seventeenth-century scholars admired and feared in equal parts, products of unrestrained and undirected talent aimed only, or mostly, at receiving applause. Although the association with miniature artworks had the potential to backfire, it was in line with Peregrini's claim that conceits were brief statements in which the author's ingenuity took center stage, as the audience wondered at the way he or she had been able to fit the parts together.[87] And while Peregrini voiced concerns that ingenuity, if noticed, was incompatible with persuasion, Minozzi's reply was plain and simple: he did not see any problem with the fact that conceits were mainly about ostentation.[88]

Recent scholarship on a variety of topics, ranging from seventeenth-century still lifes to dollhouses, has demonstrated that interest in nature's smallest creations and miniature artworks was often associated with female intellectual and artistic practice.[89] Was it only a coincidence that Ferrari placed conceits in a space often perceived as "feminine" by his contemporaries? Although it is difficult to answer that with any certainty, there is evidence that the act of conceit making was subjected to a conscious and widespread gendering. In 1644, while lecturing in Mexico City, Baltasar López argued that instead of the penetration of reality carried out by disciplines of knowledge, which López sexualized as "masculine," conceits were capable, at most, of tickling the skin.[90] For López, perceived lack of virility was all that it took to discredit conceits. It was not like that for Ferrari, who took pride in bee stings and rose thorns that bore little resemblance to swords. A few years before he published a pioneering work on ornamental agriculture and flowers, Ferrari advocated uses of ingenuity and artifice that were not primarily driven by expediency and convenience. Through conceits, he promoted a space of ambivalence that challenged conservative and misogynistic approaches to the use of intellectual talent.

Multiplicity

The sheer number and the variety of insect populations, Pliny observed, compete for attention with the artifice of their bodies and behaviors.[91] Like insects, miniature artworks also tended to be conceived in assortments. Smallness allowed them to become portable and collectible commodities, hoarded in cabinets of curiosity in which intellectual and sensory gratification embraced early modern forms of capitalism.[92]

In natural and human creations alike, multiplicity served as a counterpart to smallness. Yet, as scholars have noted in connection with cabinets and *Wunderkammern*, multiplicity and variety affected the beholder's experience of each object in a collection.[93] The same was true of conceits, which, to the horror of detractors, came in multiples. In 1644, the Jesuit Daniello Bartoli took a jab at "the modern style with conceits." He ridiculed the claim that conceits were similar to ground pearls and minced gold in "a precious mosaic of a thousand ingenious thoughts," a new version of the portrait of Pompey made up "of diamonds, rubies, sapphires, carbuncles, and pearls" once derided by Pliny.[94] Bartoli was unimpressed by the assertion that each conceit in a series was a subtle and valuable component of the whole, and decried "the mincing of periods torn into minute pieces, resulting from the multitude of so many diminutive little things, each completing its meaning and shifting to a new thought."[95] They were like sprinkles from a fountain, he thought, instead of a stream that carries the reader along with the strength of eloquence.[96]

Like fireflies swarming in the forest at dusk, conceits also favored a logic of accumulation. Fireflies and conceits not only questioned well-established hierarchies but also undermined the principle of unity that, at least since Aristotle, had governed rhetorical and poetic creation.[97] Conceits constrained the author to a sequence of brief acts of invention potentially repeated ad infinitum, to the detriment of a coherent and organic edifice. They dictated a discontinuous, at times spasmodic mode of composition that resembled the flicker of the fireflies and shaped reader expectations and enjoyment patterns by normalizing a fragmented and intermittent attention span. Writing from Madrid in 1641, António Lopes da Veiga complained that contemporary readers cared little about where the composition was leading them, intent rather on enjoying a string of brief sparks of pleasure. Once a poet à la mode, Lopes da Veiga contended that readers had become incapable of seeing the forest, focused as they were on the trees, sensitive only to the minor and transitory light of "some little conceit woven with the thread of sophistry."[98]

Even among the supporters of conceit-laden composition, the loss of a unifying principle could be felt with a touch of nostalgia. In a dialogue published in 1629 by Cristóbal Suárez de Figueroa (1571–1644), one of the characters reads from a notebook as he goes to the encounter of two friends for a walk on the hills of Posillipo, in Spanish Naples. Asked about the contents, he tells the others that he was reading works by "that monster of ingenuity, that phoenix of conceits," Luis de Góngora (1561–1627). Góngora, who had recently passed away, was for the speaker the best of Spanish poets and "the modern Martial."[99] Swiftly turning to more candid criticism, however, he avows that "all of [Góngora's] works are made of a variety of fragments. Even though they are gold pieces, it is regretful that he did not apply his superior ingenuity to a long and continued work. For without a doubt, it would have resulted in a sky incrusted with jewels."[100] Others were satisfied with a discourse composed as "a string of pearls, each part a jewel and the whole a treasure," as Manzini put it.[101] The reference to fragments as the units of appreciation in the works of "the modern Martial," as advanced by Suárez de Figueroa, seemed to confirm a new model of reading in which variety came to the rescue of smallness and its pendant multiplicity.

Conceits developed in conjunction with other successful and long-lasting anti-organicist modes of composition. Early modern collecting—including cabinets of curiosity, *Wunderkammern*, and *studioli*—built on a taste for accumulation and heterogeneity largely unaffected by sporadic attempts at imposing classification and order. Conceits also fit well with the emergence in the market of art of a class of objects created to stand alone—collectibles that need not become part of any whole save that of the collection itself. Michael W. Cole describes the advent of transportable, self-standing small bronze statues in late sixteenth-century Florence, and Michael Fried has studied a similar process regarding the appearance of autonomous and portable "gallery pictures."[102] Similar to these and other sculpture and painting forms, which became associated with a new class of collectors and a different approach to objects, so conceits were harbingers of new reading behaviors and patterns. Anticipating by several years Minozzi's imagination of discourse as a form of serialized exposure to brief and dazzling conceits, Góngora had set off a furious debate on poetic genres in

Spain. Góngora's *Soledades* (*Solitudes*, 1613), the poem that Suárez de Figueroa's character both praised and blamed as a failed attempt at forming "a long and continued work" despite over a thousand lines of poetry, had dared to imply that there could exist a long composition in verse where little happened, and where the unity of plot was not exactly absent, but merely irrelevant. What mattered was the ability to come up with ever new and surprising conceits that led the reader to experience wonder yet one more time.[103]

Multiplicity stood at the core of Gracián's *Arte de ingenio*, which contrasted the serial and discontinuous succession of conceits with a more coherent or "complex" alternative in which conceits might become fused into a whole, what Gracián labeled "the moral bond of discourse."[104] Feigning neutrality, he provided arguments both for a "free" and "unleashed" pattern and for one made to "fit."[105] Yet Gracián could not suppress a preference for freedom, which he associated with the Spanish temperament he shared with Martial, the model of conceit making born in Bilbilis, near Gracián's home city of Calatayud:

> Freedom of wit has always prevailed in Spain—not for lack of invention, but because of the nation's gravity or its choleric temperament.... Seneca's prudence never allowed him to become enslaved to the severity of a discourse, to the affectation of a scheme.... Our city of Bilbilis did not offer up monsters, as Africa did, as its tribute to the world's empress [i.e., Rome], but formed the monster of acuity and refinement. Martial arrived in Rome to pursue studies in oratory. Yet being sharp like a bird of prey, he could not bear the leash of a tongue in shackles, and flew free toward the sky in as many feints as the epigrams he wrote.[106]

Just decades before the hot and dry temperament associated with Spanish ingenuity became a target of criticism in a continent that had become wearied of conceits, Gracián's remarks proudly asserted the freedom that ancient and modern Spanish writers had in common, crystallizing in a discourse made up of a multiplicity of conceits, liberated from the constraints of unity.[107] The present state of Spanish letters, "greatly fertile of ingenious minds, and all of them discoursing with freedom," further confirmed Gracián's intuition that the Spanish temperament thrived in a style that favored juxtaposition and multiplicity over the focus on unity.[108] In fact, although the views espoused by Minozzi and Manzini were evidence that Italians also favored "freedom of wit," even there certain forms of laconism came to be perceived as Hispanicizing. Thus, in 1616,

Cardinal Federico Borromeo (1564–1631) recommended continuity and coherence of style instead of the Spanish habit of piercing the reader's intellect with incessant, self-contained conceits.[109]

The contrast between the "free" invention of more or less disjointed series of conceits and the "slavish" subjection to some form of overarching unity might strike today's reader as metaphoric and oddly hyperbolic. It made sense, however, considering the link that early modern sources were eager to point out between *ingenium* or ingenuity and the adjective *ingenuus*, meaning "freeborn" or liberal in the sense that one would oppose liberal to mechanical arts as worthy occupations for one of noble birth.[110] The ability to come up with a multiplicity of conceits without regard to a plan implied a certain aristocracy of the mind, if not social status. Citing none other than Mascardi, Mario Biagioli has argued that court culture promoted similarly casual attitudes: going from flower to flower or from gem to gem was a tactic to avoid espousing commitments that might prove dangerous or inconvenient in the long run.[111]

Yet flowers and gems are pleasing to the sight in part because each has its color and shape that render it different from the rest. It was the sum of multiplicity and variety—of animal and plant species—that gave the earth enough confidence to rival celestial bodies in the preface to Giovanni Battista Hodierna's *Opusculi* (1644).[112] Variety also made seemingly endless series of conceits tolerable and a source of pleasure, while further connecting natural and human forms of creativity.

Variety and Play

"In most authors," Gracián cautioned, "conceits are criticized not because they are too bold, but because they are all of a kind, and express the same sort of thought."[113] The multiplicity of conceits may be the index of a free mind, but it is of little use unless it is accompanied by variety. By arranging *Arte de ingenio* as a catalog of kinds or "species" of conceits, Gracián acknowledged that variety lay at the center of invention.

Most of Gracián's contemporaries, however, found it both arrogant and implausible that individual writers aspired to embody all of nature's variety. Gracián himself acknowledged that the ancients had considered conceits a matter of one's innate abilities rather than a skill that one could learn from a teacher or a book.[114] How-to manuals on conceits stood little chance of working, by some of their authors' own admission. Sarbiewski devised a method but had fears that it might curtail spontaneity.[115] Peregrini's self-defeating claim that only a buffoon would take method too seriously was likewise a strange and somewhat unexpected conclusion for a long and detailed exposition of the art.[116] Even Jesuit professor Michael Radau, who devised a method of diversifying conceits based on Sarbiewski's previous work, had to concede that "each writer will practice diligently several modalities depending on his or her genius."[117] Each writer, after all, feels inclined to the invention of certain kinds of conceits rather than others.[118] These were sobering remarks: the project of transforming writers into wellsprings of endless variety, Radau seemed to argue, was destined to fail.

Conversely, Gracián remained a staunch believer in the possibility that art, taken in the sense of method, can help individuals become as varied as nature, overcoming the limitations associated with each one's temperament. Gracián's optimism left some critics perplexed. In 1648 the Jesuit José de Ormaza scorned *Arte de ingenio* for its attempt at sharpening minds as if they were knives.[119] Yet both theology and natural philosophy left little doubt that to imagine God and nature as creative agents without recourse to multiplicity and variety was just impossible. As Thomas Aquinas once observed, even God, unable to communicate the whole of his perfection in a single creature, resorted to multiple and varied objects, like a speaker who employs a range of circumlocutions in order to express him- or herself.[120] There was hardly any transgression to which early modern rhetorical and poetic treatises would not consent in the interest of variety. In 1609, playwright Lope de Vega claimed that the amalgam of tragic and comic elements that characterized most of the plays he wrote for the stage was modeled on nature. Nature "is beautiful because it is varied," he argued, and, as illustrated by the ancient myth of the minotaur conceived by a woman and a bull, nature did not hesitate to resort to monstrosity as a means of adornment.[121] It was on behalf of Góngora's controversial practice of "weaving diverse actions" without apparent plan that in 1617 Francisco Fernández de Córdoba concluded with a similar remark: How are humans supposed to be satisfied with their individual characteristics and limitations, "when Nature itself in the attempt to look more beautiful through variety produces things that go against plan, as with monsters?"[122]

More restrained examples of nature's creativity pointed in the same direction. Flowers and seashells, for instance, taught early modern scholars to appreciate a display of shapes and colors virtually without end. In what seemed to be a clear example of purposeless heterogeneity, they saw nature obeying a ludic impulse that Pliny also identified in the physiognomy of humans. There seemed to be infinite faces, Pliny noted, made "by nature's ingenuity as toys for herself and marvels for us."[123] Gracián belonged to a culture that, drawing on

Pliny, viewed nature's recalcitrant habit of never repeating itself as one of its most distinctive and awe-striking characteristics. For the same reason, orators and poets who appeared capable of always coming up with new and different creations struck the audience as ingenious. In *Cisne de Apolo* (*Swan of Apollo*, 1602), the Jesuit Luis Alfonso de Carvallo allowed the use of obsolete verse forms only so that poets might demonstrate their ability to produce variety. The trend intensified throughout the seventeenth century, as the contests proposed in poetic tournaments tested the participants' ability to diversify their talents in different genres and verse forms, even if that meant using patterns that had come to be perceived as harsh, archaic, and even unpleasant.[124] A manuscript composed in Spain before 1630 praised Petrarch's *Canzoniere* not because of its elegance and sweetness (*suavitas*), values that its sixteenth-century commentators were likely to appreciate, but rather because its author had been able to write 365 poems on what was virtually the same subject, the poet's love for Laura, and yet "with such beautiful variety" that the result became admirable.[125] Writing in 1647, the Jesuit Juan Bautista Escardó (ca. 1581–1652) considered that a preacher's talent depended on his ability to produce variety while composing a sermon that circled a single point.[126]

The debate on variety was not circumscribed to rhetorical and poetic disciplines. At stake was the potential of humans to compete with nature's variety at the level of the individual, but also, importantly, their right to do so. Gracián's personal and professional life unfolded within Jesuit educational institutions that advocated the examination of each student's talent. Identifying individual strengths and weaknesses was a way to direct the student toward the field in which he would thrive. Collegio Romano professors at the turn of the seventeenth century and even earlier displayed little agreement regarding the procedure that might correctly assess a student's *ingenium* or inner nature. Should it be done externally, by testing his skills? Or should it perhaps be accomplished by attending to particular physical traits, or through a process of introspection? Less doubtful, however, were the nefarious consequences of "freely" choosing a career path, including frustration over one's failure to carry out scientific or artistic tasks satisfactorily, and, in the worst of cases, political sedition, religious heresy, or both. In 1613, Jesuit professor Tarquinio Galluzzi (1574–1649) argued against polymathy in a lecture he delivered in public. Describing the dangers of a multidisciplinary and multifaceted intellectual and artistic pursuit in which different disciplines become interconnected—what he called "encyclopedia"—Galluzzi promoted the ideal of a community in which individuals focus on specific and differentiated forms of expertise.[127] Variety was not a goal for the individual. It was a reward for the community whenever its members remained satisfied with the position that nature had allocated to them.

In suggesting that it was fair to remain satisfied with the limits of one's nature, even in the invention of conceits, Radau's aforementioned view that "each writer will practice diligently several modalities depending on his or her genius" cautiously yielded to the virtue of modesty advocated by Jesuit authorities. Conversely, the program set forth by Gracián would have struck the reader as a call to hubris. *Arte de ingenio* was a corrective to the nature of the individual. It aspired to instill in each member of the community of writers the ability to become multiple, creating variety in a free and expansive imitation of nature, "the great mother of beauty."[128] Gracián not only made the link between human and natural creativity explicit, but went as far as to prefer the intellect's inventiveness over that of nature: "Sameness results in limitation," he wrote. "Variety, on the other hand, is capacious, and is nobler when it causes noble objects to multiply. The stars that shine in the sky and the flowers that adorn the field seem few when compared to the subtlety of conceits that compete with one another in a fertile intellect."[129]

The association with the flowers in the field must have been ubiquitous if even a critic such as Daniello Bartoli compared strings of conceits with garlands of flowers, "which please because of variety alone."[130] It is no coincidence that Bartoli took the statement connecting flowers and variety from Pliny's *Natural History*: that was also Gracián's source. Pliny mentioned that the kind of flowers usually employed for garlands were varied because they were the product of nature's self-indulgence. Temporarily liberated from the maintenance of the order of things, nature resorted to variety for the sake of pleasure, enjoying the process of creation and reveling in its talent.[131] As Paula Findlen reminds us, the idea that some of nature's most striking and wondrous creations were the product of its playful, hedonistic mood, in contrast with more serious and purposeful instances, crystallized in the notion that those and similar phenomena were "nature's jokes" (*lusus naturae*).[132]

From opposite sides of the debate on conceits, the observations by Gracián and Bartoli suggest that flowers played a prominent role in early modern meditations on variety. Insects and any creature believed to be born through spontaneous generation elicited a similar fascination, as did life forms found in the sea, perceived as a realm of endless plasticity and imaginativeness.[133] In a book dedicated to seashells, Jesuit scholar Filippo Buonanni (1638–1723) displayed four hundred fifty shells of different shapes and colors gathered from different seas and shores. Citing Pliny, the editor's preface to Buonanni's volume claimed that the author was showing "the capricious variety of such animals, in which nature has played its game," as it is always wont to do in its smallest creations.[134]

To call nature's variety "capricious" suggested a free and exploratory behavior in pursuit of new forms. It also implied that it was without plan. Physician Huarte de San Juan once claimed that Italians called "capricious" (from *capra*) the kind of *ingenium* that, like a goat, refuses to keep to the road. "Caprice" stood in contrast with the behavior of sheep—it was hardly a coincidence that Minozzi labeled the detractors of conceits "sheep!"[135] Those endowed with goat-like ingenuity climbed up the highest mountains, approaching nature's secrets.[136] To suggest that nature too was capricious, however, had a touch of audacity. After all, it was God's plan that nature supposedly preferred to set aside every now and then. Its playful and sportive digressions resulted in products with the potential to contradict a proverb that lay at the core of early modern natural-philosophical lore, namely, that nature had done nothing in vain or that there was a reason for every choice it made.[137]

The term *capriccio*, "caprice" or "whim," occurred often in the context of drawing. In 1625, as Pietro Accolti (1579–1642) cautioned the reader that it is futile to teach children the art of painting if they lack the talent for it, he suggested a rapid and easy test: just ask them to sketch "any shape following their whim" ("qualche loro capriccio").[138] *Capriccio* was often synonymous with *ghiribizzo*, a sort of zero degree of invention that did not follow any pattern or model but contained the germ of one's talent and skills.[139] In 1626, when Tesauro was a student in Milan's Jesuit College of Brera, he delivered a panegyric of Saint Juvenal of Narni in the city of Fossano. The fact that Juvenal, Fossano's patron saint, was a native of Carthage suggested to Tesauro that he could draw on Pliny's report of the existence of unheard-of species of animals and peoples in Africa: "Some are headless, others with feet larger than their heads."[140] It was, Tesauro speculated, as though nature had behaved like a raving and delirious apprentice, neatly conflating Africa with the primitive or the unfinished, in a trope that resounded with European imperialist preconceptions as they were being shaped by recent and not-so-recent ventures. Eager to "learn" how to paint both human and animal creatures, nature had used Africa as a blank page on which to train its hesitant hand, resulting in capricious chimeras and "confused and half-sketched arabesques."[141]

It has been observed that, by the seventeenth century, the term "arabesque" had lost its earlier meaning of "Islamic ornament." "Arabesque" had become synonymous with "grotesque," a form of whimsical, nonnarrative decoration that since the turn of the sixteenth century assembled winged creatures, masks, stags, and a number of strange forms in varied patterns on the walls of public and private Italian palaces.[142] Spurred by the rediscovery of ancient mural decorations in Emperor Nero's *Domus Aurea* in Rome, Italian artists disseminated across Europe a new and fascinating—but also controversial—view of antiquity. Painted chimeras stared at individuals who had to conciliate sensory and intellectual pleasure with the arguments that Horace's *Ars poetica* (ca. 19 BCE) had flaunted against the use of forms that lacked a counterpart in nature. Grotesques remained, for Giorgio Vasari and a multitude of sixteenth-century painters, the embodiment of artistic freedom, a view that survived criticism by late sixteenth-century scholars such as Cardinal Gabriele Paleotti. Grotesques provided artists with a site to play with conventions. What is more, by featuring impossible creatures and impossible assemblages of existing ones, grotesques engaged with nature's creative powers by representing not the creations that characterized its usual and, so to speak, definitive aspect, but rather its creative and transformative powers as they could be conjectured by a perspicacious and unruly imagination. Language confirms that conceits appeared as rhetorical and poetic counterparts of grotesques. "Conceits" (*concetti*) often referred to grotesques.[143] In addition to that, in 1580 Aldrovandi referred to grotesques as *ghiribizzi*, a term that remained in vogue through the late eighteenth century to denigrate the use of conceits.[144]

Pliny's Africa embodied for Tesauro the creative and affirmative value of exception, coupled with the notion that nature once had to learn and rehearse before it settled on a more or less distinct plan. Nature the artist could display at times a playful and self-indulgent creativity, but it also had a past in which whim or caprice served as a path of exploration and search. Tesauro viewed whim (*capriccio, ghiribizzo*) not only as the drive behind the playful and unusual forms that Pliny had described as jokes of nature (*lusus naturae*) but also as a primeval and chaotic creative force that had led to nature's eventual and more sustained behavior. By mentioning nature's arabesques, Tesauro thus connected natural and human ingenuity in a new and unprecedented way: he gave to nature a history in which the past was marked by experiment and imagination, which even in the present emerged every now and then whenever it got tired of its usual course.

What if imagination was likewise a chaotic and primeval creative force, one that lay at the roots of the mind's earliest attempts at grasping the world, and now provided recreation and rest to humans every time they have enough of the strictures of logic and truth? Tesauro imagined whim or caprice—invention in its most essential and primordial state—to be at play in a conceit, a device in which human ingenuity imitated natural ingenuity not in its usual behavior, but in its furtive and playful stunts. The authors of *Logodaedalus* claim that the 1612 edition of the dictionary of the Accademia della Crusca defined *ingegno* as "the acuity used to capriciously invent anything, without a master or adviser," relegating the meaning of "individual difference" or "power of understanding" to a second place.[145] In fact, in 1654, when Tesauro eventually published *Il cannocchiale aristotelico,* he still believed that nature never appeared as powerful as when it released itself from the fetters of its usual course. Tesauro clarified that jokes of nature and jokes of art—conceits—were comparable manifestations of a form of inventiveness that takes place apart from the realm of "public usefulness," to use Pliny's term. The flowers that adorn the fields, Tesauro claimed, "are also elegant figures, and lively conceits (*vivaci argutie*) of ingenious nature. For, just as we call the poet's conceits 'flowers,' so we call the flowers of nature 'conceits.'"[146] It is thus unnecessary to ask why the muses are friends with dawn, he concluded: for it is the time of day when nature itself jokes and speaks "with a thousand subtle and ingenious conceits."[147]

The flowers in the field represent a case in which nature has gone far beyond what usefulness demands. Tesauro playfully suggested that the fact that tradition had named tropes and rhetorical figures "flowers" was far from anecdotal. It was simply one element in a comprehensive approach to ingenuity that imagined conceits as sites in which theology and natural philosophy encountered rhetoric and poetics. *Il cannocchiale aristotelico* depicted God, angels, nature, and humans as comparable creators of conceits differing in degree rather than kind and ranked according to decreasing ingenuity. In a reversal of the convention that depicted the poet as a divine or godlike creator, Tesauro imagined God to be a superior imitator of "the poet and the ingenious speaker." The evidence of God's ingenuity was in Scripture, where he avoided sounding trivial in order to arouse veneration.[148] Just below God, Tesauro placed oracles and visions—conceits that angels and spiritual creatures address to humans on God's behalf. Then, just above the lowest tier occupied by humans, there was nature, with its variety, its monsters, and its prodigies, including comets. Nature—Tesauro described it as God's intellect "adapted to the matter that he fabricated at the beginning, positioned for the continuation of the universe . . . both creating intellect and created matter"—stirs wonder in the philosophers with creations that are "ingenious beyond belief."[149] Most of the time, nature considers "public usefulness," seeing to the persistence of the whole by following its usual ways without calling anyone's attention. However, on occasion nature indulges in creations that are useless but pleasant "for the pomp of ingenuity and to show itself capable of conceits and humor."[150] While Pliny had justified nature's seemingly useless performances as a form of solipsistic pleasure, meant only secondarily as a spectacle, Tesauro's nature, with a focus oriented to theatrical display, boasted of its creativity before the eyes of humans.

The continuity between natural and human conceit making manifested itself in the form of competition and challenge. The perception of nature's creations as acts of playfulness pushed writers to join the game. The way Tesauro saw it, it was as though they simply could not help reacting to nature's ingenuity. For instance, faced with the variety of flowers—nature's conceits—Tesauro rushed to repay in the same currency: "It is," he contended, "as though the rising sun, to make a sky of the earth, shook the stars down from the sky."[151] Exploiting the similitude between a flowery meadow and a starry sky, Tesauro suggested that nature had used flowers to transform the earth into heaven. Tesauro's account of a playful, even humorous nature determined the way he understood conceits, which he often called "jokes" (*scherzi*) as a way of countering the suspicion that conceits were merely instances of sophistry.[152] As jokes, conceits aimed to disengage the audience from serious reasoning, yet without seeking to instill false beliefs as sophists do.[153] As early as 1640, Radau had labeled conceits *lusus ingenii*, or "games of ingenuity," and Martial himself called the epigrams he wrote *lusus*.[154]

Conceits belonged to a broad and diversified sphere of competition between natural and human ingenuity that also included the creation of automatons, similarly labeled as jokes and playful recreations of nature.[155] The collection that Tesauro's former student Manfredo Settala gathered in Milan brings the point home neatly. The catalogs of Settala's collection, prepared in 1664 and 1666 by physicians Paolo Maria Terzago (d. 1695) and Pietro Francesco Scarabelli, are representative of what happens when writers get carried away by enthusiasm at nature's talent and artifice.[156] The one published in Latin by Terzago in 1664 devotes a chapter to Settala's amber lumps and inclusions. For instance, the third item was a piece with two frogs seemingly trapped in pine resin by accident—unbeknownst to Terzago, it was a forgery crafted by the ingenuity of early modern artisans. It had belonged to Cassiano dal Pozzo (1588–1657), secretary to Cardinal Francesco Barberini and a member of the Accademia dei Lincei (Figure 3.3).[157]

As he approached the item, Terzago must have felt haunted by a double anxiety and pressure: nature's ingenuity was conspicuous in the artful and jewel-like nature of the piece, with its ability to freeze in time and transform "into stone" two creatures once teeming with life. In addition to nature, there was

Figure 3.3. Attributed to Vincenzo Leonardi. Amber with two frogs that once belonged to Cassiano Dal Pozzo. 150 × 206 mm. RL 25481. Royal Collection Trust / © His Majesty King Charles III 2023.

Martial's precedent too. Martial wrote a small but well-defined group of three compositions dedicated to amber inclusions like the one Terzago now had before him. Epigram 4.32 argued that nature had transformed a bee accidentally ensnared in poplar resin into a jewel, its color reminiscent of the honey for which the insect once toiled so hard. Then, Epigram 4.59 portrayed a viper suddenly trapped in the slow and viscous element, teaching Egyptian Queen Cleopatra a lesson of humility from its sumptuous, lucid tomb. Epigram 6.15, the story of a trapped, dead ant, continued the meditation on amber's ability to dignify even the vilest of creatures.[158] Martial's three epigrams accompanied early modern discussions of amber inclusions as a shadow does the body. From Cardano to Francis Bacon, scholars seemed plainly incapable of describing amber's preservative properties without either quoting one of the three epigrams or vying with Martial through endless forms of variation.[159]

For the contemporaries of Terzago, amber inclusions were representative of nature when it leaves its usual course in favor of a playful and artificial mode. Nature had created a jewel like those made by humans. It had transformed unfortunate beasts into immortal, embodied monuments by overcoming the laws that preside over the corruption of matter. Martial's conceit was a different but comparable feat of ingenuity, which bestowed form on amorphous acts of mere happenstance and chance through the similitude between the colors of amber and honey and by transforming fossilized poplar resin into a queen's tomb. Terzago must have felt both daunted and exalted as he entered the competition with nature and Martial. In the frontispiece to the catalog, he had boasted mastery of the laconic style, and now the third item in Settala's collection of amber inclusions gave him the opportunity to prove it. He addressed nature's joke through a conceit, a rhetorical and poetic joke that built on Martial's precedent without being servile. "Neptune," Terzago wrote in reference to the custom of gathering amber in Baltic shoals described by Tacitus, "enveloped in amber two frogs that croaked irritatingly. For this we owe amber no small thanks."[160] Terzago imagined the sea god sparing the spectators a headache by silencing two irksome, noisy frogs inside a prison of gleaming, translucid amber.

Two years later, when Scarabelli published Terzago's catalog in the vernacular, he decided to go all the way to craft a new conceit instead of simply translating the one by Terzago. Now the third item consisted of "a piece enclosing two frogs, more boisterous than ever in the silence of death." Scarabelli understood that variety was critical as evidence of ingenuity, but also that there was more to it than just inventing a new and different conceit about the same object or topic. Why not invent one that was also more challenging and complicated?

In *Arte de ingenio,* Gracián ranked conceits according to the extent to which they defied expectation: when the difficulty was greater, so was the wonder. Far-fetched similarities challenge the reader, but not nearly as much as paradoxes, let alone the kind of seemingly incongruent statements that became understandable only following clarification.[161] Faced with Terzago's well-crafted but moderate conceit, Scarabelli added a component of paradox in order to outdo the author he was in charge of translating. He insinuated that killing the creatures had failed to do the trick. Muted by death, they seemed condemned to try in vain to emit a sound for all eternity, grimacing through the amber's transparent and polished surface in a way that felt disturbing.

Through the invention of a new conceit on the same object or topic, Scarabelli successfully met the challenge posed to him by the ingenuity of nature, Martial, and Terzago. Somewhat unconvinced, however, Scarabelli went back to the piece of amber, adding a detail that had escaped his predecessor ("a fairly large drop of water") and a new conceit based on it ("lest such creatures might miss their swamp").[162] It is in the nature of competition, however, that Scarabelli's triumph over Terzago's ingenuity was not destined to last. In a biography of Settala published in 1681, Alfredo Bernardo de Yrissarri reacted to the detail first remarked by Scarabelli. He described two frogs swimming "in a drop of water," which, the way he saw it, "represents a spacious lagoon."[163] A sliver in the amber that resembled solidified water inspired Scarabelli, and later Yrissarri, to compare the lump of fossilized resin with the natural habitat of the creatures, first conceived of as a swamp and then, with a small but noticeable modification, as a lagoon.

Terzago, Scarabelli, and Yrissarri understood that in order to emulate and do justice to nature's ingenuity, it was imperative that the writer deviate too from conventional means of expression. From a consumer's perspective, the dissemination of catalogs such as those authored by Terzago and Scarabelli, and the possibility that individuals might approach the collection exclusively on the page—or perhaps by browsing the catalog with the objects in sight—established some kind of continuity between the objects and the conceits.[164] The proximity to one another was informed by a lively and playful rivalry between alternative forms of creativity. At the intersection between nature and language, the catalogs commissioned to publicize seventeenth-century collections embody the link between nature's ingenuity and that of orators and poets. Scarabelli's unrestrained impulse to compete on three fronts—with nature, with Martial, and with his predecessor Terzago—drew on the views of Tesauro, who, as Settala's former teacher, is likely to have been familiar with the collector's repertory of items encased in amber. In the catalog, Scarabelli praised Tesauro for the invention of a conceit that variated and outdid Martial's Epigram 4.32, about a bee enclosed in amber.[165]

In *Il cannocchiale aristotelico,* Tesauro used Epigram 4.32 to teach would-be inventors of conceits how to "joke" in almost infinite ways. In its most sophisticated and dignified version, "to joke" (*scherzare*) meant for Tesauro a meticulous process leading to the invention of countless conceits about the same topic. Tesauro's method came to fruition in twenty-five pages dedicated to Martial's Epigram 4.32.[166] He began by excerpting and recording any materials that might be of use in future conceits. Then, to illustrate the method's application, Tesauro imagined a group in amiable conversation. One might easily picture them after dinner, walking along a river, or in a garden as they passed a summer's afternoon: "Imagine that, in the middle of a casual conversation, news arrived that *a bee has died in the amber.* Someone might follow Martial and say: *She was able to choose a death worthy of her.* Another one: *Not even Queen Cleopatra had such a noble tomb.* And then yet another: *She is more precious dead than alive.* And thus other thoughts may come up without a change of subject."[167]

Starting with a theme, by which Tesauro understood "the topic that either someone else has proposed to us, or that we have proposed to ourselves, upon which we may build ingenious and conceit-like jokes," Tesauro imagined a contest among individuals, a way to "arouse one's ingenuity and make it abundant with conceits."[168] One of the exercises that Tesauro believed was "full of spirit" was "the one that awakens two ingenious intellects to persist on a subject with a row of propositions, answers, replies, blows, and counterblows, as in a duel of ingenuity."[169]

Terzago and Scarabelli participated in the duel of ingenious blows and counterblows described by Tesauro. Their conceits, however, would lose all of their theological and natural-philosophical significance if they became dissociated from their character as mere replies to nature's initial challenge, which by encasing creatures in amber had extended a provocative and pleasant invitation to Terzago and Scarabelli—and to Martial many centuries before them—to join in the process of invention.

Conclusion

Fireflies made Minozzi think of a discontinuous and inorganic pattern, based on juxtaposition, spontaneity, and the cultivation of endless variety. Pliny's observation that nature reveled in never generating the same creature twice fascinated Minozzi's contemporaries, who imagined the invention of a multiplicity of different conceits as the counterpart to nature's playful and sportive mood, embodied in its "jokes," ranging from monsters to the countless shapes and colors adopted by flowers and seashells. Conceits were also jokes, Minozzi argued, yet there was nothing vile about joking. "God himself," he clarified citing a verse from Scripture, "who aimed to persuade each and every human to admire him, created the universe as a joke, *Ludens in orbe terrarum.*"[170]

Even Scripture, Minozzi discovered, came to the aid of the scholar willing to integrate theology, natural philosophy, and the arts of language and reason, placing conceits in a general and comprehensive framework of ingenuity. In 1990, Paula Findlen cogently argued for the need to engage with the notion of *lusus naturae* in order to determine its impact on the development of different views about nature but also further to illuminate "the relation between science and poetics."[171] The notion of *lusus* illuminates and expands Gracián's contrast between the invention of conceits with "freedom" in unfettered multiplicity and the form of "slavishness" that he associated with a coherent and organic composition. There was, for Gracián and for Minozzi, a correspondence between freedom from unity in a discourse or poem and freedom from usefulness and purpose. Discontinuity and play were two different but interconnected paths for both natural and human ingenuity to lay their claim to the nobility that some believed to be at work in the etymology that connected *ingenium* with *ingenuus*—noble, highly born. Both notions embodied the right to a pleasant, purposeless art.

Yet there was more to the ecology of natural and human ingenuity than smallness, multiplicity, and variety. Another set of paradoxical values reinforced the association of fireflies and conceits, tying together contemporaneous developments in natural philosophy, rhetoric, and poetics. The belief that insects, including fireflies (*pace* Scaliger), multiplied through spontaneous generation connected fireflies with a form of propagation that epitomized nature's boundless creativity, and which offered a suggestive, alternative model to explain the invention of conceits. Last, the weak but enchanting light of fireflies, with its mysterious origin and purpose, corresponded to a conceit's ambivalent and precarious but suggestive relation to truth. That is the topic of chapter 4.

CHAPTER FOUR

The Life of Conceits

Juggling, Magic, and Alchemy

In 1623, poet Tommaso Stigliani (1573–1651) published a poem in praise of Rome's most famous sculptor, Gian Lorenzo Bernini:

> Let the torch of Prometheus, good Lorenzo,
> And the head of Medusa
> Yield to your chisel—the torch
> When it grants life to bodies without sense;
> The head when it makes stone out of limbs
> Full of soul. For each statue of yours
> Is so vivacious, and my stupor is such
> When I look at them, that
> They seem to have soul, and I to be a stone;
> They seem to be alive, and I, a sculpture.[1]

Like Prometheus, who instilled life in statues made of clay, Bernini (1598–1680) made marble breathe.[2] He created vivacious sculptures seemingly endowed with soul, the outcome of a creative endeavor brimming with similar audacity and hubris to that which once brought ruin to the titan.

Powerless to threaten the distinction between the forms of nature, which appeared "active, and as if alive" to early modern Aristotelians, and those of art, which were "as if inert and dead" and without effective force, Bernini's works also lacked the inner principle of movement that only nature can instill in its creations.[3] Yet they were capable of paralyzing Stigliani, struck by a blend of awe and Medusa-like terror. In a similar vein, the engineer Bernardino Baldi claimed in 1589 that the motion of lifeless automatons caused living and thinking humans to remain motionless with stupefaction.[4] The claim that Bernini's sculptures could freeze the beholder must have appeared only moderately extravagant to a reader educated in the language of early modern art history and criticism. In 1550, Giorgio Vasari reported that painter Francesco Francia (1482–1518) was so struck on seeing Raphael's *Saint Cecilia* at the church of San Giovanni in Monte in Bologna—"not painted, but alive"—that he soon became ill and died.[5] Ancient Greek epigrams containing descriptions or εκφράσεις of artworks reveled in the tenuous line between the viewer's awareness that a sculpture was inert and the illusion that it was on the verge of speaking and moving. A proverbial case was that of the cow sculpted by Myron in the fifth century BCE. Poets were unanimous that the statue must have soul, a fact that would also explain why a calf tried to drink from her udders and a bull became attracted to it.[6]

Seventeenth-century scholars imagined that the rhetorical and poetic ingenuity displayed by Stigliani stirred reactions comparable to the experiences provoked by lifelike sculptures and automatons. Using conceits—arguments mediated by tropes—Stigliani described a set of metaphorical rather than actual transformations that, in a manner of speaking, brought artifacts to life while rendering the beholder virtually lifeless. Conceits did not aim to pass for true any more than the sculptor and the engineer intended their creations to pass for actual living beings. They merely sought to place the reader in a liminal and transitory state, between the illusion of truth and a skeptical and inquisitive response to the awareness that the use of figurative language removed the safety net of "natural" and sound argument.[7]

Artifice and the appearance of truth were not the sole aspects in which conceits resembled lifelike sculptures and automatons. I argue that, just as Stigliani's contemporaries expected that sculptures should look as if they were alive and breathing, they expected as much from conceits. In fact, artifice, with its component of untruth, and lifelikeness were two crucial and—as I show—interconnected characteristics that transformed conceits into rhetorical and poetic counterparts to automatons as well as lifelike sculptures and paintings.

The connection between falseness and lifelikeness or vivacity might appear paradoxical. In fact, it was pervasive. Countless breathing and animated sculptures and canvases paraded through *Le Veneri* (*The Venuses*), a volume of poetry published in 1633 by Antonio Bruni (1593–1635). The book contained a number of compositions in praise of artworks, including Bernini's group of Apollo grasping for the nymph Daphne created for Cardinal Borghese's villa on Rome's Pincian Hill.[8] Bruni claimed the inspiration of the celestial as much as the earthly Venus, yet the latter seems to have exerted a particularly powerful charm on him. Bruni described her as the embodiment of generation: flowers pullulate beneath her feet as she seduces everyone with a girdle that, as stated in Homer's *Iliad*, is the emblem of "sweetness, pleasure, charm, persuasion, deception, and painted enchantment."[9] Bruni struggled to suppress his fascination with a model of inventiveness that amalgamated liveliness and deceit, two qualities similarly enmeshed with one another in the discussion of conceits found in Pier Francesco Minozzi's *Sfogamenti d'ingegno* (1641). Rather predictably, Minozzi also called conceits "Venuses" (*Veneri*).[10]

As I discussed in chapter 3, Minozzi transformed a real or fictitious encounter with clouds of fireflies outside the gates of Ferrara into a vindication of the rhetorical and poetic craze for conceits. Owing to a controversy renewed by Agostino Mascardi and Matteo Peregrini between 1636 and 1639, fireflies had become code for conceits, as comparable sources of a weak and false "light." Although the analogy was intended for the discredit of both conceits and fireflies, Minozzi rehabilitated both by transforming the insects into heralds of a new approach to nature focused on smallness, vivacity, and the pursuit of novelty through artifice, all of them qualities that conceits were supposed to embody in the field of style. With their swift and capricious swirl, fireflies appeared to Minozzi as "living and innocent fireworks of nocturnal joy." They were "animated little flames," moving incessantly and teeming with life. Something in the behavior of fireflies, constantly alert and responsive, struck a chord with the traveler, even more so owing to their imminent and premature death. In the analogy with conceits, the death of fireflies corresponded to the trope's untruth and artifice, which the reader would soon perceive as he or she unraveled its striking yet merely imaginative linkages.[11]

Seventeenth-century scholars were adamant that conceits, like fireflies, were endowed with life. Minozzi described detractors of conceits as "dead wits, who live without spirit" and are incapable of perceiving life in other people's writings.[12] With a keen sensibility to nomenclature, Peregrini titled his treatise *Delle acutezze, che altrimenti spiriti, vivezze, e concetti, volgarmente si appellano* (1639). All but the first of Peregrini's four names for conceits placed them directly within the sphere of life. "Sparks of life" (*vivezze*), "spirits" (*spiriti*), and "conceits" (*concetti*) were all apt names because they pointed to "I know not what of vital" that Peregrini perceived to be at work in the invention of the rhetorical and poetic devices.[13] Three years later, the Jesuit Baltasar Gracián claimed that there are individuals so subtle that they give birth to "living bodies" with "conceits as soul." Less delicate minds produce "corpses that lie down in dusty tombs, a banquet for moths."[14]

The claim that conceits were, in a way, living bodies was met with skepticism and even disdain by critics of later periods. In 1899, Benedetto Croce dismissed Gracián's repeated claims that to use conceits is to write "with soul" (*con alma*) as just lyrical expressions. In what amounted to the first study in modern times about seventeenth-century writings on the rhetorical and poetic devices, Croce condemned the debate over the life of conceits to oblivion.[15] Yet art historian Fredrika Jacobs has argued that to disregard the recurrent allusion to "living," almost "speaking," paintings as mere topoi devoid of any specific or technical meaning is to overlook early modern natural philosophers' and physicians' investigation of "aliveness as a physiological and psychological condition of being."[16] Art theorists and critics of the period used a lexicon complicated by conflicting notions of vitality, in which the ideas on what made a body alive in Aristotelian natural philosophy coexisted with mechanical, magical, and alchemical debates over the possibility of creating life or its simulacrum artificially.[17] I contend that to dismiss, as Croce once did, the references to life, lifelikeness, and vivacity that permeated writings on conceits similarly compromises any attempt at understanding the place of the rhetorical and poetic devices in broader debates about artifice and its relationship to nature.

Life was at the center of increasing interest in early seventeenth-century creative and intellectual activity. To repeatedly stress that conceits were endowed with life was thus of necessity a meaningful gesture. Life placed even the smallest of insects above celestial bodies. In 1602, naturalist Ulisse Aldrovandi drew on the authority of Augustine to contend that a fly, because it was endowed with soul, provided the intellect with stronger evidence of God's greatness than any of the stars.[18] The textbooks prepared by the Jesuits at the University of Coimbra during the same years argued as much from the angle of Aristotelian-Thomist natural philosophy.[19] Yet if Minozzi's embrace of the analogy with fireflies sought to capitalize on the phenomenon of life, the question then follows, how were conceits alive? The quest for answers, in turn, will lead us through a subset of interrelated problems: What were the conditions required in a conceit's invention—or "conception"—for it to come to life? How did a conceit's aliveness impact its audience or readership?

In the present chapter, I argue that seventeenth-century scholars accounted for the life of conceits by deeming it a product of artifice. Resorting to nonnatural or "artificial" modes of conception, the intellect brought forth creations that were brimming with life, even though—or precisely because—they deviated from logical and truth-oriented forms of reasoning. Explanations of this kind went a long way to reinforce the analogy between fireflies and conceits. While the insects' precarious and ephemeral existence signaled vitality in the eyes of Minozzi, the light they radiated acted as a proxy for artifice by evoking forms of causality that defied nature's usual ways. The light of fireflies thus recalled the arts of the magician, the alchemist, and the engineer of automatons, accomplices of the inventor of conceits in the pursuit of a new and ambivalent entanglement between natural and human ingenuity, culminating in the quest to create life.

I take as a point of departure the paradox inherent in the association between the life and the artifice of conceits. How is it, one might ask, that humans need to resort to artificial, unnatural language—rather than logical and therefore "natural" ways of meaning—to replicate life, which is nature's greatest achievement? The answer placed conceits in a position to deal a fatal blow to the distinction between nature and art, and that between scientific knowledge and creativity.

Opus Vitale

If Minozzi was right to maintain that seventeenth-century Bologna was a garden of ingenuity, that was in part thanks to Giovanni Battista Manzini.[20] In 1652, *Delle meteore rettoriche* (*On Rhetorical Meteors*) made good on Manzini's earlier promise to articulate the principles underlying the idiosyncratic, conceit-laden style that had made him popular across different languages and literary markets.[21] As he sought to justify why he wrote the way he did, Manzini interrogated the life of conceits with remarkable breadth of perspective.

Manzini's approach to the entanglement between natural and human ingenuity exploited the old and well-known analogy between nature and the mind as generative agents. The parallel that made the author of a book, a device, or a discovery its parent was common since Greek and Roman antiquity.[22] In a similar vein, *conceptus* was the term for "embryo" in early modern medicine but also designated what physician Juan Huarte de San Juan called the mind's offspring, its grasp of the object.[23] The problem, Manzini—and Peregrini before him—was quick to realize, lay with the ambiguity of the terms *concetto* in Italian and *concepto* in Spanish, which could refer to a concept as much as a conceit.[24] In a concept, Manzini averred, the mind limits itself to abstract reality in a way that is both truthful and accurate, creating a lifeless creature apt only for scientific demonstration. Conversely, a rhetorical and poetic conceit has, at best, a circuitous and mediated relationship with reality, but there is something lively and vital about it that is missing in a concept.

What, then, determined whether the mind would beget concepts or conceits? Manzini's answer hinged on a distinction between two kinds of intellectual conception, both construed by analogy with nature. Concepts were products of natural intellectual conception, modeled by analogy with Aristotle's account of sexual reproduction. In *Generation of Animals*, Aristotle claimed that the temperament of the male is warmer than that of the female. As a result, he can turn digested food into the seed, which is subtler than the menstrual blood contributed by the female.[25] The seed originating in the father imposes its form on the menstrual blood, which gives the embryo only its matter. This is why the child resembles the father except when the seed fails to entirely conquer the menstrual blood.[26] Manzini argued that, in concepts, the seed, corresponding to sensory input, gives rise to concepts by imposing its form on the subject's passive intellect, functioning as the equivalent of the menstrual blood. He clarified that concepts are merely products of "generation"; in them, the intellect "conceives" but "like a mother" who becomes "impregnated" by sensory input.[27] Building on Aristotle's gendered, misogynistic natural-philosophical perspective on sexual reproduction, which relegated the female to a passive role, Manzini viewed the process of natural intellectual conception favored by truth-oriented disciplines as a case in which the intellect adopts the role of the female and surrenders to sensory input. It is a passive, self-effacing modality in which the intellect leaves a mark only in case of failure, when, being insufficiently subtle, it makes it difficult for the seed of sensory input to impose its form, resulting in a concept that deviates from reality as nature does when it begets monsters.[28] Natural intellectual conception is also incapable of producing life.

Others around the same years also sought to articulate the intellect's agency and authorship—or lack thereof—in terms of gender. In 1639, the Jesuit Baltasar López told the audience at Mexico City's College of Saint Peter and Saint Paul, "Someone's ingenuity should be considered excellent only when it bears fruit by itself and gives birth and delivers what it has created."[29] Speaking at the same institution in 1631, López had resorted to the ancient Greco-Roman parallel between ingenuity and a soil capable of bearing fruits.[30] Now, eight years had passed and López felt the need to clarify that as long as the soil merely brings forth the seed without adding anything to the product, it is not truly a parent to it. Using Manzini's terms, it is not "conception" but merely "generation." "It belongs to a masculine *ingenium* . . . to bring forth its creation with the mind's force and without borrowing from outside," López argued. Conversely, "it befits a sluggish *ingenium*, and one that is so-to-speak soft and feminine to become always impregnated through someone else's force, and to grow without bringing forth anything new or fresh except when it is born from someone else's diligence, discovered by someone else's trouble and work."[31]

López disclosed for his audience of male students and male civil and religious authorities a view that opposed "masculine" inventiveness to "feminine" memorization. Early modern educational and medical sources often praised ingenuity at the expense of memory, a passive and less esteemed faculty than imagination and the intellect.[32] Those who merely convey notions that they have found somewhere else exert their memory, but not their ingenuity, the only faculty that truly "brings forth out of itself, thanks to its inner virtue, creations that fructify because of inborn fertility."[33] Early modern physicians might have explained similarly active and passive behaviors according to differences in temperature. Just as Aristotle had argued that the female, unable to concoct seed, could digest and refine the food only to the stage of the menstrual blood, Huarte de San Juan linked excellence in memory to a colder temperament than the one required for superior intellectual and imaginative qualities. The members of the audience faced a choice between becoming parents to creations born from inside them or merely being "nurses" (*nutricem, ac gerulam*) to those of someone else.[34]

For López, ingenuity should be masculine, yet he constantly compared it to the soil, which traditionally occupied the place of the female in the analogy between sexual reproduction and agriculture. Was he incurring contradiction or depicting ingenuity as androgynous? Instead of that, the claim that ingenuity alone "brings forth out of itself, thanks to its inner virtue, creations that fructify because of inborn fertility" suggests a model of generation without a second parent, one that transcends the sexual binary in favor of a single and self-sufficient thinking and creating subject, reminiscent of precedents in ancient myth that were conveniently related to the sphere of ingenuity. Athena, for instance, was born from the head of Zeus after he had swallowed Metis, the goddess of wile and craft who was already pregnant from him. When Hesiod (fl. ca. 700 BCE) narrated the birth of Athena—who, like Metis, was associated with craft and wile—he followed with the revenge of Zeus's wife Hera. Without having intercourse, through parthenogenesis, Hera gave birth to Hephaistos, "expert with his skilled hands beyond all of Sky's descendants."[35] It is certainly intriguing that both Athena and Hephaistos, two divinities embodying forms of mental and manual ingenuity closely linked to one another in the ancient and the early modern imaginary, originated from nonsexual reproduction.

In Manzini's view, artificial intellectual conception "conceived" creatures endowed with life—that is, rhetorical and poetic conceits—by relying likewise on a single parent, and being similarly oblivious to the binary between male and female. Instead of passively adopting the form of the seed, representing sensory input, artificial intellectual conception was, Manzini wrote, "entirely a work of the intellect, which, as a creator rather than a generator, operates without the species' concourse or aid. And these are the rhetorical and poetic conceits. In procreating them by means of fantasy, matter provides only the occasion, while they are born spontaneously from the fertility of a soil in which native fecundity germinates even without a seed and without a farmer."[36] Artificial intellectual conception followed in the footsteps of nature because "artifice" here meant "nature's artifice." Manzini was eager to find alternatives to sexual reproduction without entirely leaving behind nature and its authority.

Likening the abstraction of concepts to sexual reproduction had been Manzini's way of vilifying the intellect's passive and self-effacing behavior when it merely seeks resemblance to the seed. Sexual reproduction created a society structured around parent-child relations, and parent-child resemblance had long been a simile for rhetorical and poetic imitation, which Manzini considered "one of the most unfortunate and useless instruments that a great intellect may employ to shape its style."[37] With its cult of similarity, sexual reproduction was nature's exemplar of imitation, and a model for a servile, slavish intellect incapable of bringing forth life.

The opposite was true of artificial intellectual conception. Impatient to liberate conceits from the cast of parent-child relations, Manzini associated the rhetorical and poetic devices with a form of creativity that enabled nature to bring to life creatures that he and his contemporaries believed propagated without sexual reproduction. When Neapolitan natural philosopher Giambattista Della Porta argued in 1589 that fireflies were born from dew, he was assigning them to the order of life forms that were born through "spontaneous generation," a process that, depending on the author, could be characterized as simply mysterious or as a potentially undisciplined and riotous outburst of fertility.[38] Before such scholars as Francesco Redi (1626–1697) and Jan Swammerdam (1637–1680) ruled out that animals could be born from anything but the seed, it was universally acknowledged that a number of plants and animals emerged spontaneously from either inorganic or decaying matter, without parents or intercourse: bees and flies sprang forth from carcasses of mammals, while moths and butterflies arose from the matter of caterpillars as it decomposed.[39]

Nature adopted a protean, plastic mode marked by constant transformation as soon as it left the relatively predictable and repetitive trappings of sexual reproduction aside. The belief that individuals of a species could be born from a substance of a different kind unshackled animal propagation from the laws of resemblance in favor of virtually unrestrained and boundless creativity.[40] To maintain the appearance of order, natural philosophers went back to Augustine. Convinced that the world was the result of a single act of creation, Augustine claimed that God had sowed "seed principles" (*rationes seminales*) that, under the right conditions, became actualized as living beings: they were part of God's creation back at the beginning, even if they came into existence only now.[41] Augustine's expedient meant that spontaneous generation was not entirely spontaneous. Insects and lizards may seem to be born out of decaying matter or mud; in truth, they originate from seed principles containing a resemblance of the resulting insect or lizard. In different guises, the notion of *rationes seminales*, which went back to Stoic natural philosophy, enabled Athanasius Kircher and a host of early modern scholars to tame spontaneous generation and, at least to a certain extent, subject it to a plan (Figure 4.1).[42] Such efforts were nonetheless insufficient to extinguish a belief that remained enduringly seductive to generations of poets, artists, and natural philosophers.

Conceits were merely the latest addition to the catalog of small creatures that escaped sexual reproduction. Manzini's claim that conceits were born "spontaneously" evoked a model that did not require the concourse of both the male and the female parent. Just as the earth, a swamp, or decaying organic matter deposited in them begot creatures that did not resemble the substance from which they seemingly originated, in a conceit the intellect created a product that diverged creatively from the stimuli conveyed by sensory input. Making spontaneous generation a model for the invention of conceits aligned with a renewed interest in life at small scale. It meant that conceits became integrated in a realm governed by protean and metamorphic powers that resisted explanation through the principles of seventeenth-century natural philosophy. It also fit with Manzini's aversion to similarity in rhetorical and poetic creation.[43] Just as Gracián considered the adherence to a plan to be a sign of slavishness unworthy of the nobility of *ingenium*, so Manzini maintained that surrendering one's liberty to tread in a predecessor's footsteps was utterly servile, worthy of a mechanical rather than a liberal profession of the kind that should be reserved to high-born minds.

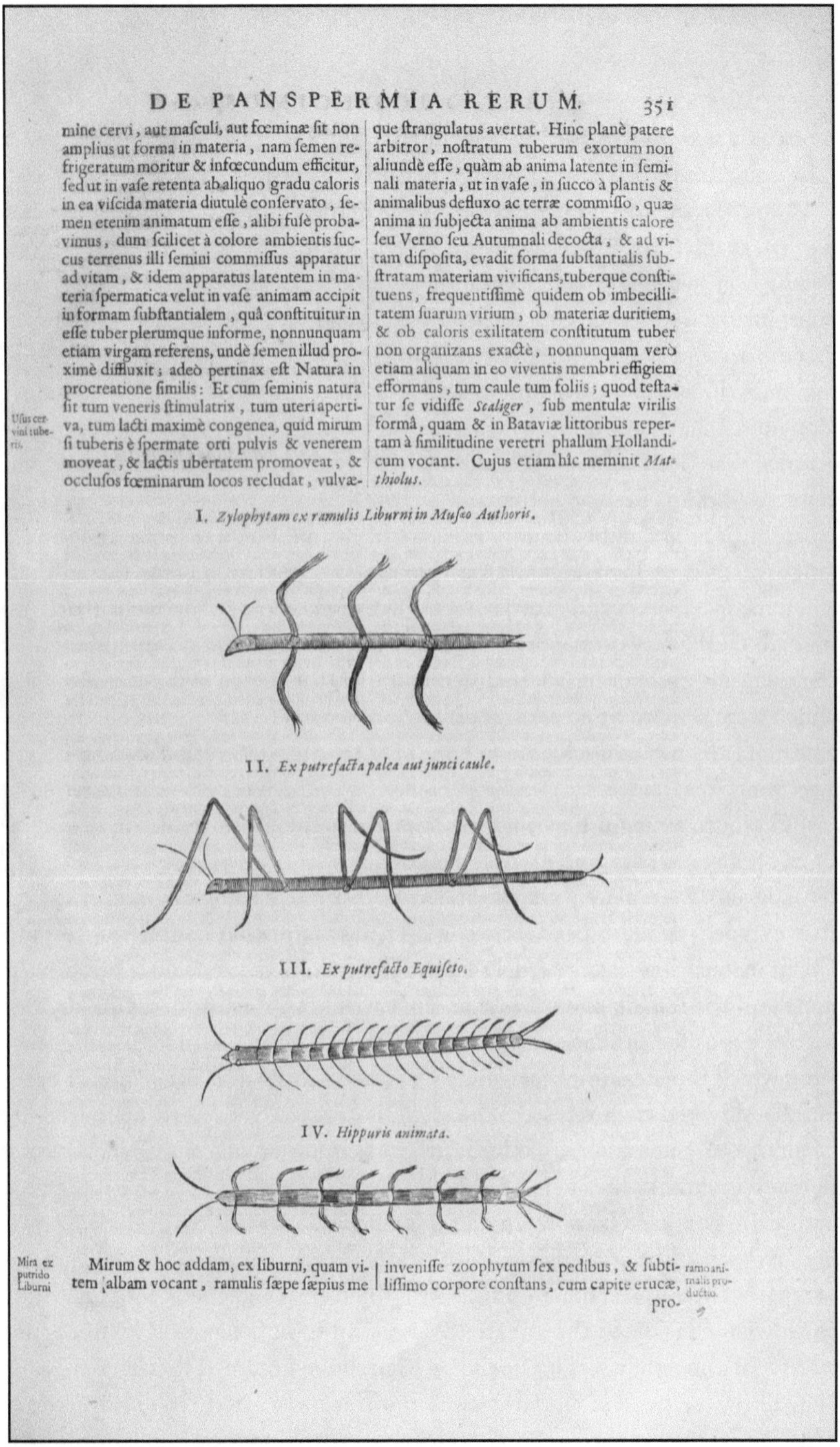

DE PANSPERMIA RERUM. 351

mine cervi, aut maſculi, aut fœminæ ſit non amplius ut forma in materia, nam ſemen refrigeratum moritur & infœcundum efficitur, ſed ut in vaſe retenta ab aliquo gradu caloris in ea viſcida materia diutulè conſervato, ſemen etenim animatum eſſe, alibi fuſè probavimus, dum ſcilicet à colore ambientis ſuccus terrenus illi ſemini commiſſus apparatur ad vitam, & idem apparatus latentem in materia ſpermatica velut in vaſe animam accipit in formam ſubſtantialem, quâ conſtituitur in eſſe tuber plerumque informe, nonnunquam etiam virgam referens, undè ſemen illud proximè diffluxit; adeò pertinax eſt Natura in procreatione ſimilis: Et cum ſeminis natura ſit tum veneris ſtimulatrix, tum uteri apertiva, tum lacti maximè congenea, quid mirum ſi tuberis è ſpermate orti pulvis & venerem moveat, & lactis ubertatem promoveat, & occluſos fœminarum locos recludat, vulvæque ſtrangulatus avertat. Hinc planè patere arbitror, noſtratum tuberum exortum non aliundè eſſe, quàm ab anima latente in ſeminali materia, ut in vaſe, in ſucco à plantis & animalibus defluxo ac terræ commiſſo, quæ anima in ſubjecta anima ab ambientis calore ſeu Verno ſeu Autumnali decocta, & ad vitam diſpoſita, evadit forma ſubſtantialis ſubſtratam materiam vivificans, tuberque conſtituens, frequentiſſimè quidem ob imbecillitatem ſuarum virium, ob materiæ duritiem, & ob caloris exilitatem conſtitutum tuber non organizans exactè, nonnunquam verò etiam aliquam in eo viventis membri effigiem efformans, tum caule tum foliis; quod teſtatur ſe vidiſſe *Scaliger*, ſub mentulæ virilis formâ, quam & in Bataviæ littoribus repertam à ſimilitudine veretri phallum Hollandicum vocant. Cujus etiam hîc meminit *Matthiolus*.

Uſus cervini tuberis.

I. *Zylophytam ex ramulis Liburni in Muſæo Authoris.*

II. *Ex putrefacta palea aut junci caule.*

III. *Ex putrefacto Equiſeto.*

IV. *Hippuris animata.*

Mira ex putrido Liburni

Mirum & hoc addam, ex liburni, quam vitem albam vocant, ramulis ſæpe ſæpius me inveniſſe zoophytum ſex pedibus, & ſubtiliſſimo corpore conſtans, cum capite erucæ, pro-

ramo animalis productio.

Figure 4.1. Insects born through spontaneous generation. Athanasius Kircher, *Athanasii Kircheri e Soc. Iesu Mundi Subterranei Tomus II.us In V libros digestus* . . . (Amsterdam: Johannes Janssonius van Waesberge and Elizaeus Weyerstraten, 1665), 2:35. Courtesy of The Linda Hall Library of Science, Engineering, and Technology.

The choice to model a form of invention self-styled as free and noble on the process of spontaneous generation associated with insects, toads, and a range of creatures traditionally relegated to the margins of natural philosophy is admittedly surprising. It is evidence of the tortuous but innovative paths followed by seventeenth-century individuals eager to picture the entanglement of natural and human operations in alternative ways. William R. Newman claims that medieval and early modern alchemists modeled their recipes for the creation of life not on "natural" sexual reproduction but on "artificial," spontaneous generation.[44] Newman's suggestion is not exactly a precedent for Manzini's views but confirms that writings on conceits shared with other scholarly and artistic disciplines the belief that spontaneous generation was the product of what one may call nature's "artificial" and "artisanal" persona.

Manzini's ideal of the entanglement between natural and human ingenuity was nowhere more clearly embodied than in the fertility of a soil capable, to all appearances, of spontaneously bringing forth countless and varied creatures without consideration of origin or parenthood.[45] Ingenuity resisted any attempt to be contained within any limits or plan. It was "all sulfur" (*tutto di zolfo*), always pursuing new roads. It behaved like fire, which, after it is detonated in a mine, "never returns through the path it took first but, instead of that, tears a mountain open just to find a new way."[46] The comparison between ingenuity and fire was conventional to a degree. Manzini had called artificial intellectual conception the work of fantasy or imagination, a faculty that, as Huarte de San Juan had underscored in the late sixteenth century, was associated with heat since antiquity.[47] Citing Huarte de San Juan in 1602, Luis Alfonso de Carvallo warned that only a brain with three degrees of heat, signaling the predominance of yellow bile or cholera, made a good poet.[48] The fact that poetry comes easier during the summer or when individuals fall in love, Carvallo argued, confirmed the link between heat and creativity.[49]

In nature, the generation of life required "vital heat." According to Aristotle's *Generation of Animals*, the seed was a foamy and bubbly mixture of water and pneuma or spirit (πνεῦμα in Greek, *spiritus* in Latin). Pneuma was not one of the four elements but was akin to the ether of which celestial bodies are made. Although Aristotle tentatively described pneuma as a kind of "hot air," unlike fire, pneuma possessed vital heat, also called "innate heat" in medical and natural-philosophical works, with the power to inform bodies, fill them with movement and life, and keep them alive until it wears out.[50] Aristotle borrowed the term *pepsis* (πέψις, "digestion") from Hippocratic medicine and then generalized its use for any process involving vital heat, including the ripening of fruit and spontaneous generation. Pneuma served as the substrate of any process initiated by vital heat and also served "as a formative agent" in living beings.[51]

Even though vital heat was not just heat, for centuries natural philosophers and alchemists operated within the framework of analogies between vital heat and heat in general. The torch of Prometheus, Emanuele Tesauro had claimed in 1626, instilled life into statues because it had been kindled in the flame of the sun, the source of vital heat that caused insects to arise by spontaneous generation and created metals under the earth.[52] Kircher conceived the earth's interior as a set of furnaces, and alchemists worked on the premise that artificial and natural processes were in a way analogous.[53] Some even spoke of the individual's vital heat as "a lamp of life."[54] Just as alchemists sought to bridge the divide between art's lifeless fire and vital heat as it existed in nature, discussions of rhetoric, poetics, and visual arts also resorted to heat as a creative and active principle with life-giving powers, as was the case in Manzini's analogy between spontaneous generation and the invention of conceits by the heat of the author's imagination.[55]

Yet even if we concede, as Manzini probably intended us to do, that the life of conceits was merely a certain feeling or impression that it managed to stir, we still wonder, Why is artificial intellectual conception, with its analogy to spontaneous generation, capable of obtaining that certain feeling and impression? Manzini's answer was as dramatic as it was opaque. "Because it is *opus vitale*," he claimed, a conceit "is the noblest among the intellect's acts. Elevated in some way to a certain I know not what of the divine, it has appropriated for itself the name 'conceit,' which in principle was generic and pertained to both ways of conceiving."[56] The notion that a conceit is *opus vitale*, a substance animated by operations similar to those found in animals and plants, was metaphoric, just like the claim that a conceit was the spirit that inspired life in the body of style.[57] A detour through *Delle acutezze*, a work in which Peregrini also acknowledged that a conceit has "I know not what of vital," might help disentangle the web of Manzini's vitality- and soul-related allusions. Manzini's contention that the artifice of a conceit denotes the involvement of imagination aligns with Peregrini's views, which distinguished "artificial" from "natural" reasoning through the presence in the former of figurative language, a product of imagination.[58] Peregrini claimed that the artifice of a conceit enabled the reader to detect the writer's ingenuity and thus to revel in the encounter with a peer.

Peregrini's hypothesis that the reader's *ingenium* recognizes the author's *ingenium* as it makes sense of a conceit suggests a new approach to Manzini's *Delle meteore rettoriche*, published thirteen years later. Manzini claimed that the artificial is more congenial to humans than the natural because we are curious, eager to be carried away by wonder and its movement of restless inquiry.[59] A conceit, the product of artificial intellectual conception, instilled in the mind of the audience the principle of self-motion that served as one of the soul's defining features in Aristotelian natural philosophy.[60] A conceit deserved the name of "soul" because it triggered a movement of inquiry leading the audience through different avenues, seeking to make sense of facets of reality connected by the author's imagination in a process likened to spontaneous generation. Without artifice, as Manzini claimed in a twist of the argument, elocution lacks "life, and with it the immortality that is often the only reason for writing."[61] Manzini seemed to be arguing that a conceit resurrects its author as many times as the mind of listeners or readers becomes animated by the desire to make sense of it, a process experienced as a sort of encounter with the intellect from which it originated.

Assuming that the life of a conceit lay in its ability to instill motion in the mind, then it comes as little surprise that seventeenth-century writings on the topic exploited the link between rhetorical invention and the automatons that often served as heuristic models for animal life in the works of René Descartes and other proponents of "mechanical philosophy."[62] The parallel between the life of a conceit and that of self-moving machines emerged in 1654, in a passage of *Il cannocchiale aristotelico* in which Tesauro dubbed a conceit's essence or *argutezza* a vital spirit. *Argutezza* is like a miracle, he wrote, which brings it about that "mute things speak, insensible ones become alive, and those that are dead return to life. When tombs, marbles, [and] statues receive voice, spirit, and movement from this enchantress of minds, they argue ingeniously with ingenious individuals. In sum, the only dead thing is the one that is not enlivened by *argutezza*."[63] A conceit opened a conversation between the author and the reader because the latter, observing the artifice at work and being ingenious him- or herself, rejoiced in the encounter with the author's ingenuity. Intriguingly, Tesauro excerpted the previous passage from a letter sent by Cassiodorus Senator (d. ca. 483 CE), the secretary of Ostrogoth King Theodoric, to the mathematician and philosopher Boethius. The passage was a eulogy of mechanics, the field of applied mathematics that was responsible for the creation of automatons.[64] By repurposing the passage, Tesauro transformed the art of devising self-moving and lifelike machines into a figure for the invention of conceits.

Through a tortuous but productive dialogue that engaged rhetorical and poetic theory, mathematics, and natural philosophy, seventeenth-century writings linked the invention of conceits to the generation of insects, toads, and reptiles as well as to the creation of automatons. The aspiration to create life, nature's greatest achievement, was invariably part and parcel of what prompted orators and poets to resort to imagination and artifice in the imitation of nature. That was not without risk.

A conceit's soul and vivacity, Tesauro surmised, was evidence of its inventor's godlike, superhuman character.[65] The claim, echoed by Manzini's remark that devising a conceit elevated the intellect to "a certain I know not what of the divine," gestured to a component of Frankensteinian hubris that was never lost on detractors of conceits. In 1608, Jesuit professor Famiano Strada told Collegio Romano students that the poet was like a god in that he or she brought forth something that was not there before.[66] Yet the commentary on Aristotle's *Physics* published in 1592 by the Jesuits of Coimbra left little doubt that only God could bring forth things from nothing; art only rearranged the components of God's creation, and its products lacked the principle of inner movement that we call life and that is a prerogative of nature's creations.[67]

Writers, however, did not appear to be godlike because they created vivacious, lifelike conceits. In fact, the reverse was true: conceits appeared to be alive because the audience noticed in them the presence of artifice and, through it, the author's simultaneous attempt and failure at becoming more than human. The challenge to one's limits, captured in a conceit's unstable and precarious argument, suggested to the audience that the author was up to something grand. Conceits walked a thin line between the awareness of what made humans mortal and limited and the possibility of using imagination and artifice to play God.

Manzini knew that placing one's hopes of invention beyond art's inert, lifeless tricks could appear as evidence of hubris and a form of rebellious dissatisfaction. However, developing further the link between ingenuity (*ingenium*) and the adjective for "noble" or "high-born" (*ingenuus*), and drawing on another term with the root *gen-*, he argued that the writer needed to be magnanimous or *generoso*, as well. Manzini maintained that both natural and human ingenuity extended to "all that can be made."[68] He inveighed against any mind "that does not dare and try anything," even if he was not oblivious to the fact that audacity and the drive to push boundaries forward stood dangerously close to transgression and excess.[69] Without due care, the pursuit of excellence might turn the writer into a new Icarus, unable to stay the course between the earth and the sun.[70]

I will show that the danger of rising too high and falling in the process was not accidental to conceit making but lay at the core of what rendered conceits potentially alive. The instability inherent in a conceit's pairing of disparate realities rendered the feat of its invention conspicuous and the ingenuity and godlike character of its author apparent enough to evoke the kind of vitality that both supporters and detractors attributed to conceits. What kindled life in the mind of the audience was the writer's performative, aspirational attempt at creating something new by using dissonance to push the boundaries of acceptability near to a breaking point.

Risk Management

Silkworms came front and center as Minozzi celebrated Bologna's reputation as a hub of audacious creativity.[71] The insects conjured the hubris associated with the inventor of conceits, comparable at small scale with such mythical characters as Icarus and Phaeton. In silk factories, animals and human ingenuity partnered with one another to weave stupefying works that redounded to the name of the city no less than Manzini's conceits and the laconic style of Virgilio Malvezzi.[72] Minozzi inspected the operations of such small but wondrous creatures as he delved into the capabilities and the limitations of human art and ingenuity.[73] If fireflies were code for conceits, silkworms embodied ingenuity in absolute terms.

Silk manufacture started to thrive in Bologna and other cities in northern Italy starting in the late twelfth century. With time, the introduction of mulberry trees and the culture of domesticated silkworms (*Bombyx mori*) in the city's hinterland reduced its dependence on imports of raw silk. By the sixteenth century, northern Italy had become a center for technological innovation in the industry, and Minozzi's visit paid tribute to its centrality in Bologna's economy.[74] Marcello Malpighi (1628–1694), one of the city's most famous scholars, wrote a tract on silkworms in 1669 in fulfillment of a request from London's Royal Society. Malpighi opened *De Bombyce* (*On the Silkworm*) with Pliny's claim, now familiar to us, that nature is nowhere more present than "in its smallest creations."[75]

Closely tied to ingenuity, sericulture was one of the inventions featured in a series of prints titled *Nova reperta* (*New Discoveries*, ca. 1600), engraved after late sixteenth-century designs by Joannes Stradanus (Figure 4.2).[76] The inventions illustrated by Stradanus—which included the printing press, spectacles for reading, and the use of *guaiacum* against syphilis—were central to the early modern self-consciousness of what made the present modern and superior

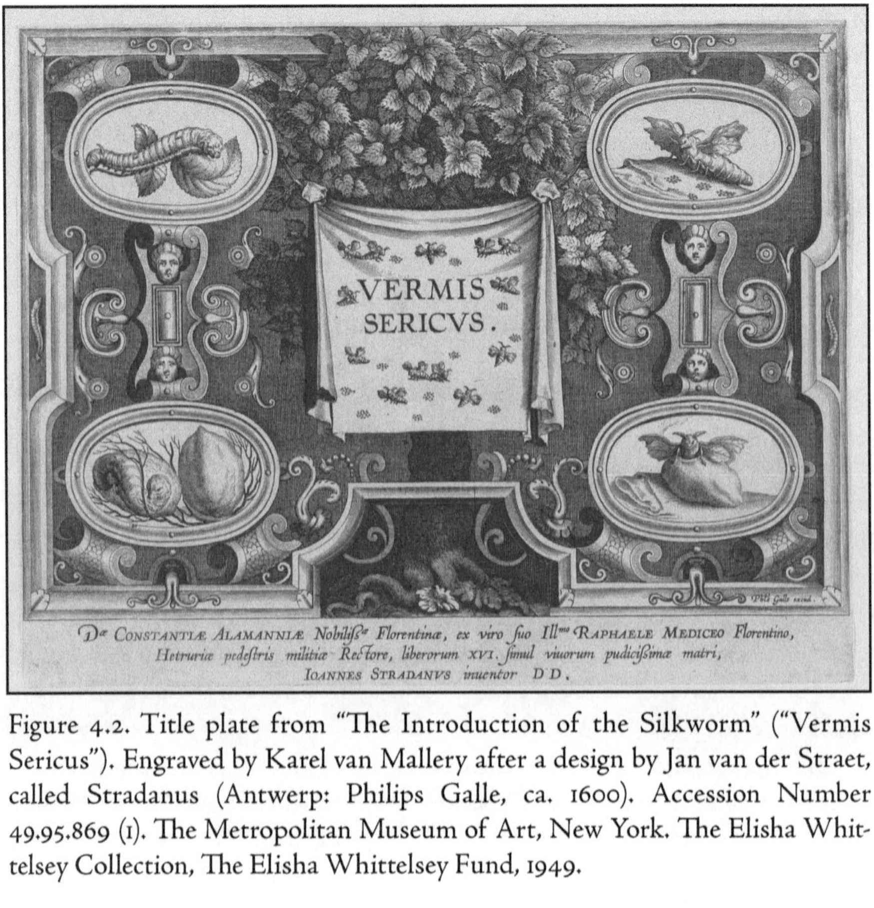

Figure 4.2. Title plate from "The Introduction of the Silkworm" ("Vermis Sericus"). Engraved by Karel van Mallery after a design by Jan van der Straet, called Stradanus (Antwerp: Philips Galle, ca. 1600). Accession Number 49.95.869 (1). The Metropolitan Museum of Art, New York. The Elisha Whittelsey Collection, The Elisha Whittelsey Fund, 1949.

to antiquity. Without a doubt, the likes of Minozzi and Manzini would not have hesitated to complete the list with conceits, invented or at least carried to perfection in the age of subtlety that was the seventeenth century.

Novelty, however, remained likely to provoke ambivalent reactions. Gunpowder certainly topped the list of inventions that it was elegant to celebrate with one or two caveats, but a cloud of mistrust also hung over less controversial items. In fact, mistrust of novelty was more than just the dark side of invention's early modern mythology; one could say that mistrust lay at the basis of novelty's ability to fascinate. The cult of the inventor's personality and the dispute over priority that nourished early modern works such as Polydore Vergil's *De inventoribus rerum* (*On the Inventors of Things*, 1499) could extend to the inventor of a single metaphor, as philosopher and translator Alessandro Piccolomini (1508–1579) argued in 1572.[77]

Secrecy was likewise central to the early modern culture of ingenuity. A closed door served as the emblem of the guild of silk workers or *Arte della seta* in Florence, and secrecy was also paramount among the ethical and social concerns related to invention with which Minozzi peppered the celebration of sericulture as a mark of ingenuity. At one point, he recalled the story of Ugolino Menzani, publicly condemned for exporting equipment for throwing and twisting silk from Bologna to Venice. While Menzani escaped bodily punishment, a painting showing him "hung upside down" remained on public display in Bologna for over a century.[78] He was a traitor to his country. Art secrets were valuable. Some were so dangerous that they should never have left the mind that created them, as suggested by another story transmitted by Minozzi. In the city of Danzig, "one could see a most ingenious artifice. It ensured that several looms in a room, as many as one might desire, could weave every fabric and every cloth by themselves. They were programmed and adjusted to work twenty-four hours a day like the machinery of a clock. But, seeing that so many people lived off the art of weaving and that now they would die of hunger, the city's magistrate forbade the invention, and its author was secretly drowned."[79] The source of the anecdote was a book by Olivetan monk Secondo Lancellotti (1583–1643) in defense of modern ingenuity.[80] Minozzi's account of a few days spent in Bologna connected conceits with art secrets and transformations in labor through a description that was deeply permeated by the period's simultaneous fascination with and mistrust of ingenuity.[81] Not coincidentally, the discussion of silk was ensconced between the two sections of the book dedicated to conceits and fireflies, underscoring the early modern role of silkworms as symbols of the concern with authorship and hubris that also haunted the debate over conceits—including their aliveness or vivacity.

The silkworm seemed to long for isolation. Some believed that the insect pulled fibers from its substance in order to bury itself in a cocoon and die in it, until a moth would emerge from its lifeless and decaying remains.[82] The silkworm's solipsistic drive, indicating a maker of boundaries, bore negative connotations, as when the narrator in Rodrigo Fernández de Ribera's *Los anteojos de mejor vista* (ca. 1625) compared a character's refusal to revise his opinion with the insect's pursuit of solitude in its cocoon.[83] Even when the silkworm stood for creative autonomy and independence, it suggested a model of invention that was as heroic as it was dangerous. Its perils stood out in Petrarch's likely fictitious letter to Tommaso Caloiro da Messina (ca. 1350–1351). Asked his opinion on invention, Petrarch reminded Tommaso of Seneca's advice that one should become like the bee, which fabricates honey out of the nectar it has gathered from flowers.[84] Through digestion, the bee claims authorship over the honey like a writer who transforms what he or she has read into something that is essentially new.

Of course, Petrarch timidly added, one could also think of a different, more audacious standard of invention. The silkworm provided a model for a poet who creates without relying on what others have done first. For the majority, however, Petrarch sternly warned, self-sufficiency was not without danger. We should weigh our talent carefully and be aware of our limitations. Imitating the silkworm could lead us to pursuits that lie beyond and above us. Petrarch advised Tommaso to use moderation (*modestia*) and follow the bee, setting aside the temptation of fashioning himself as a creator ex nihilo.

In the years leading up to the works by Minozzi and Manzini, scholars gradually took exception to Petrarch's recommendation. Lodovico Castelvetro, a loud advocate of the silkworm's daring and self-sufficient creativity, advised using one's ingenuity without following others. Prefiguring Manzini's distaste for imitation, Castelvetro voiced his opinion in a controversial but greatly influential commentary on Aristotle's *Poetics* (1570) that situated novelty at the heart of the poet's duty. By endorsing the silkworm over the bee, Castelvetro subscribed to a bold but ambivalent model of authorship destined to grow increasingly popular in subsequent decades.[85] The debate over creative autonomy that pitted the silkworm against the bee was not free from ethical and political ramifications. Petrarch's advice to know one's limits, inspired by Horace's *Ars poetica*, coalesced with a moral and psychological injunction to know oneself (*Nosce te ipsum*) and often implicated another creature traditionally linked to antisocial behavior and hubris: the spider.[86] The spider competed with the silkworm as a symbol of isolation and aloofness. In *Novum Organum* (1620), Francis Bacon compared the spider, the bee, and the ant as contrasting figures in the pursuit of truths about nature: "Those who have handled sciences have been either men of experiment or men of dogmas. The men of experiment are like the ant; they only collect and use: the reasoners resemble spiders who make cobwebs out of their own substance. But the bee takes a middle course; it gathers its material from the flowers of the garden and of the field, but transforms and digests it by a power of its own."[87] Anticipating Manzini's view of a conceit created without the concourse of "the seed," the spider became the symbol of any writer self-styled as modern.[88]

Always erecting boundaries, the spider proved captivating and upsetting at the same time. Indeed, its self-sufficiency became dangerous for two reasons. First, its disregard for interpersonal dependence threatened social cohesion, as remarked by Jesuit orations delivered in the first half of the seventeenth century, which, as I have mentioned, made the point that talent should be spread across the community.[89] Self-sufficiency also conjured disrespect for hierarchies, as illustrated in the most notorious appearance of the spider in Greco-Roman myth, the punishment of the weaver Arachne when she dared to boast that her skills and art owed nothing to the goddess Athena, as narrated in Ovid's *Metamorphoses* (8 CE).[90] Heralding Manzini's vindication of talent-based nobility (magnanimity or *generosità*), Ovid highlighted that Arachne's excellence had to do with her ability and not her birth.[91]

Yet self-sufficiency imperiled not only the community but also the subject who boasted it. Compared to the silkworm and the spider, which gave away part of themselves without return, the bee, which merely carried out digestion, was not only less arrogant but also more prudent. Even Minozzi had to admit that the invention of conceits took a toll on the author. He conveyed the conversion of the author's substance into his or her creation through the metaphor of the alchemist's alembic. In a period in which, as Mario Biagioli has remarked, nobles and prelates had distilleries in their palaces and residences, it comes as little surprise that the alembic, which purified a substance through the action of heat, stood for intellective and creative processes.[92] The metaphor of "brain distillation" was greatly popular. A curious inkwell, crafted in Italy around 1600, makes the point with humor by depicting a man placing his head inside the alembic as he declares, "I distill my brain and am totally happy" (Figure 4.3).[93]

In a tract "On the Distillation of the Human Brain," published posthumously in 1628, scholar and writer Vincenzo Gramigna (1580–1627) explained that, in distillation, heat isolates a subtle spirit or quintessence from the flowers in the alembic, imitating a process found across a variety of natural and human activities, both manual and intellectual. For instance, Gramigna perceived the formation of rain from water surfaces and the transformation into gold of less valuable metals as manifestations of the alembic's mode of operation in nature. However, the creation of miniature artworks and subtle ideas, he claimed, resulted likewise from placing one's brain in the alembic. As was the case with Manzini's analogy between the invention of conceits and spontaneous generation, for Gramigna, the heat of the brain and that of the sun performed analogous roles.[94] Eleven years after the publication of Gramigna's reflection, Minozzi wrote that in Bologna, as in a sort of alchemical Parnassus, "distilled quintessences of the intellect" flew down instead of the stream from the fountain of Aganippe.[95] Elaborating on the alembic as a trope, Minozzi explained that inventors of conceits stood in opposition to antiquarians and all

Figure 4.3. *Inkstand with A Madman Distilling His Brains.* Italian, probably Urbino ca. 1600. Maiolica (tin-glazed earthenware). 24.1 × 21 × 14.3 cm. Accession Number 04.9.15. The Metropolitan Museum of Art, New York. Rogers Fund, 1904.

those who live off the merits of a prodigious memory. Unlike them, "those who know how to compose sweat quite a bit," he claimed. "As they distill their intellects in ingenious inventions, they become weak, tired, and consume themselves up."[96]

Minozzi failed to observe that consumption affected creator and creation alike, a fact that might help elucidate the connection between self-sufficiency, hubris, and the life of conceits. Brain distillation resulted in conceits so subtle and thin that they often risked becoming a form of nonviable offspring. That was precisely the point of the Jesuit López, when in 1644 he argued that conceits did not penetrate like swords but tickled like needles, unable to persist beyond the first or second blow without either breaking or becoming blunt.[97]

Spiderwebs symbolized subtlety carried to the extreme. They were admirable but useless, as Bacon insinuated in 1605 in *The Advancement of Learning*: "The wit and mind of man . . . if it work upon itself, as the spider worketh his web, then it is endless, and brings forth indeed cobwebs of learning, admirable for the fineness of thread and work, but of no substance or profit."[98] According to Diogenes Laertius (fl. third century CE), Stoic philosopher Ariston of Chios (fl. ca. 260 BCE) compared spider threads to the sophisticated but useless arguments crafted in the schools of dialectic.[99] Ariston's analogy would inspire one of the emblems included by the Jesuit Antoine de Bourgogne (d. 1657) in the collection *Mundi lapis lydius* (*Touchstone of the World*, 1639). Bearing the inscription "Subtle, [therefore] useless" (*Subtilis, inutilis*), Bourgogne's emblem and its commentary compared the graceful but sturdy textiles produced in Cambrai, Holland, and Brabant to spider threads, concluding that "fineness (*tenuitas*) ultimately diminishes strength."[100] As Raphaële Garrod has summarized, "The emblem defines subtlety as an extreme performance of ingenuity which ultimately renders it inane."[101] Spiderwebs were the victims of the same slender and delicate nature that made them admirable. The examples of Icarus and Phaeton evoked by Manzini provided a tragic and heroic counterpart to the subtlety of spiderwebs, emblems high and low of the risk of self-destruction perpetually looming over the maker of wonders. Far from accidental, the risk of self-destruction was central to the life of conceits and informed each stage of artificial intellectual conception, in which imagination linked different realities to one another via that wondrous but unreliable instrument, figurative language, the product of imagination.

The cult of fragility was at the center of the biting but perceptive criticism of conceits formulated in 1644 by the Jesuit Daniello Bartoli. Inventors of conceits, he claimed, "toil producing conceits that most of the time are disconcerting . . . glass manufactures worked with the tip of a lantern, which, no sooner than you touch, or even see, them, they break to pieces—and yet the more fragile, the more beautiful they are; *even more, it is their fragility that makes them precious*."[102] On the margin, a note directed the reader to Bartoli's source, supposedly Plutarch but actually Pliny's discussion of a mineral endowed with glass-like qualities called fluorspar or murrine ware (*murrina*), "things which their mere fragility rendered costly." Describing it, Pliny concluded pithily, "It came to be deemed the proof of wealth, the true glory of luxury, to possess something that might be absolutely destroyed in a moment."[103] Always ambivalent about luxury, Pliny was less than enthusiastic that fragility, a quality that detracted from the usefulness of goods, should become the measure of value.[104]

A similar apprehension informed Bartoli's criticism of conceits, which fools appreciate in proportion to how close they are to perishing. Conceits were rendered precious by a stability so precarious that it was always teetering on collapse, the result of walking a fine line between deficiency and excess. Their success—or failure—lay with their ability to advance a connection rare enough to strike as wondrous yet somehow plausible; their subtlety denoted the ability to penetrate through the fissures of reality beyond what others could perceive, but it also referred to the form of the conceit itself, which struck the audience as fragile. A conceit should make sense and yet be on the brink of not making sense, of breaking "its point," similar to spiderwebs that arouse wonder because their threads seem to contain the bare minimum of matter needed to subsist.[105]

The analogy between glass-like minerals and conceits implied a form of pleasure based on fragility and risk, while also pointing to broader ethical and social implications. Glass served as a metaphor for honor, particularly in reference to female chastity, which even the slightest suspicion might compromise and "break."[106] Conceits and honor were similarly glass-like, and the strategies of risk management that assured the writer's success also applied to the courtier. The fact that Peregrini, Manzini, and Gracián authored writings on both conceits and life at court renders the parallel especially persuasive. Between 1637 and 1647, Gracián laid out the strategies best suited to help the courtier increase his social capital or *caudal*, his value in the eyes of others.[107] At stake in *caudal* could be wealth (how much he was believed to have), learning (how much he was believed to know), or social networking (on how many and what kind of people he was believed to count). The courtier should give interlocutors the impression that he possessed more than they could fathom, conveniently clouding his assets through the appearance of depth. Gracián called the risk thus incurred *empeño*, a term suggesting that the courtier's *caudal* or social capital has been placed "in pawn." Risk grew in proportion to the expectations that he was able to arouse in others. The size of a bet, measured in the fraction of one's social capital put on the table, determined the value of any transaction.

The courtier's *caudal* was always volatile and defined by fragility. In the grip of competition, it was prudent to avoid unnecessary danger. Yet while Gracián advised individuals to "avoid entangling themselves in obligation" ("huir los empeños"), he knew only too well that the courtier needed to put himself on the line if he wanted to survive, let alone thrive.[108] The courtier's prudence evoked in the title of Gracián's last work on the topic, *Oráculo manual y arte de prudencia*, was mainly about managing risk. The same held true of the invention of conceits. Orators and poets arouse wonder through the use of dissonance, which constrains the audience to make sense of arguments involving figurative language. If they play too safely, using trite or low-demanding tropes, they might end up arousing little to no dissonance in the audience, similar to the courtier who never promises anything to anyone. Conversely, if they dare too much through astonishing, far-fetched metaphors, the audience might end up defrauded of too high and unrealistic expectations if the argument fails to make sense. Between deficiency and excess, the inventor of conceits has little room to maneuver.

Different stakes, however, rendered the parallel between the courtier and the inventor of conceits only partial. Facing the risk of losing face, the courtier was sparing with his capital where the orator and the poet could be calculating yet bold. Conceits reproduced the pattern of "pawning" under altered conditions: inventors created a problem that they themselves solved.[109] Even though Gracián compared the intellect with the ancient hero Theseus finding a way out of King Minos's labyrinth, and with Odysseus escaping Polyphemus through ruse, the inventor of conceits created the problem with a solution first in mind. Rather than Theseus or Odysseus, Daedalus, who created the labyrinth, and Homer, who put Odysseus in trouble and then helped him escape, were his models.

To lead the audience out of a problem by pointing the solution to them: this is what conceits brought about by using figurative language in a way that was both striking and highly fitting. In doing so, Gracián claimed, a conceit displayed "soul of subtlety," which was contingent on evading the danger.[110] It was the conceit's fragility that elicited from the audience a feeling of aliveness and vivaciousness. As Jesuit theologian Francisco de Toledo first argued in 1574, the soul, which is the form of the body, holds together everything that makes life in the individual possible—and turns it into *opus vitale*, in Manzini's terms—including accidents, organs, temperament, and disposition. Then, "when [the soul] leaves, all gets dispersed and corrupted."[111] Like the soul, so the argument went, the trope in a conceit held different realities together in a way rendered evident by a strained, unstable character, always threatening to dissolve.

Risk informed a model of rhetorical and poetic invention based on dissonance and wonder. In 1622, former Jesuit Agostino Mascardi disparagingly compared inventors of conceits to tightrope artists who "delight the spectators inasmuch as just a slip of their feet threatens to cause them to fall."[112] For

Mascardi, the whole was not worth it: "To have one's foot always suspended," he claimed, "removes any delight and produces horror."[113] The characters of *Philosophía antigua poética* (1596), by physician and poetic theorist Alonso López Pinciano (ca. 1547–1627), compare narrative and dramatic plots—with their unstable and precarious balance between verisimilar and wondrous events—with tightrope art. A juggler named Buratín has just arrived in Madrid. To the amazement of the people, he performs every manner of movement and dance on the tightrope using different shoes, even stilts. What Buratín does is possible but not verisimilar, one of the characters claims. He has seen it, but the mind finds it hard to believe.[114] Conceits produce the same reaction.

Pliny's discussion of luxury shaped the debate over conceits through two different but interrelated ideas. The first, connecting fragility and value, prompted Bartoli to compare conceits to glass-like minerals, esteemed because they are always on the brink of destruction. Yet Pliny also disparaged the search for goods that it is costly and even dangerous to obtain, a consideration reminiscent of the writer's self-consumption at the alembic imagined by Minozzi and the peril of his or her stunts on the tightrope, in Mascardi's analogy—even though these were, of course, metaphoric and hyperbolic comparisons.[115] Both in Pliny's assessment of the culture of luxury and in the literature about conceits, simultaneous danger to the creator and his or her creation results in both value and aesthetic pleasure. As in the case of a spider that spins its threads out of its own matter, the appearance of life that characterized conceits was dearly bought, costing the author part of his or her own substance. The life of conceits derived from the vitality that the author gave away in favor of his or her creation, but also its attendant precariousness and risk of contingencies. The author's fragility and that of his or her conceits were two sides of a materialistic conceptualization of creativity and, specifically, of rhetorical and poetic invention.

The mention of tightrope art, however, was far from isolated. It pointed to a link between the life of conceits, the arts of the fair, the range of disciplines that early modern individuals categorized under the rubric of magic, and, last but not least, alchemy.

Praestigia, Natural Magic, and Alchemy

The analogy with tightrope art, which served Mascardi to condemn conceits as puerile, dated to at least the fourth century CE, when Jerome had compared the rhetoricians who juggle with likelihood to the sleight-of-hand of a mountebank.[116] Mentions of the fair, its professions, and activities permeated seventeenth-century writings on conceits. In 1644, López angrily inveighed against the inventors of conceits, calling them jugglers and mountebanks who did not mind debasing themselves for the sake of entertaining a popular and hardly discerning audience. Appropriating a passage from Seneca's *Letters to Lucilius* originally targeted at sophistry, López compared the inventor of conceits to a swindler who, "with his mouth open like a warehouse, vomits a multitude of variegated ribbons, now a purple one, then green, blue and yellow."[117]

From a different perspective, the fair became a positive and empowering symbol of conceit making in Tesauro's account of *versability*, the faculty of the mind in charge of finding connections. Addressing the kind of criticism that López had directed against conceits, Tesauro repurposed the analogy between the juggler and the inventor of conceits as he described the intellect's ability to link via figurative language the circumstances it has discovered "inside" the topic: "*Versability* compares quickly all the circumstances among themselves or with the subject; it ties them up or divides them; it aggrandizes or diminishes them . . . and hints to the one through the other, as the juggler does with cups and balls. . . . And this is what metaphor does, the mother of poems, symbols, and devices. And those excel in ingenuity that get to know and pair circumstances that are more remote between themselves."[118] Tesauro likened intellectual to manual ingenuity by comparing the transformation of reality via metaphor with the juggler's ability to make things appear in places in which they previously did not seem to be.

Through the embrace of carnivalesque skills and activities, Tesauro suggested that the intellect had switched to a playful, not entirely trustworthy mode of argumentation in which tropes made one thing pass for another, augmenting and diminishing them just as the juggler in cup-and-ball routines resorted to quickness and dexterity of hand to deceive the audience.[119] Undisturbed by the prospect of contaminating the invention of conceits with a manual and mechanical art, Tesauro—who did not hesitate to include deceit as one of the components of metaphor—saw fit to place conceits in a space marked by fraud.[120] Vera Keller explains that in 1610 German physician and alchemist Johann Ernst Burggrav "counteract[ed] the charge that alchemy was mere prestidigitation by championing prestidigitation itself as a form of ingenuity."[121] Tesauro likewise engaged in a vindication of conceits that involved a reassessment of the entire sphere of ingenuity, as well as dignifying any manual and mechanical counterparts of conceit making.

In fact, a reassessment of juggling was already underway in the context of early modern magical theory and practice. In 1599, Jesuit theologian Martín Del

Río assigned juggling to a category of magic that he labeled as *praestigiatrix*, which fabricates deceptions or *praestigia*. Although Del Río remarked that jugglers were creatures of the fair who pretended to make incantations with "the agility of hands and feet," Ann-Sophie Barwich has observed that the magic of the fair also involved "early forms of specialist knowledge in mathematics (riddles), physics and astronomy (optical effects), chemistry (transmutation), and theatrical display (mechanical props)."[122] Tesauro's enthusiasm for juggling as a term of comparison thus made entire sense within the early modern natural-philosophical and mathematical culture. As noted by Stuart Clark, Francis Bacon included "The History of Jugglers and Mountebanks" (*Historia praestigiatorum et circulatorum*) among the materials that would serve as a basis for "a natural and experimental history." According to Clark, in Bacon's program "the feats of jugglers were used to demonstrate the sympathy and antipathy of plants and the powers of imagination."[123] The juggler's ingenuity denoted intimacy with nature.

Allusions to the sympathy and antipathy of plants were even more common in connection with natural magic, a discipline that, according to Del Río, "frequents the intimate natures of things."[124] Rumor had it that the earliest practitioner of natural magic, in ancient Iran, was Zoroaster, a character destined to enjoy a fleeting but suggestive association with conceits in Minozzi's *Sfogamenti*.[125] Halfway through the digression on fireflies, the author directed the reader to the works of Count Carlo Della Lengueglia, described as "seedbeds of charms" that captivate the mind. Sporting a style brimming with conceits, the count was "a new Zoroaster and a composer of magic full of spirit," prefiguring Tesauro's claim that conceits hold the mind spellbound.[126]

In a veiled but evocative way, Minozzi suggested that natural magic was a bridge to understanding the entanglement of natural and human ingenuity embodied in the parallel between fireflies and conceits. Natural magic professed that by understanding the causes through which nature brings about its effects, one could replicate nature and control it.[127] Del Río called natural magic "a more accurate knowledge of the secrets of nature, by means of which, having observed the course and influence of the skies and stars, and the sympathies and antipathies of singular things, some things are applied to others in consideration of time, place, and mode, and some wondrous things are done in this way, which seem deceitful and miraculous to those who are ignorant of the causes."[128]

Magic had always been regarded with suspicion, despite attempts by medieval scholars to identify a benign and respectable, "natural" version of it.[129] William of Auvergne (d. 1249) observed that while the presence of fire in a substance caused its tendency to rise, a number of phenomena resisted explanation by the presence of one or more of the four elements, suggesting a causality defined by "concealed forces and powers"—what later authors would call "occult" causes or properties—which it was possible for one with knowledge of nature to manipulate.[130] The reference to a deeper logic than the one "visible" to Aristotelian physics resurfaced in Thomas Aquinas's letter *De operationibus occultis naturae* (*On Nature's Occult Operations*).[131] Despite the climate of institutional persecution that witnessed the publication of the infamous *Malleus Malleficarum* (*The Hammer of Witches*, 1486–1487), fifteenth-century scholars such as Marsilio Ficino (1433–1499) and Giovanni Pico Della Mirandola elevated the status of the discipline by laying claim to its ancient pedigree.[132] Natural magic became increasingly specialized as a study of nature's occult causes and operations: "the practical side of the knowledge of nature," according to Pico's formulation.[133]

The relevance of natural magic for the period's material and experimental culture only intensified in the sixteenth century. In 1558, Della Porta described natural magic as knowledge of the entirety of nature and defined it in the following terms: "A master of great power that flows forth secret mysteries and the qualities and properties of hidden things. It bestows freely upon the cognition of the whole of nature, and teaches (knowing as it does the agreement and disagreement that exists among things) how to separate and unite them through a mutual and fit application of one unto another, so that it may bring forth works that people call miracles, and which surpass every admiration or grasp of a human mind."[134] Both natural and black magic—which differed in that the latter involved the aid of demons—operated by "knitting together causes and virtues of things rarely conjoined," as Lorraine Daston has put it.[135] The magician made use of properties that were less evident, and accordingly more wondrous, to the intellect. Della Porta, who left a recipe for crushing fireflies in the confection of glow-in-the-dark ink, argued that natural magic made use of occult properties that nature had concealed from the majority.[136] Paramount among them were agreements and disagreements ("sympathies" and "antipathies") such as those mentioned, for instance, in the title of Juan Eusebio Nieremberg's *Oculta filosofia de la sympatia, y antipatia de las cosas* (1633). Sympathies and antipathies structured the world in a covert and mysterious way, predisposing things to seek and escape one another in manners that were not evident to the eye and the mind.[137] Della Porta's views found lasting echoes not only in best-selling books of secrets but also in university lectures delivered by natural philosophy professors, such as Nieremberg himself.

By calling Della Lengueglia "a new Zoroaster," Minozzi aimed to elevate conceits from the realm of the fair, with its mechanical, deceptive arts that so charmed Tesauro, to the status of rhetorical and poetic counterparts of a discipline that asserted knowledge and control of nature. Even though the opposition between *praestigia* and natural magic was less than clear-cut for some of Minozzi's contemporaries—who often perceived natural magic as closer to a vain and empty spectacle than to the earnest and meticulous knowledge of nature it purported to be—Minozzi's likening of conceits to natural magic was hardly idiosyncratic.[138] Late sixteenth- and seventeenth-century writings seemed to suggest that both were similar in operation. Conceits, for one, aroused wonder in the audience. Del Río identified wonder as also being the effect of natural magic, a coincidence corroborated by Minozzi's suggestion that conceits "cast a spell" on the intellect.[139] Yet conceits also resembled magic in terms of invention. Della Porta's claim that magic operated the separation and the union of things "through a mutual and fit application of one unto another" found a direct echo in Peregrini's argument that a conceit delighted the audience through "a mutual and greatly fit arrangement" between the terms, crafted by the author's ingenuity.[140] For Gracián, a conceit was a product of intellectual sharpness or acuity, which discovered within reality "some conformity or proportion" between different things, worthier in proportion to its concealed or occult character.[141] Just as natural magicians used their familiarity with the sympathies and antipathies that tie nature's creations to one another in order to bring about wondrous and surprising effects, so orators and poets, "having penetrated [the topic] greatly and successfully," exploit the connections, concealed or occult from the minds of the majority, that they discover among its circumstances in order to shock the intellect of the audience with wonder at unheard-of but strikingly suitable pairings.[142]

Minozzi's account allowed for two different yet complementary explanations for the life of conceits. The first, drawing on magic, suggested that the author's imagination aroused amazement at connections that had remained concealed or occult, casting a spell on the mind and stirring it with the restless, inquisitive movement of wonder. Minozzi called conceits "spirits" (*spiriti*), and "spirited" (*spiritosi*) the writers who excelled at conceits.[143] Moreover, Minozzi's description of conceits as "little spirits" (*spiritelli*) capitalized on the ambiguity between the term's medical and natural-philosophical meaning and its demonic counterpart.[144] Della Lengueglia's conceits or spirits could refer to the subtle and airy substances that carry vital functions as they mediate between soul and body, but they could also be taken by small, breathing, and moving preternatural creatures taking hold of the reader's mind.[145] Minozzi's hesitation to decide whether conceits "gave" him soul or "stole" it from him underscored the simultaneous and conflicting presence of natural-philosophical and demonic meanings in the term "spirit."[146] In any event, by alluding to Della Lengueglia's writings as "seedbeds" (*seminari*) of charms, Minozzi sought to highlight a connection between conceits and the sphere of life.[147]

The second of Minozzi's explanations for the life of conceits involved alchemy, and foreshadowed Tesauro's claim that conceits made one thing pass for another, as a form of transmutation. Drawing on the analogy between conceit making and distillation, seeing conceits as alchemy had the potential to shed further light on a materialistic, interdisciplinary approach to early modern invention. As I mentioned earlier, distillation applies heat to a substance to liberate its spirit from its matter.[148] Conceits were called "spiriti" because, in creating them, the heat of imagination operated like the alembic that separates the subtler and pure occult or concealed correspondences—which, in the analogy, served as counterparts to spirit—from a thick and impure material layer of appearances.

In early modern medicine, spirits were the soul's instrument, through which it communicated with the body, but also that which kept the body alive.[149] Early modern natural-philosophical and medical works linked spirit to life, either as the cause of generation or as a subtle vapor responsible for the performance of vital functions. Metaphorically, writers and plastic artists also aspired to isolate spirit's enlivening and ensouling properties to instill them in conceits, paintings, and statues. Commenting on Agnolo Bronzino's panel of *Pygmalion and Galatea* (ca. 1530), Vasari described the sculptor imploring Venus so that the statue he had sculpted "receiving the spirit, might come to life" (*ricevendo lo spirito s'aviva*).[150] The passage on *Pygmalion* followed Vasari's discussion of Jacopo da Pontormo's now-lost *Raising of Lazarus*. In it, the flesh was painted in such a way that one could tell the parts that had been reinvigorated from those "where the spirit had not reached yet" (*doue non era ancora lo spirito arriuato*).[151] Spirit instills life into both the sculptor's marble and the flesh of Lazarus. Similarly, language informed by pneuma or spirit gives rise to rhetorical and poetic conceits, which are, in a way, alive.[152]

In Minozzi's view, the author's imagination distilled spirit and its life principle as he or she pondered on the topic being discussed. In doing so, orators and poets replicated the quest of early modern alchemists such as Petrus Severinus (Peder Sørensen, ca. 1540–1602). Severinus popularized the application of chemistry to medicine advanced by German-Swiss physician Paracelsus (Theophrastus

Bombastus von Hohenheim, 1493–1541). In 1571, he maintained that one could distill a quintessence or vital principle from the body. As Hiro Hirai has explained, for Severinus the substance that operated as vital principle in nature was a type of salt.[153] Similarly, in 1665, Kircher called salt "the true and genuine element of the earth."[154] He argued that inside the folds of common salt there was "spiritus sulphureo-salino-merculialis," "a seminal force" that was responsible for the generation of all things through its participation of sulfur, salt, and mercury, the three primordial elements in nature, or *tria prima*, according to Paracelsus.[155] Salt thus understood was nothing but a quintessence, the fifth element that Kircher identified with Aristotle's pneuma or spirit, which "fertilizes, vivifies, nourishes, [and] preserves everything."[156] "The seed," Kircher concluded, "could be called animated salt."[157] Even the term "alchemy," Kircher speculated, meant in origin "melting of salt" (τοῦ ἁλὸς χυμίαν . . . *salis fusionem*).[158]

The insights of Severinus, Kircher, and other mid-seventeenth-century practitioners of alchemy intent on isolating spirit and its life principle are highly valuable to establishing connections across a variety of disciplines. Without them, one would try in vain to understand notions vaguely sketched in rhetorical and poetic writings from the period that become suddenly meaningful in light of aspirations expressed in works from seemingly remote and supposedly unrelated fields. According to Kircher, in the middle of a discussion on healing waters, Hippocrates had claimed that "salt endowed with the virtue of sulfur" (*sulphurea vi*) was "divine" (τὸ θεῖον, *id est, quid divinum*).[159] Could it be a coincidence that Manzini, who claimed that ingenuity is "all sulfur" (*tutto di zolfo*) and that it instills life in conceits, also claimed that, in doing so, it becomes elevated to "a certain I know not what of the divine"? The case of Manzini presents us with a rhetorical and poetic theorist who, without necessarily being himself a practitioner of alchemy, was educated in a culture in which ideas about the presence of spirit and its life principle in salt were pervasive and played a role as he, Minozzi, and others sought to figure out how conceits could be "alive."[160]

Kircher's account of the alchemy of salt includes further links to early modern views on the life of conceits. According to Kircher, "in dunghills, stables, sepulchers, latrines, and excrements," salt carries out the process of fermentation that keeps nature's economy afloat by transforming decay into the beginning of new life.[161] Kircher's catalog includes ecosystems favorable to spontaneous generation, and thus germane to the invention of conceits. Could fermentation be, like salt, connected with conceits in the early modern culture of ingenuity?

In a set of notebooks gathered by the young Giovanni Maria Lancisi (1654–1720) starting in the year 1672, the future physician to three popes collected a variety of opinions on topics relevant for the profession, arranged in alphabetical order. Lancisi recorded that fermentation was "the internal movement of the parts of which [a body] is composed," a view that he quoted from a diatribe on fermentation published by Thomas Willis (1621–1675) in 1659.[162] Willis remarked that Aristotelian natural philosophers traditionally paid little attention to fermentation, a process that was well-known to artisans who made bread, wine, or beer. A number of authors had turned to the topic recently, however, recognizing at last what artisans had always known, namely, that matter is seemingly permeated with life and a form of inner activity and movement.[163]

By pairing salt with fermentation, Kircher brings us back to the claim by astronomer Johannes Kepler, discussed in chapter 1, that dissonance in music is like salt or vinegar in a banquet, which serves as "the yeast" that "ferments" what otherwise would be a whole made up of inert and static consonance.[164] Dissonance also "ferments" conceits by endowing them with inner activity and movement through the unstable and precarious equilibrium, based on figurative language, which keeps the whole together.

The notion of inquiry as movement had Greco-Roman origins. Thought was, for Hippocrates, "the soul's promenading," as Huarte recalled.[165] As it moves, the soul longs to attain the truth for which it "naturally" exists, just as fire and earth tend to rise and then descend. In 1599, Ginés Rocamora y Torrano commended mathematics on the grounds that its demonstrations and theorems allow the mind to rest in certainty.[166] Contrary to mathematics, conceits keep the reader's mind in constant, endless motion: they are filled with inner movement and activity because the yeast or *fermentum* of dissonance keeps the reader's mind perpetually wondering how to make sense of the linkages that they advance through tropes, preventing it from resting in a certain and satisfactory truth.

As Jesuit Maciej K. Sarbiewski suggested in 1626, conceits exist in a troubled union, which he labeled *concordia discors*.[167] Although Sarbiewski did not make any connection between conceits and matter explicit, the tag *concordia discors* evoked not only the claim by Empedocles that the world results from the clash between Love and Strife but also its influence on Aristotle.[168] For Aristotle, heterogeneous bodies are made up of fire, air, water, and earth, mixed in unstable and precarious equilibrium and always ready to devolve into its constituent parts. Vital heat keeps heterogeneous bodies in place and prevents them from disintegrating, with fire ascending, earth descending, and each striving to find rest in the place where it naturally belongs to the detriment of the whole. Conceits likewise argue in seemingly different and irreconcilable directions.

Only the heat of the author's and the reader's imagination makes a transient, almost inapprehensible union possible and, in a way, enduring.[169] Sarbiewski's allusion to *concordia discors* evoked, albeit indirectly, a link between conceits and matter's inner activity and movement.

If dissonance was the yeast or *fermentum* in harmony, then the analogy tying music, rhetoric, and poetics to nature verified itself also from the perspective of alchemy. For instance, Kircher maintained that fermentation, "the cause of generation and corruption in all things," occurs in heterogeneous bodies through the agitation and the separation in them of particles of sulfur, salt, and mercury.[170] What remains from Aristotle's natural philosophy is the notion that movement permeates matter because heterogeneous bodies are made up of components with different, irreconcilable tendencies.

I come now to the last of the links in the chain connecting the life of matter with that of conceits. Halfway into a discussion of spontaneous generation, Kircher recounted a story from *Commentarii de sale* (*Commentaries on Salt*, 1572, revised in 1579), a work by Spanish humanist Bernardino Gómez Miedes (1520–1589) started in mid-sixteenth-century Rome. Gómez Miedes argued that salt is "the common cause, or even better, occasion" through which things come to be.[171] Prefiguring Kircher's argument on salt and generation as much as Manzini's language about conceits, Gómez Miedes claimed that salt deserves to be called "divine" because it is reminiscent of spirit and its life principle. Yet the parallel with Kircher ended there. For Gómez Miedes, in fact, it was salt's ability to prevent putrefaction that justified its claim to divinity: "Salt exerts a form of power that one might call divine, and plays the role of soul, because it is like a certain spirit that, while occupying a body, constrains it to subsist."[172] Salt encapsulated the link between nature's ingenuity, embodied by spirit and its life principle, and the mind's ingenuity. To confirm the analogy, Gómez Miedes dedicated book 4 of *Commentaries* to what he called "festive or humorous salt" (*de sale geniali siue iocoso*), that is, conceits.[173] In a work exclusively devoted to salt that predated the seventeenth-century craze for conceits by several decades, Gómez Miedes thus confirmed the continuity between natural and human ingenuity centered on salt to which Minozzi would allude in 1639, when he compared Bologna's cured meats (*salumi*) to the conceits of the writers in the city.[174]

From natural magic to alchemy, early modern disciplines aimed at investigating and manipulating nature informed—and perhaps were informed by—rhetorical and poetic reflections on conceits. Regardless of how each scholar envisioned the process that Manzini called artificial intellectual conception, they all agreed that its product, conceits, brimmed with "life" resulting from imagination's ability to artificially excavate, or distill, or perhaps ferment, the concealed or occult connections underlying reality, and to build with it a product marked by unstable and precarious unity, always arousing wonder and never inert.

Few phenomena were more apt at exemplifying nature's occult qualities or properties than the light of fireflies, the third and most important ingredient justifying the analogy between the insects and conceits. Minozzi's observations about bioluminescence provide us with a case study to question, and even disprove, one of the main tenets of the literature about conceits as formulated by some of its main exponents, such as Tesauro. I am referring to the assumption that for a conceit to succeed—and to appear alive—it needs to be false. By denying that the link between conceit making and falseness was necessary, I show that the search for natural-philosophical truths and the invention of conceits were entangled in highly productive and, so far, unsuspected ways.

A New Philosophy

"If it were for me," the traveler wrote in allusion to fireflies, "I would call them relics of the sun already set." Faced with the impatient and nervous pulse of the insects, Minozzi paid like with like as he suddenly burst into a string of conceits, the form of plenty of which he was capable. He called the insects "eyes of the summer that, with episodes of light and shadow, are now open, now shut; small and fiery Arguses that transform the blind mole of the night into a lynx, the earth's flying stars, winged compliments, wandering bolts, small flames endowed with soul, sparkling atoms and, in sum, living fireworks of nocturnal joy."[175]

Excessive as they might seem, these could be described as relatively effortless conceits. They built on the reader's ability to picture similarities between fireflies and certain natural and artificial phenomena. Admittedly, they still required basic knowledge of Greco-Roman culture, including hundred-eyed Argus, the mole's proverbial blindness, and the lynx's ability to see through solid bodies. Limited familiarity with ancient myth and the emblematics of nature—often derived from Aristotle's works of natural history and, to a greater extent, Pliny—provided access to a string of conceits that, in a nutshell, told the reader that clouds of eyelike fireflies turned the darkness into light.[176]

Not entirely satisfied, Minozzi put on the vest of the natural philosopher and, with conceits as instruments of choice, tried—or pretended to try—to examine the operations that rendered the light of fireflies so intriguing. In doing so, and as he ventured into the study of nature, Minozzi confronted head-on the fraught and controversial issue that lay at the origins of the debate between Mascardi, Peregrini, and himself: the relationship of conceits to truth.

In fireflies, Minozzi observed that "with a new philosophy, movement acts as the cause of light, rather than of heat."[177] Though he had mentioned the house of Aldrovandi in Bologna, it is unclear whether Minozzi had read the zoologist's work on insects.[178] Believing, along with Pliny, that fireflies glowed only as they flew—Aldrovandi had denied as much in 1602—Minozzi built a conceit based on what he saw as a departure from nature's usual course.[179] In fireflies, from Minozzi's perspective, the connection between movement and heat that constituted a pillar of Aristotelian natural philosophy appeared to have been replaced by a link between movement and light—or at least Minozzi believed that he could argue as much by replacing a relationship of simultaneity with one of causality via a trope, as conceits were wont to do: fireflies glow when they move, therefore, they glow because they move. On a humorous yet earnest note, Minozzi concluded that fireflies posed a challenge to available explanations. They heralded "a new philosophy" that called for new attitudes in response to phenomena that did not fit within the expectations of how nature behaves "always or for the most part," as Aristotle described the scope of science.[180]

The mention of a new philosophy was ambiguous. On the one hand, it signaled Minozzi's aim to arouse wonder and steal a smile from the reader. Like fireflies, conceits introduced "a new philosophy" by operating through alternative forms of causality. Minozzi's claim that in fireflies "movement acts as the cause of light, rather than of heat" imitated the search for cause-and-effect relationships characteristic of the student of nature, while simultaneously captivating the mind with a seemingly paradoxical and inconclusive argument. On the other hand, Minozzi's exploration of causality also leads the reader along a path in which, as I demonstrate, the distinction between the pursuit of knowledge and that of beauty—between natural philosophy and conceits—becomes compromised. This is so because, as one ventures into the investigation of new, previously unknown relationships of cause and effect, truth and falseness become more difficult to discern without qualification in a terrain that is, of necessity, plagued with uncertainty. Let us consider the following example.

Minozzi claimed that, in fireflies, "one may see how, with a strange generation, light is born of shadow, movement begotten by stillness, and fire by the humors of the night."[181] Minozzi identified a paradox. Experience and elementary physics teach us that fire, not shadow, is the cause of light. Likewise, they also tell us that the movement of one body, not stillness, sets another body in motion. Last, they assure us that it would run against expectation if fire were born of a damp or watery substance. At first, it all seems to be a fanciful, far-fetched argument, a conceit exploiting a series of opposites. The writer has transformed the darkness that served as a backdrop to fireflies into the womb from which they have sprung. A still and humid night has likewise become the origin of fireflies, not only the deafening and quiet scene in which they cheerfully fly. It is tempting to take Minozzi's remarks as a parody of the quest for causes characteristic of natural philosophy. Just as Aldrovandi and his contemporaries set out to pry into yet-unknown layers of nature's behavior, the inventor of conceits basked in the exploration of alternative modes of making sense of reality by positing new and striking relationships between different objects. He and the natural philosopher made use of the same faculty, imagination, as they examined and compared circumstances on the lookout for new explanations, which ought to appear probable in the case of the natural philosopher, but merely figurative in that of the inventor of conceits.[182]

Although some of Minozzi's seventeenth-century readers probably took the statement as a string of conceits, it is likely that the author had a true or probable natural-philosophical explanation in mind, resulting in a case that effectively eroded the division between the heuristic and the aesthetic use of imagination. Let us imagine for a moment that Minozzi's allusion to "a strange generation" referred, as it probably did, to the mechanism of spontaneous generation that natural philosophers associated with the reproduction of insects, including fireflies and a lower world of so-called imperfect creatures. If that were so, then the series of paradoxes making up Minozzi's conceits would be true to reality rather than mere figments of the writer's imagination: Shadow would stand for the hot and humid soil from which fireflies emerged, the allusion to stillness rather than movement might suggest that the insects have not been generated from prior forms of life through sexual reproduction, and the claim that fireflies have been born of "the humors of the night" would pay tribute to Della Porta's claim that they were born from dew.[183]

Nature itself facilitated the encounter between conceit making and natural philosophy by displaying in its behavior a penchant for paradox comparable to that of the writer, who merely chronicled the event's strangeness. Minozzi's conceits thus highlighted the relationship between early modern views of natural and human ingenuity, and with it the question that lies at the core of the present and previous chapters: How do the operations of conceits relate to nature's operations?[184] To take Minozzi's conceits seriously—or at least, according to a notion dear to the period, as *a serious joke*—is the only path to move beyond a crude, black-and-white opposition between truth-oriented and beauty-oriented disciplines, respectively embodied by natural philosophy and conceit making. By focusing on nature's rare and wondrous phenomena, such as the luminescence of fireflies, it becomes possible to connect natural and human ingenuity in terms that, rather than simply judging whether specific conceits are true or not, might project onto nature's workings a similar degree of ambivalence and doubt.

When early modern Aristotelian natural philosophers based a claim on experience, they were launching "an implicit appeal to what the reader, as a person with normal faculties, has seen or might see or else an explicit appeal to an authority."[185] Experience was a labile and variable notion that rendered the novelty of specific objects of natural-philosophical pursuit less than clear-cut. Comets remained a rarity even for those who became accustomed to waking up to them almost every night in the fall of 1618. Fireflies too came back every summer. It was hard to call them "monstrous" in any sense of the term. As poet Antonio Telesio (1482–1534) once wrote, fireflies were familiar enough to double as children's playmates in the regions they inhabited.[186] Yet considering that known species of light-bearing animals could be counted with the fingers of a single hand and that instances of "cold light" in nature were extremely rare, fireflies remained fairly wondrous, if common, creatures.[187]

Similarly entrenched between wonder and familiarity, a range of more or less earnest, experience-related conceits resisted the label of falseness. Unlike fallacious conceits, which aroused wonder by violating logic, experience-related conceits struck as novel and paradoxical just because they dealt with realities that could be entirely true—or not—but were both unknown and counterintuitive. Without lying, they merely exploited the audience's lack of omniscience. Experience-based conceits seemed to refute Tesauro's claim that conceits were all about knowing how to lie well.[188] The notion of experience was murky, nuanced, and—despite the claims of universality that Aristotelian natural philosophy attached to it—context-specific, relying as it did on blurred facts, dubious testimonies, and, least dependable of all, personal knowledge. By capitalizing on what remained unknown to the audience, experience-related conceits aspired to exploit the gray area between rhetorical invention and natural philosophy and to arouse wonder not only without lying but also in the hope of discovering new truths. The invention of experience-related conceits coalesced with natural philosophy, especially in connection with the kind of rare particulars—cold light included—that lay at the core of early modern investigations of nature.[189] Whenever the aim was to elucidate phenomena that did not conform to the principles of Aristotelian elementary physics, any attempt at attributing a cause to them remained tentative, and any opinion merely probable. In such circumstances, imagination's quest for explanations in natural-philosophical investigations looked remarkably similar, if not identical, to its counterpart in the realm of conceit making.[190]

Another instance from Minozzi's travel narrative illustrates the notion of conceits that do not lie and even tell the truth. What could be more absurd, Minozzi seems to imply, than to imagine fireflies replacing lamps, for instance at one's study table? The proverb "To take fireflies for lamps," which lay at the roots of the analogy between conceit and fireflies, poked fun at even the suggestion of it.[191] Yet even the proverb's appeal to shared experience was subject to debate. In a group of three madrigals, first printed in 1623 with the poem on Bernini already discussed, Stigliani lamented that the summer had slipped away. It was already too late to catch fireflies and place them in a jar. The poet's cat, with its shimmering eyes, had escaped, and now Stigliani could use the light of the insects to continue his work through the night. With rising oil prices, refilling his lamp was out of question.[192] Fireflies and cats' eyes were usual suspects in early modern catalogs of cold light. In 1612, Alessandro Tassoni (1565–1635) listed them with fish scales, gems, and certain kinds of rotting wood and bones.[193] Stigliani wished to appear facetious, and his use of humor was not lost on Minozzi: "Stigliani makes me laugh when he summons [fireflies] to light up for him, and replace the eyes of his cat, 'the ancient lamp,' as he calls it, 'of his study room.' And he indeed seems willing to sell us fireflies for lamps when he pretends that they might serve him as sources of light."[194]

By suggesting far-fetched possibilities, conceits could amuse readers disposed to take joy in the author's harmless attempt at deceiving them. Yet the truth, or lack thereof, of Stigliani's conceit hinged on the reader's experience and was not as easy to determine as Minozzi wants us to believe. Although the notion that light-bearing creatures might replace candles in a writer's study might sound hyperbolic, it could have taken inspiration from sober descriptions found in widely circulating New World natural histories and chronicles. And if there was ever a poet who was familiar with publications on New World matters, it was Stigliani. In preparation for composing the epic poem *Nuovo Mondo* (1617), he must have read Gonzalo Fernández de Oviedo's *La Historia General de Las Indias* (*General History of the Indies*, 1535), which in the section on the island of Hispaniola mentioned a species of light-bearing creature, the fire beetle *cocuyo*.[195] With four or five of them, Oviedo claimed, one could walk through the fields at night or wage war. The natives kept *cocuyos* captive to light up their homes at night. Just one beetle, or even better two of them, made it possible to read and write in the dark.[196] A few years after Oviedo, Dominican friar Bartolomé de Las Casas (1484–1566) recalled that he had prayed the matins from a breviary printed in small type with the aid of light emitted by *cocuyos*.[197]

Transplanted from a Caribbean to a European context, Stigliani's conceit transformed Oviedo's description from a curious but everyday wonder of nature into a contrived and helplessly hyperbolic suggestion—unless we take seriously Cardano's claim that he had used Italian fireflies to read and write, or the poem by Andrew Marvell "The Mower and the Glowworm," which imagined the nightingale rehearsing its song by the light of fireflies, which later showed the lost and wandering mowers a path through the night.[198] Indeed, to capture fireflies in a jar was less than extraordinary in parts of Europe. The poet's imagination took in earnest—and, in the case of Marvell, compounded with further fiction—reports originating in scholarly and, at least in principle, nonpoetic fields of knowledge. The poems by Stigliani and Marvell thus exemplify the entanglement between conceits and truth, in which the ability of rhetorical invention to kindle life in the mind via connections found by imagination depends on a nuanced spectrum based on individual and highly variable forms of experience. The faint and vacillating boundaries between the familiar and the new that enabled fireflies to arouse wonder every summer also explained why experience-related conceits could appear extraordinary even if they advanced claims that were less than new or far-fetched.

To further illustrate the overlap between natural philosophy and conceit making, one might turn to Minozzi's observation that fireflies "seem to be carrying both their funerals and their torches."[199] Natural philosophy was a quest for causes, and those that denoted purpose, or "final causes," were central to a view of the world in which nothing occurred in vain.[200] To be sure, to suggest that the light of fireflies was meant for use in the insects' funeral seems more representative of what Jesuit professor Michael Radau celebrated in 1640 as the attribution of "ingenious," rather than true, final causes.[201] What could be more apt for creating "a new philosophy," a playful, alternative nature, than to persuade a mind predisposed to fiction of false but to some extent acceptable causes?

A different approach is also possible, though. If the study of rare particulars, including cold light, that became central to early modern natural philosophy implied the study of a playful and ingenious nature, could it be that, on occasion, imagination's "ingenious" causes hit the mark of truth? Just as Stigliani aroused wonder while describing actually existing traditions, is it possible that, in a field characterized by merely probable answers, the use of imagination by natural philosophers and inventors of conceits produced causal accounts with similar chances of capturing something true?

With apparent disregard for accuracy, Minozzi confronted a question destined to intrigue scholars both before and after him, namely, why did nature give light to fireflies? Aldrovandi offered two concurrent and complementary explanations (Figure 4.4). It could be, he argued, that light enabled the creatures to find their way in the dark, just as they, in turn, helped Marvell's mowers, and as *cocuyos* helped the natives of Hispaniola.[202] Or, perhaps, the insects glowed when they felt harassed.[203] The second explanation might have been suggested to Aldrovandi by Telesio's above-mentioned poem. A re-creation of the poet's childhood, it described Telesio's awe before the fields swarming with fireflies. At "a frail age" in which one is still "ignorant of things," the young Antonio would avoid contact with the insects for fear that they might burn him.[204] Elaborating on Telesio's memories of fear, Kircher imagined "the torches" of fireflies to be actually shields or deterrents that the small and helpless creatures could use to persuade their enemies, as they also did Telesio, that they would be burned by merely touching them.[205]

Natural philosophy and poetry found themselves in a complicated but often productive collaboration. As Kircher was staying in Malta in May 1637, he saw firsthand a behavior that was slightly at odds with what Telesio had imagined convincingly enough to persuade first Aldrovandi and then himself. Surrounded by fireflies, Kircher observed that their lights seemed to go off and on in the proximity of a hostile or a friendly presence, respectively.[206] Then, Kircher thought he had observed something else: "It was especially when a multiplicity of fireflies, or lamps, were placed together," he waxed lyrical, "that, as though they were exulting in light's ambitious glory, they most used the superb ornaments of their light-bearing liquid. You would say that they longed to be looked upon."[207]

Kircher's suggestion that fireflies glowed in the presence of friends, and notably of fellow fireflies, soon evolved into a second conclusion: the insects seemed to vie for attention and take pride in their beauty and their ability to arouse wonder in the beholder. How serious Kircher was about his claim is a matter of interpretation. He worked within a tradition in which, it has been argued, it was acceptable to "report and accept a wide range of probable opinions" for a single problem without always concluding in favor of one view or another.[208] Or it could be something else. Kircher's argument was marked conditionally, as a fictitious, even adventurous hypothesis. It was a conceit that imagined that

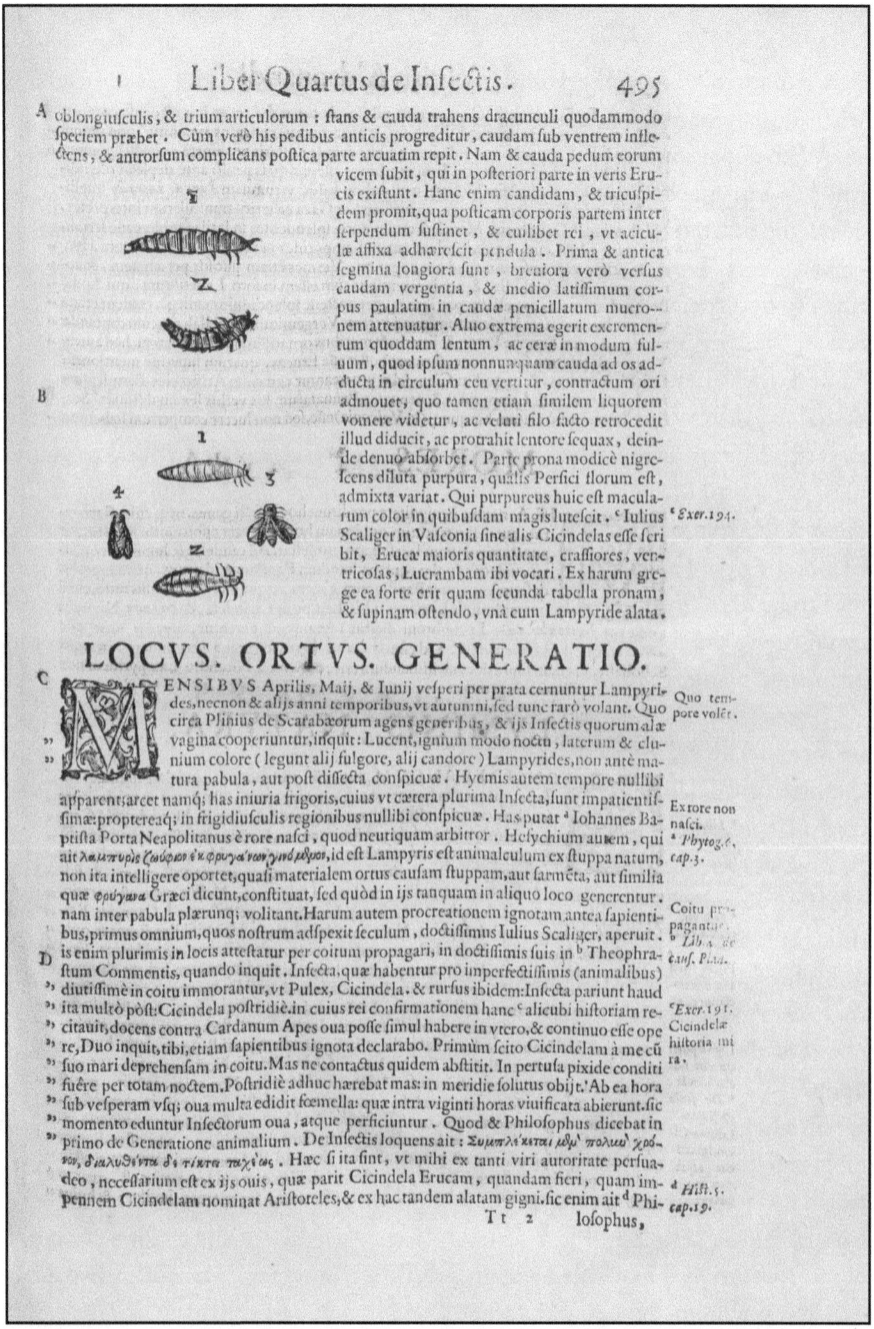

I Liber Quartus de Insectis. 495

A oblongiusculis, & trium articulorum: stans & cauda trahens dracunculi quodammodo speciem præbet. Cùm verò his pedibus anticis progreditur, caudam sub ventrem inflectens, & antrorsum complicans postica parte arcuatim repit. Nam & cauda pedum eorum vicem subit, qui in posteriori parte in veris Erucis existunt. Hanc enim candidam, & tricuspidem promit, qua posticam corporis partem inter serpendum sustinet, & cuilibet rei, vt aciculæ affixa adhærescit pendula. Prima & antica segmina longiora sunt, breuiora verò versus caudam vergentia, & medio latissimum corpus paulatim in caudæ penicillatum mucronem attenuatur. Aluo extrema egerit excrementum quoddam lentum, ac ceræ in modum fuluum, quod ipsum nonnunquam cauda ad os adducta in circulum ceu vertitur, contractum ori

B admouet, quo tamen etiam similem liquorem vomere videtur, ac veluti filo facto retrocedit illud diducit, ac protrahit lentore sequax, deinde denuo absorbet. Parte prona modicè nigrescens diluta purpura, qualis Persici florum est, admixta variat. Qui purpureus huic est macularum color in quibusdam magis lutescit. [c] Iulius Scaliger in Vasconia sine alis Cicindelas esse scribit, Erucæ maioris quantitate, crassiores, ventricosas, Lucrambam ibi vocari. Ex harum grege ea forte erit quam secunda tabella pronam, & supinam ostendo, vnà cum Lampyride alata. [c] *Exer. 194.*

1 2 1 3 4 2

LOCVS. ORTVS. GENERATIO.

C MENSIBVS Aprilis, Maij, & Iunij vesperi per prata cernuntur Lampyrides, necnon & alijs anni temporibus, vt autumni, sed tunc rarò volant. Quo circa Plinius de Scarabæorum agens generibus, & ijs Insectis quorum alæ vagina cooperiuntur, inquit: Lucent, ignium modo noctu, laterum & clunium colore (legunt alij fulgore, alij candore) Lampyrides, non antè matura pabula, aut post dissecta conspicuæ. Hyemis autem tempore nullibi apparent; arcet namq; has iniuria frigoris, cuius vt cætera plurima Insecta, sunt impatientissimæ: proptereaq; in frigidiusculis regionibus nullibi conspicuæ. Has putat [a] Iohannes Baptista Porta Neapolitanus è rore nasci, quod neutiquam arbitror. Hesychium autem, qui ait λαμπυρὶς ζωύφιον ἐκ φρυγάνων γινόμενον, id est Lampyris est animalculum ex stuppa natum, non ita intelligere oportet, quasi materialem ortus causam stuppam, aut sarmēta, aut similia quæ φρύγανα Græci dicunt, constituat, sed quòd in ijs tanquam in aliquo loco generentur. nam inter pabula plærunq; volitant. Harum autem procreationem ignotam antea sapientibus, primus omnium, quos nostrum adspexit seculum, doctissimus Iulius Scaliger, aperuit.

D is enim plurimis in locis attestatur per coitum propagari, in doctissimis suis in [b] Theophrastum Commentis, quando inquit. Insecta, quæ habentur pro imperfectissimis (animalibus) diutissimè in coitu immorantur, vt Pulex, Cicindela. & rursus ibidem: Insecta pariunt haud ita multò pòst: Cicindela postridiè. in cuius rei confirmationem hanc [c] alicubi historiam recitauit, docens contra Cardanum Apes oua posse simul habere in vtero, & continuo esse opere, Duo inquit, tibi, etiam sapientibus ignota declarabo. Primùm scito Cicindelam à me cū suo mari deprehensam in coitu. Mas ne contactus quidem abstitit. In pertusa pixide conditi fuêre per totam noctem. Postridiè adhuc hærebat mas: in meridie solutus obijt. Ab ea hora sub vesperam vsq; oua multa edidit fœmella: quæ intra viginti horas viuificata abierunt. sic momento eduntur Insectorum oua, atque perficiuntur. Quod & Philosophus dicebat in primo de Generatione animalium. De Insectis loquens ait: Συμπλέκεται μὲν πολὺν χρόνον, διαλυθέντα δὲ τίκτει ταχέως. Hæc si ita sint, vt mihi ex tanti viri autoritate persuadeo, necessarium est ex ijs ouis, quæ parit Cicindela Erucam, quandam fieri, quam impennem Cicindelam nominat Aristoteles, & ex hac tandem alatam gigni. sic enim ait [d] Philosophus,

Quo tempore volēt.

Ex rore non nasci.

[a] *Phytog. 6. cap. 3.*

Coitu propagantur.

[b] *Lib. 2. de caus. Plan.*

[c] *Exer. 191.*

Cicindelæ historia mira.

[d] *Hist. 5. cap. 19.*

Tt 2

Figure 4.4. Fireflies. Ulisse Aldrovandi, *De animalibus insectis libri septem*... (Bologna: Giovanni Battista Bellagamba, 1602), 495. Courtesy of The Linda Hall Library of Science, Engineering, and Technology.

fireflies were animals endowed with intellect and will, and thus with seemingly human passions and sins. Yet at the height of Kircher's "poetic" imagination, he came remarkably close to the consensus shared by today's entomologists and bioluminescence scientists that fireflies use light not only to scare predators, as he and Aldrovandi believed based on Telesio's poem, but also to render themselves conspicuous in front of potential mates.[209] As he searched for causes, Kircher confirmed a point advanced decades ago by Lorraine Daston, and more recently by Juan Pimentel's study of the controversies surrounding the unearthing of megatherium bones in late eighteenth-century South America: imagination, the faculty involved in conceit making and poetic activity more generally, was also central to its more "serious" and "respectable" counterpart, scientific discovery. Guesswork and playful speculation, as the example of Kircher reveals, occupied the frequent and unexpected boundaries between the investigation of nature and rhetorical and poetic invention.[210]

Conclusion

Physician Huarte de San Juan described intellectual conception in the sciences and the arts as a process marked by trial and error. To grasp reality with one's mind, he claimed, "it is necessary to draw a thousand lines in the air, compose many models, and eventually put one's hands to work." And even then, Huarte was forced to admit, most notions come out wrong.[211] In conceits likewise, imagination acts partially blindfolded as it aims to take hold of a reality that it does not know yet or entirely.

To find a middle term ("un medio") connecting the ends of a syllogism, but also to find some solution or means, was Gracián's choice expression to define conceit making.[212] It was also a pertinent label for the back-and-forth movement between natural-philosophical and rhetorical and poetic imagination that informed Minozzi's and Kircher's investigations of cold light and fireflies. Their approach was truthful to the nature of invention as a tool of problem solving that connects remote ideas through missing links. Telesio's poem suggested to both Aldrovandi and Kircher a plausible explanation for why fireflies produced light, one that could be cast in terms acceptable in a work of natural philosophy. Then Kircher added a new explanation, a conceit requiring the reader to accept less than strict forms of causality. By suggesting that fireflies lit up in search of attention, Kircher ignored a crucial admonition in Bourgogne's criticism of subtlety: to know something is to know its proximate causes. Far-fetched and metaphoric causes were evidence of the author's ingenuity, so the argument went, but failed to convey knowledge.[213] Yet the fact that time has proven Kircher's conceit not entirely ill-advised suggests that the border separating *scientia* from rhetoric and poetics was, to say the least, murky. In a work devoted to the physics of light, Kircher's search for probable causes stood in a strange and productive promiscuity alongside far-fetched and playful explanations, which proved to be at least partially adequate in accounting for nature's processes. To demonstrate that this was so because nature appeared to Kircher's contemporaries as similarly playful has been one of the aims of the present book.

Epilogue

Mere months after Orazio Grassi's lecture on the three comets of 1618, mathematician Charles Malapert (1581–1630) climbed to the lectern at the Jesuit College of Douay. To judge by his tone, those were difficult times: "You would easily say that telescopes are war engines pointed at the sky, which caused more than little harm to the strength of its orbits, so far undefeated. . . . And what would you say of comets? We recently saw one of them ramming with great vehemence against the solid walls of the Peripatetic sky."[1] Observations made with the aid of telescopes corroborated that the orbits of Mercury and Venus required new, alternative models to describe the universe. Knowledge of comets likewise threatened to end long-held beliefs. The previous winter, Grassi had demonstrated that comets were celestial bodies describing trajectories that did not match the supposedly circular, homocentric orbits of planets. As Malapert humorously argued, comets were bound to shatter the solid spheres that some still believed carried the planets as they moved through the heavens.[2] Comets, he implied, would hit them like rams battering the walls of a city, but telescopes, which enhanced the power of the eye, were similarly destructive.

Like Grassi, Malapert had a taste for conceits. It mattered little that some of the novelties that he evoked were relatively old. Even as the specifics faded from the audience's memory, they certainly retained the violence of discovery and savored the shock that always accompanies the new and unexpected.

Comets and telescopes were a strange, heterogeneous pair. How did one of nature's most spectacular phenomena relate to a device that was human-made, one that, in the ten years since its invention, had been variously fashioned as a tool for military survey and as the instrument enabling astronomers to scrutinize the secrets of the sky? For Malapert at least, both comets and telescopes challenged received views and habits. Seventeenth-century scholars of rhetoric and poetics pointed to comets and telescopes as analogs of conceits because conceits likewise wrought havoc with the mind's views and habits, which they short-circuited through the use of tropes.

In 1609, Claudio Achillini dubbed conceits "poetry's comets." Comets departed from nature's usual course, represented by the regular, predictable positions of planets and stars. They constrained the likes of Malapert to contemplate the possibility that the sky was perhaps different from what they previously believed and aroused wonder at the open, unsolved mystery that they put in front of their senses and intellect. Conceits similarly departed from logic in a process conducive to new and alternative realities that similarly resulted in wonder. Telescopes, in turn, transformed appearances by means of the artifice of optics, just as conceits resorted to the artifice of metaphor to connect objects unrelated from the point of view of logic and to depict a new and different reality.

Even so, comets did more than shatter familiar habits and ways of looking at nature. In a way, wholeheartedly pursuing comets as objects of study meant embracing a sensibility that natural philosophers since antiquity had sought to repress. It meant overcoming the reluctance to provide humans with unrestricted amounts of novelty and artifice, the only spice that effectively awakened their sensory and intellectual appetite, and from which the students of nature had been kept at a distance as a dangerous and empty frivolity. Rhetoricians and poets promoted conceits by likening them to the kind of phenomena that early modern natural philosophers had agreed to indulge in, objects that charmed the mind with their seemingly capricious and unpredictable behavior.

Another instance of the alignment between early modern natural-philosophical developments and discussions of style occurred in connection with insects. At some point around 1600, insects began to receive increased attention as objects of natural-philosophical research and even started to become prominent in still-life paintings. Too small to be fully appreciated by the naked eye, insects resisted traditional canons of beauty. Beyond that, they defied basic principles of identity. It was believed that, unlike larger animals and humans, insects came into existence through spontaneous generation from decaying matter or mud, not sexual reproduction from like to like. Difficult to see and unpredictable in their origins and lineage, insects held the key to otherwise inaccessible aspects of nature precisely for the reasons that made them insignificant according to conventional views.

It was only a matter of time before conceits became associated with insects. As individuals came to feel overwhelmed by the ability of orators and poets to create a multiplicity of endlessly different conceits, often disconnected from each other, they were reminded of the seemingly chaotic and endless resourcefulness displayed by God when he works at small scale. Conceits also resembled insects in the way that they came into existence. According to Giovanni Battista Manzini, conceits did not result from converting sensory stimuli into a notion that resembled them, a process that he compared to sexual reproduction, in which the offspring is, to a greater or lesser extent, its parent's likeness. Instead, imagination begot conceits through a process that Manzini viewed as being akin to spontaneous generation, in which the outcome remained unpredictable and always surprising.

Manzini's analysis of conceits as products of spontaneous generation emerged in the wake of a controversy that raged between 1636 and 1639. In it, fireflies became code for conceits. Like conceits, the small creatures that glow in the summer at dusk were believed, at least by most scholars, to be products of spontaneous generation. Yet it was the light in the tail of the insects that made the analogy with conceits so appealing. Just as conceits sparkled with a brief but enchanting appearance of truth, one that vanished as the intellect realized that they were products of metaphor and fiction, fireflies also possessed a charming but precarious light. Conceits resonated with a domain of nature that went against the grain of theologically and metaphysically acceptable values, including identity and permanence, and instead thrived in ephemerality, instability, and spontaneity. Although those in pursuit of steady sources of light could choose the sun and the stars over fireflies, which are useless in comparison, fireflies were more suggestive to individuals who, in the age of subtlety, had grown sensitive to nature's mysteries. In fireflies and the like, they saw greater opportunities for building knowledge by focusing on objects that could not be taken for granted and that held value not despite their precarious and obscure existence but because of it.

By analyzing rhetorical and poetic writings alongside works of natural philosophy on topics ranging from astronomy to entomology, *The Age of Subtlety* has unearthed several links between the seventeenth-century craze for conceits and contemporary transformations in scientific sensibilities. Indeed, it has emerged that early modern natural-philosophical research and controversies over conceits shared numerous aspirations, including the belief that exception harbors greater promise for research and explanatory potential than the norm. Added to that, the two fields shared anxieties about the impossibility of truly knowing anything without resorting to imagination and, above all, about the unlikelihood that humans would ever be able to remove themselves, and artifice, from the ways in which they understood nature.

All the clues about the shared ecologies of natural and human ingenuity were there right from the start for anyone who would pay attention to them. They adopted the form of complex and startling analogies between natural-philosophical topics and conceits. If there is anything daring about this book, then it is its resolve to treat the claim that conceits were poetry's comets—or, for that matter, style's fireflies—as if those were not merely ways of speaking, but rather—as is almost always the case with ways of speaking—meant with a purpose. That might explain why the analogies in question thrived for years while circulating through different individuals and settings, and while surviving episodes of minor or intense disagreement between the individuals who resorted to them. A case in point was Pier Francesco Minozzi's reaction to the analogy between fireflies and conceits, first advanced by Agostino Mascardi as a means to disparage conceits and repurposed by Minozzi to achieve the opposite of Mascardi's original intent. Fireflies became emblematic of a new way of looking at, and relating to, reality, one in which conceits also played a part and that can be characterized by its embrace of wonder, artifice, and fascination with the precarious and unstable manifestations of life at small scale. The analogy between fireflies and conceits persisted, and both its significance and its implications for the relationship between natural and human ingenuity transcended the disagreements among the participants who had set the debate in motion in the first place.

By paying close attention to a series of ideologically resilient intersections between stylistic fashion and the investigation of nature, relevant continuities have emerged, and with them a picture of the broader intellectual, scientific, artistic, and literary culture that witnessed the rise of the craze for conceits. By setting aside today's disciplinary boundaries, it is possible to not only obtain a deeper understanding of a rhetorical and poetic fashion that otherwise risks striking twenty-first-century readers as extravagant and even empty but also suggest that the controversy regarding conceits had repercussions across a variety of intellectual and cultural fields. At least some of the concerns that haunted the likes of Maciej K. Sarbiewski, Matteo Peregrini, Baltasar

Gracián, Minozzi, Manzini, and Emanuele Tesauro also lurked in the minds of the students of nature who peered at comets through telescopes and those who quietly held their breath as they examined the behavior of a swirl of fireflies. Their doubts still resound in their printed and manuscript works, their lectures, and their private correspondence, and speak loud enough for us to hear them.

Notes

Introduction

1. Scheiner, *Sol ellipticus*, 1. See also Reeves and Van Helden, *On Sunspots*; Van Helden, "Galileo"; Biagioli, *Galileo's Instruments*, 135–217.

2. Locher and Scheiner, *Disquisitiones mathematicae*, 67.

3. Reeves and Van Helden, *On Sunspots*, 319–320.

4. Scheiner, *Sol ellipticus*, fol. A2r–v. The title page of *Sol ellipticus* indicated that the book should serve "strenae loco." On Kepler's gift, see Aït-Touati, *Fictions*, 1–2.

5. Biagioli, *Galileo, Courtier*, 103–157.

6. Scheiner, *Refractiones Coelestes*, fol. A3v.

7. Scheiner, *Sol ellipticus*, fol. A2r.

8. Scheiner, *Sol ellipticus*, fol. A2v.

9. Bacon, *The Works*, 4:411; Lewis, *Language*, 6–7; Werlin, "Francis Bacon."

10. It is common to take conceits to be tropes or, more specifically, metaphors. Mazzeo, "Metaphysical Poetry," 222; *Merriam-Webster.com Dictionary*, s.v. conceit. Yet seventeenth-century writers on conceits, Baltasar Gracián chief among them, insisted that conceits were intellectual acts and tropes merely the mind's "instruments" for connecting different terms into a new reality. Gracián, *Arte*, 180–181.

11. Seventeenth-century writers consistently paired plots and conceits. Gracián described narrative plots as a specific kind of conceit, while Emanuele Tesauro likened the way readers react to character recognitions that prove them wrong about someone's identity to the way readers respond to the component of deception in conceits. Gracián, *Arte*, 287; Tesauro, *Il cannocchiale aristotelico*, 687–688.

12. Tesauro, *Il cannocchiale aristotelico*, 551–552.

13. "Vna Vipera vera, laqual di repente ti morde & auuelena: & vna Vipera dipinta, che par ti voglia mordere, & pur ti piace." Tesauro, *Il cannocchiale aristotelico*, 589. All translations are mine unless stated otherwise.

14. Reeves and Van Helden, *On Sunspots*, 54. "Erubuit scilicet & proripuit sese, ne suas intueremur nuptias." Scheiner, *Tres Epistolae*, fol. A4v.

15. Scheiner, *Sol ellipticus*, fol. A2v.

16. "Incubanti, *ouum quoddam*." Scheiner, *Refractiones Coelestes*, fol. A2r, emphasis original.

17. Scheiner, *Refractiones Coelestes*, fol. B1r.

18. Tesauro's early musings on conceits are partly lost. Zanardi, "Sulla genesi (I)"; Zanardi, "Sulla Genesi (II)"; Doglio, "Introduzione."

19. In 1646, Gracián's patron, Vincencio Juan de Lastanosa, accused Peregrini of plagiarizing *Arte de ingenio*, even though Gracián's treatise was published only in 1642, three years after Peregrini's *Delle acutezze*. Peregrini replied four years later, suggesting that *Arte de ingenio* was in fact a translation of *Delle acutezze*. Gracián, *El Discreto*, 157; Peregrini, *I fonti*, 21. For Pallavicino's criticism of Peregrini, see Sforza Pallavicino, *Considerazioni*, 141–146.

20. For a distinction between *acutum* and *argutum*, see Sarbiewski, *Wykłady*, 30.

21. Tuve, *Elizabethan*; Mazzeo, "Metaphysical Poetry"; Brodsky, "Donne"; Smith, *Metaphysical Wit*.

22. On Italian *concettismo*, see Praz, *Studi*; Croce, *Tre momenti*, 6–220; Raimondi, *Anatomie secentesche*, 1–41; Zanardi, "La metafora"; Carminati, "Alcune considerazioni." The term *concettismo* appeared in the title of the earliest treatment of writings on conceits in modern times. See Croce, *I trattatisti italiani*. On Spanish *conceptismo*, see Ayala Martínez, "Introducción"; Pozuelo Yvancos, "La Agudeza"; Grady, "Rhetoric." Mercedes Blanco's *Les rhétoriques de la pointe* (1992), the most detailed summary of the topic so far, bore the subtitle *Baltasar Gracián et le Conceptisme en Europe*.

23. See, for instance, Gracián, *Arte*, 136–137, 142, 162, 170; Ormaza, *Censura*, 24, 64, 70; Minozzi, *Sfogamenti*, 265; Manzini, *Delle meteore rettoriche*, 123; Tesauro, *Il cannocchiale aristotelico*, 153, 500, 542. *Concettoso/conceptuoso* also described the intellect of a writer who, for better or for worse, brimmed with conceits. Gracián, *Arte*, 287; Ormaza, *Censura*, 84; Bartoli, *Dell'huomo*, 399, 410.

24. Smith, *Body*; Long, *Artisan/Practitioners*. On matter and materials in early modern art, see Cole, "Cult"; Anderson, Dunlop, and Smith, *Matter*; Klein and Spary, *Materials and Expertise*; Rublack, "Matter."

25. Smith, "Between Nature." See also the Making and Knowing Project, directed by Pamela H. Smith (https://www.makingandknowing.org/).

26. On language as the matter of the inventor of conceits, see Gracián, *Arte*, 409.

27. Gracián, *Arte*, 133, 221.

28. Gracián, *Arte*, 426.

29. On using nature's laws against nature, see Del Monte, *Mechanicorum liber*, fol. +2v. On natural versus artificial language, see Peregrini, *Delle acutezze*, 42–43.

30. "Penitiora rimatur, intima permeat, & sese totum in rerum medullas insinuat." López, *Oratio*, fol. 2v. See also Ripa, *Iconologia*, 10.

31. Cardano, *Hieronymi Cardani*, 1. On objective and subjective subtlety in, respectively, Cardano's thought and medieval philosophy, see Mori, "Mathematical Subtleties," 13–14; Aubert, "'Subtile.'" For more on subtlety, see chapter 1.

32. Fracastoro, *Scritti inediti*, 176. On Fracastoro and contagion, see Nutton, "Seeds"; Nutton, "Reception." On the language of plague applied to the dissemination of conceits in Jesuit Quito, Ecuador, see Jouanen, *Historia*, 1:156. On contagion and painting, see Weststeijn, "'Painting's Enchanting Poison.'"

33. Lewis, "Francis Bacon," 113–116; Pons, "Ingenium"; Marr et al., *Logodaedalus*. See also Marr, "Pregnant Wit."

34. On wit as "superficial cleverness," see Goodey, *History*, 52. On conceits and generation, see chapter 4.

35. Marr et al., *Logodaedalus*, 19–21.

36. Estienne, *Dictionarivm*, s.v. ingenium. See also Oosterhoff, "Ingenious Materials"; Marr et al., *Logodaedalus*, 37. In the first century CE, Pliny the Elder advised that the place to plant vines be chosen "according to the nature of the soil, the ingenuity of the place" ("ad soli naturam, ad loci ingenium"). *Natural History*, 5:14–15, §17.2.

37. Attention to inborn talent played a prominent role in Pier Paolo Vergerio's *De ingenuis moribus et liberalibus adulescentiae studiis* (*The Character and Studies Befitting a Freeborn Youth*, before 1402) and Juan Luis Vives's *De tradendis disciplinis* (*On Education*, 1531). See Kallendorf, *Humanist Educational Treatises*, 56–59; Vives, *On Education*, 62. Huarte de San Juan's *Examen* reinforced the trend by making the assessment of a student's inborn talent the key to a successful education and, indirectly, a stable and prosperous state. Huarte de San Juan, *Examen*, esp. 149–156.

38. In the first century, rhetorician Marcus F. Quintilian listed *ingenium* among the aspects that the orator cannot learn. Quintilian, *Orator's Education*, 5:326–327, §10.2. In 1623, the Accademia della Crusca dictionary quoted Francesco da Buti's fourteenth-century commentary on Dante's *Divine Comedy* on the distinction between *ingegno*, which is "man's natural understanding" ("lo naturale intendimento, che l'huomo ha"), and *arte*, "which instructs man with rules and teachings" ("quella, che ammaestra l'huomo con regole, e ammaestramenti"). Crusca, *Vocabolario* (1623), s.v. ingegno.

39. Interestingly, Gracián believed differently that the intellect made use of imagination in the invention of conceits. *Arte*, 137–140.

40. Manzini, *Delle meteore rettoriche*, 115–117. On "concept" and "conceit," see Parker, "Concept."

41. On the internal senses, see Park, "Organic Soul"; Casini, "Renaissance Theories." On imagination, see Tuominen, "Ancient Theories"; De Boer, "Imagination," 11–16; Rusu, "Imagination"; Lyons, *Before Imagination*.

42. "οὐδέποτε νοεῖ ἄνευ φαντάσματος ἡ ψυχή," Aristotle, *On the Soul*, 176–177, §3.7, 431a17. See De Boer, "Imagination"; Rusu, "Imagination," 2.

43. On imagination between body and mind, see Rusu, "Imagination," 1; Lyons, *Before Imagination*, 1–4; De Boer, "Imagination," 11–15.

44. Pascal, *Pensées* 41, in *Oeuvres complètes*, 2:551. Quoted in Cassin, Apter, Lezra, and Wood, *Dictionary*, s.v. imagination.

45. On imagination and fantasy, see Rusu, "Imagination," 1; Guest, "Plato's Stag Goats."

46. Quoted in Casini, "Renaissance Theories," 149. The original reads, "Phantasia est mirifice expedita & libera: quicquid collibitum est, fingit, refingit, componit, deuincit, dissoluit, res disiunctissimas connectit, coniunctissimas autem longissime separat. Itaque nisi regatur, & cohibeatur à ratione, haud secus animum percellit ac perturbat, quàm procella mare." Vives, *De anima et vita*, 32–33. For Vives on emotions and the mind, see Casini, "Emotions."

47. Johnson, *Lives*, 1:29.

48. Daston, "Fear," esp. 81–85. See also Ingham, *Medieval New*, 8.

49. Recent studies have approached the early modern culture of ingenuity through the study of lexica and the role of *ingenium* in a variety of natural-philosophical and artistic disciplines. Marr et al., *Logodaedalus*; Oosterhoff, Marcaida López, and Marr, *Ingenuity*.

Scholars have been taking long-due steps toward filling the gaps in the history of *ingenium* by bringing together not only the mechanical and the liberal arts but also the work of European learned men and that of a multiplicity of individuals of different gender and social statuses who produced epoch-making contributions, often in colonial settings. See Bigelow and Cruz, "Ingenios," 521.

50. "Impropriamente si dice ingegno quel delle mani, ma deesi chiamare attitudine." Crusca, *Vocabolario* (1623), s.v. ingegno. The authors of *Logodaedalus* have argued that the Accademia della Crusca dictionary tended to underrepresent the mechanical arts, lending priority to poetic and scholarly sources. Marr et al., *Logodaedalus*, 64.

51. Keller, "Charlatan's Ingenuity," 181. On conceits as juggling, see chapter 4.

52. For both of Manetti's works, see Manetti, *Vita*. See also Ascoli, "History."

53. Marr et al., *Logodaedalus*, 47. Charles du Fresne's *Glossarium mediae et infimae latinitatis* (1678), attempting to document medieval language, defines *ingenium* as "the intention behind an evil deed, which a court would assess to determine guilt," as well as a war machine or device. Quoted in Marr et al., *Logodaedalus*, 47. On medieval usages of *ingenium* versus *machina*, see Hamesse, "L'apport." The 1623 edition of the Accademia della Crusca dictionary listed "deceit, cunning, stratagem" ("inganno, astuzia, stratagemma") among the meanings of *ingegno*. Crusca, *Vocabolario* (1623), s.v. ingegno.

54. Micheli, *Le origini*, 9.

55. "Ritrovare qualche via, o astutia." Vittori, *Tesoro*, s.v. maquinar. The authors of *Logodaedalus* have pointed out that the prominence of engineering in early modern dictionary definitions of ingenuity guaranteed "a complex interplay between liberal and mechanical senses of creative potency . . . , while feeding the 'deceitful' or 'tricksy' aspects long ascribed to poets and artisans alike." Marr et al., *Logodaedalus*, 59.

56. Ingham, *Medieval New*, 8.

57. Huarte de San Juan, *Examen*, 452–453.

58. Huarte de San Juan, *Examen*, 395–396. In fact, seventeenth-century works of politics codified in aphorisms in the manner of Hippocrates argued for the analogy between the physician's back-and-forth between theory and practice and that of the politician. See Álamos de Barrientos, *Tácito español*, fol. †8r–v.

59. The first occurrence of *machina* in Latin, in fact, had been in Plautus, with the meaning of "ingenious solution." Simeoni, "Machina," 147. In 1617, John Minsheu translated *maquinar* neutrally as "to devise," *Ductor*, s.v. maquinar. In 1611, Sebastián de Covarrubias also defined "machine" as both a mechanical device and a plan in someone's mind. Covarrubias Orozco, *Tesoro*, s.v. maquina.

60. On μῆτις, see Detienne and Vernant, *Les ruses*.

61. "Escogitare un congegno, un apparato, un mezzo che permetta di superare una determinata difficoltà pratica e compiere operazioni altrimenti impossibili." Simeoni, "Machina," 150. Vitruvius also claimed that a machine was the product of "summa sollertia." *On Architecture*, 2:318–319, §10.9.1. See also Micheli, *Le origini*, 9–10. For the connection between ingenuity and engineering, see Schiefsky, "Art"; Marr et al., *Logodaedalus*, passim; Ingham, *Medieval New*, 8; Vázquez Manassero, *El "yngenio" en palacio*.

62. "Vulgarmente llamamos ingenio vna fuerça natural de entendimiento inuestigadora de lo que por razon y discurso se puede alcançar en todo genero de ciencias, diciplinas, artes liberales, y mecanicas, sutilezas, inuenciones, y engaños: y assí llamamos ingeniero

al que fabrica maquinas para defenderse del enemigo, y ofenderle: ingenioso, el que tiene sutil y delgado ingenio." Covarrubias Orozco, *Tesoro*, s.v. ingenio.

63. "Qualquiera cosa que se fabrica con entendimiento, y facilita el executar lo que con fuerças era dificultoso y costoso." Covarrubias Orozco, *Tesoro*, s.v. ingenio.

64. Gracián, *Arte*, 287.

65. "Ay laberintos del discurso que el mental Theseo, con el precioso ovillo de una acertada salida, mide y vence." Gracián, *Arte*, 283. The notion that natural philosophers navigated nature's labyrinth as if they were Theseus—or Ariadne—was commonplace. See Mulerius, *Institutionum astronomicarum*, 26; Kircher, *Mundus subterraneus*, 2:326; Keller, "Charlatan's Ingenuity," 179.

66. On Gracián's use of "hallar un medio," see Gracián, *Arte*, 283, 300.

67. "Sitiada la inteligencia de una perplexidad y cogidos todos los passos . . . socorrida de su prontitud, halla repentina salida." Gracián, *Arte*, 283.

68. Gracián, *Arte*, 283.

69. Castelvetro, *Poetica*, fol. 119r–v.

70. "Ristretto dentro da certi termini da quali non è licito vscire." Castelvetro, *Poetica*, fol. 119v.

71. "La barba era spezzata & mancheuole." Castelvetro, *Poetica*, fol. 119v.

72. "Per quella parte che le era restata intorno al mento mostrauasiche [*sic*] la barba se fosse stata intera douesse giugnere secondo proportione infino al belico, & non dimeno la punta della barba si vedeua rimasa nella sommità del petto non passare piu oltre. Diche ognuno si marauigliaua, ne alcuno si poteua imaginare come stesse quella barba quando era intera. Solo Michel Angelo Bonarotti scoltore di rarissimo ingegno che si trouaua presente stato alquanto sopra se comprese come la cosa stesse, & disse Rechimisi creta, la quale recata formò con quella la parte della barba che mancaua tanto grande quanto si richiedeua alla proportione della rimasa, & aggiungendouela la tirò infino al belico poi annodatala con vn groppo fece vedere chiaramente, che la punta della barba formata da lui batteua alla sommità del petto in quel luogo medesimo nel quale era la punta della barba rotta." Castelvetro, *Poetica*, fol. 119v.

73. Castelvetro, *Poetica*, fol. 119v.

74. Smith, "Between Nature," 49–55; Smith, *Body*; Roberts and Schaffer, "Preface." See also Bredekamp, *Galileo's Thinking Hand*. On artisans and knowledge of nature, see also Long, *Artisan/Practitioners*.

75. Smith, *Body*, 8.

76. Oosterhoff, "Ingenious Materials"; also Smith and Lores-Chavez, "Counterfeiting Materials," 31–32; Smith, "Between Nature," 55–56. On alchemy, see Rampling, "Spirited Matter," 14. For the idea that the alchemist's ingenuity partners with that of nature, see also Oosterhoff, "Ingenious Materials." See also chapter 4.

77. Rusu, "Deceiving," 35.

78. Muir, *Culture Wars*, 8, and esp. 79. On skepticism in early modern Europe, see especially Popkin, *History*; Dooley, *Social History*.

79. "Come sembra che i sudditi facilmente si vestano delle inclinazioni e de' costumi de' loro Signori, gli Italiani divennero per così dire Spagnuoli." Tiraboschi, *Storia*, 2:21. On Tiraboschi, see Zucchi, "Tiraboschi, Girolamo." For a similar notion, see Bettinelli, *Dell'entusiasmo*, 304.

80. Rapin, *Reflexions sur l'usage*, 148; Rapin, *Reflexions sur la Poétique*, 184; Thouvenin, "Une histoire littéraire," 224. See also Blanco, *Les rhétoriques*, 68–70.

81. Andrés, *Dell'origine*, 1:i

82. Six years later, in 1773, Pope Clement XIV suppressed the Society of Jesus. This means that some of the scholars whom I label as "Jesuits" were, at some point in the story, former Jesuits. On Spanish Jesuits in eighteenth-century Italy, see Batllori, *La cultura hispano-italiana*; Guasti, *L'esilio italiano*; Guasti, *Juan Andrés*; Arato, "I gesuiti spagnoli."

83. See Lampillas, *Saggio storico-apologetico*, 2.4:192–194 and passim.

84. Rodríguez Mohedano and Rodríguez Mohedano, *Historia Literaria*, 7:239–241.

85. Published first in Italian, Andrés's letter appeared in Spanish translation four years later. See Andrés, *Carta*.

86. Andrés, *Lettera*, 9–11.

87. Andrés, *Lettera*, 6–7. The objection advanced by Andrés was echoed a few years later in Rodríguez Mohedano and Rodríguez Mohedano, *Historia Literaria*, 7:243–245.

88. "È una certa sottigliezza, e misteriosa sublimità che si discosta dalla naturalezza, e rende oscure le composizioni." Andrés, *Lettera*, 44. The term "subtlety" was more often than not invested with ambivalence and tainted by suspicion. Referring to a thin and delicate imagination, subtlety was often associated with dubious or negative characters such as tricksters, lowlifes, and *pícaros* in seventeenth-century Spanish literature. See, for instance, the title of Alonso Jerónimo de Salas Barbadillo's *El subtil cordobés Pedro de Urdemalas* (1620). See also Salas Barbadillo, *La hija de Celestina*, 84. On subtlety's negative connotations, see also Bruni, "Semantica."

89. "Ghiribizzi, ed in sottigliezze insussistenti." Andrés, *Dell'origine*, 5:529.

90. Daston, "What Can Be," 49.

91. Aristotle, *Metaphysics*, 1:302–305, §6.2, 1027a20–21. See also Daston and Park, *Wonders*, 120; Daston, *Against Nature*, 25–28.

92. Nieremberg, *Oculta filosofía*, fols. 113r–114r.

93. Quoted in Daston, "Curiosity," 391.

94. Tesauro, *Il cannocchiale aristotelico*, 91–146.

95. Vega, *Arte nuevo*, 312–313, ll. 174–180.

96. See Weeks, "Francis Bacon." "A monster," Lorraine Daston writes about early modern natural-philosophical views, "might be regarded as nature's error or as its finest workmanship." "Nature of Nature," 158.

97. Daston, "Cold Light," 21.

98. Findlen, "Jokes"; Findlen, "Ludic Postscript."

99. Mulsow, "Idolatry," 698–699.

100. On nature's moral and legal authority, see Daston, *Against Nature*; Lehoux, "Laws."

101. Croce, *I trattatisti italiani*, 1. Similarly, Croce's disciple Joel Elias Spingarn described seventeenth-century writings on conceits as obscure and decadent. Spingarn, *Critical Essays*, 1:xxii.

102. "Des manifestations inférieures de l'intelligence." Coster, *Baltasar Gracián*, 289–290. This was a paraphrase of a passage in Peregrini's *Delle acutezze*, but the ambivalence of the source was entirely lost on Coster.

103. Astrain, *Historia*, 5:110.

104. Van Hook, "Concupiscence," 25. See Morpurgo Tagliabue, "Aristotelismo e Barocco"; Donato, "Tesauro's Poetics"; Croce, *Tre momenti*; Tuve, *Elizabethan*; Mazzeo, "Metaphysical Poetry."

105. Vélez-Posada, "Handsteine," 52.

106. See Hermogenes, *On Types of Style*, 23.

107. González de Salas, *Nueva idea*, 2:576. On the porous boundaries between literal and figurative language, see Quintilian, *Orator's Education*; Cummings, "Metalepsis."

108. González de Salas, *Nueva idea*, 2:576.

109. Russo, "Artistic Humanity." See also Hanß, "New World Feathers."

110. Gracián, *Arte*, 138.

111. "Más es menester para tratar con un solo hombre en estos tiempos que con todo un pueblo en los pasados." Gracián, *Oráculo manual*, 101.

112. Peregrini, *Delle acutezze*, 8–9.

113. Peregrini, *Delle acutezze*, 66. Actually, the notion that the audience could feel a pleasure connected with self-love was common in Greco-Roman rhetoric. See Aristotle, *Art of Rhetoric*, 126–127, §11.1, 1371b25–27; Quintilian, *Orator's Education*, 3:338–339, §8.2; Aristotle, Longinus, and Demetrius, *Aristotle*, 178–179.

114. Sforza Pallavicino, *Considerazioni*, 136–137. However, Pallavicino's professor Famiano Strada had argued in a lecture delivered in Rome and first published in 1617 that conceits, at least the way they were cultivated in the present, were a recent invention and had been unknown to Cicero. Strada, *Prolusiones Academicae*, 168–169.

115. On secrecy and crafts, see Long, *Openness*; Long, "Openness of Knowledge"; Belfanti, "Guilds."

116. Langbein, *Commentatio*, lxxv–lxxviii. For a discussion of Sarbiewski's exchange on conceits, see chapter 1.

117. "El primogénito de la agudeza." Gracián, *Arte*, 163.

118. On Gracián, conceits, and late medieval poetry, see Gracián, *Agudeza*, 1:264–273, 284–285.

119. Gracián, *Arte*, 133–135.

120. On Galileo's attempts at construing his invention of the telescope as a product of reflection, see Biagioli, *Inventing Invention*, 9; Shea and Bascelli, *Galileo's Sidereus Nuncius*, 19.

121. "Che gl'Ingegni moderni . . . siano di gran lunga superiori à gli Antichi." Minozzi, *Sfogamenti*, 257–258.

122. "Ma più di due volte dieci, con applauso tale, c'hà tirato quasi affatto questo secolo à mutar maniera di scriuere." Manzini, *Lettera apologetica*, 23–24. Minozzi agreed with Manzini. Minozzi, *Sfogamenti*, 260. See also Russo, "Introduzione," 9–10.

123. Manzini, *Lettera apologetica*, 27–28.

124. Peregrini's treatise had only two editions, which appeared within a few months of one another. Gracián published *Arte de ingenio* in 1642. It was followed in 1648 by a considerably altered and enlarged edition titled *Agudeza y arte de ingenio*, often reprinted as part of the author's complete works. Pallavicino's *Considerazioni sopra lo stile e il dialogo* appeared twice, in 1646 and 1647. In 1662, it was published again, in a revised and augmented version, as *Trattato dello stile e del dialogo*. Tesauro's *Il cannocchiale aristotelico*, published in 1654 and again, in updated and extended form, in 1663 and 1670, had appeared in twenty

editions by 1702. By the mid-eighteenth century, it boasted three editions in languages other than Italian (Latin, 1698 and 1714; and Spanish, 1741). See Cutrì, "Leggere," 11; Cutrì, "Storia editoriale."

125. "Condimenti sì nobili si conuertono in cibi à chi hà lo stomaco . . . proporzionato a digerire la sostanza del zuccaro, e del mele." Minozzi, *Sfogamenti*, 270.

126. Minozzi's elevation of sugar to a main dish finds echo in a remark by Sidney Mintz, who claimed that modernity entailed the end of the opposition between complex carbohydrates that served as "center foods" and "edge foods," including sugar, which were consumed in smaller amounts. Mintz, *Sweetness*, 11–14.

127. For the criticism of "such a delicate century," see, for instance, Mascardi, *Orationi*, 325–326; Bartoli, *Dell'huomo*, 400–401; Rapin, *Reflexions sur la Poétique*, 69. I further explore Mascardi's ideas in chapters 1 and 3; for Bartoli, see chapter 4. On spirit, see Hirai, "Spirit."

128. "Il Mondo và raffinandosi d'ogni hora più, e particolarmente nell'opere dell'ingegno. Molte cose che furon già rare, son hora vulgari." Manzini, *Delle meteore rettoriche*, 55.

129. Gal and Chen-Morris, *Baroque Science*, 6–7.

130. Dear, "Mixed Mathematics," 149. See also Herrera, *Institución*, 5r–17v; Baldi, *De gli automati*, fol. 4r–v.

131. "Scritto in lingua matematica, e i caratteri son triangoli, cerchi, ed altre figure geometriche." Galilei, *Il saggiatore*, 119. See Palmerino, "Reading." On the debate between mathematicians and natural philosophers, see, for instance, Mori, "Mathematical Subtleties"; Carolino, "Cristoforo Borri"; Schöttler, "From Causes to Relations."

132. Peregrini, *Delle acutezze*, fol. a6r–v. See chapter 3.

133. On literature and science in Italy and Spain, see, for instance, Reeves, *Galileo's Glassworks*; Aït-Touati, *Fictions*; García Santo-Tomás, *Refracted Muse*. On science and the visual arts, see Reeves, *Painting*; Marr, *Between Raphael*; Marcaida, *Arte*.

134. Hanafi, *Monster*, 187–217; Snyder, "Art." See also Marr, "Mirror"; Gómez López, "Encounter"; Arena, "Omnis."

135. I do not aim to offer a comprehensive exposition of seventeenth-century writings on conceits, a world that has been reconstructed in detail by Mercedes Blanco and Jon R. Snyder. See Blanco, *Les rhétoriques*; Snyder, *L'estetica*. Blanco and Snyder culminated over a century of studies of specific seventeenth-century theorists of conceits. Among the early studies of Gracián's ideas on conceits, see Sarmiento, "Gracián's 'Agudeza'"; May, "Gracián's Idea." On the intersections between Italian and Spanish writings on conceits, in addition to Franco Croce's *Tre Momenti* and the contributions by Blanco and Snyder, see also García Berrio, *España*; Woods, *Gracián*. On Gracián and Tesauro, respectively, see Hidalgo-Serna, *El pensamiento ingenioso*; Frare, *Per istraforo*. Scholars of English metaphysical poetry often delve into Italian- and Spanish-language theorists as well. See Tuve, *Elizabethan*; Mazzeo, "Metaphysical Poetry"; Smith, *Metaphysical Wit*; Van Hook, "Concupiscence." For a detailed bibliography, see Frare, *Per istraforo*, 13–21. For a recent and insightful contribution, see Snyder, "Strokes."

136. The notion of a radical rupture with the past evoked by the term "Scientific Revolution" has fallen out of favor, in the wake of the reappraisal of the role of Aristotelian natural philosophy and disciplines such as natural magic, alchemy, hermeticism, and Paracelsianism in the development of "the new science." The role played by religious institu-

tions such as the Society of Jesus as contributors, and not only antagonists, has been widely admitted, and the place of women, artisans, and nonwhite, non-European individuals as agents of scientific and technological innovation has gained increasing scholarly attention. Dear, *Revolutionizing*; Shapin, *Scientific Revolution*. See Daston, "Cold Light," 17; Rossi, *La nascita*, xiv–xvii.

137. "Como estrella errante, no tiene casa fixa." Gracián, *Arte*, 143.

138. On the blind spots that result from abiding by present disciplinary boundaries rather than those that were in place in the early modern period, see Grafton, "Humanism."

Chapter 1 Poetry's Comets

1. Silva y Figueroa, *Comentarios*, 2:341–342; Kapoor, "Nūr Ud-Dīn Jahāngīr"; Kirwitzer, *Observationes*.

2. Parker, *Global Crisis*, xxvi–xxvii, 11–12.

3. The four orations are extant in manuscript form. Varia de Cometis, Jesuit Collection (Ges.), Biblioteca Nazionale Centrale di Roma, Ms. 458. See Baldini, *Legem*, 251–281, 565–582. Girolamo Nappi's *Annals* of the Collegio Romano, which cover the history of the institution until the author's death in 1648, mention Grassi's oration in addition to several poems composed by a variety of professors. See Parte Seconda delli Annali del Seminario Romano . . . , APUG Collection, Archivio della Pontificia Università Gregoriana, Ms. 2801, 685–686. Both Nappi and private correspondence confirm that Grassi, one of the pupils of Cristoph Grienberger, was the author of the lecture. Anonymous and pseudonymous publication were widespread practices in the Society of Jesus; for instance, Baltasar Gracián published his 1642 and 1648 treatises on conceits under someone else's name. See Manning, "Publication Protocol."

4. Grassi, *Libra astronomica*, 3. For the translation, see Drake and O'Malley, *Controversy*, 69. For Grassi's biography, see Preti and Ercolino, "Grassi."

5. "Postremos pedes." Grassi, *De tribus cometis*, 5.

6. Drake and O'Malley, *Controversy*, 6. "Nulla somni cura, nullo algentis Aquilonis timore, cogerentur." Grassi, *De tribus cometis*, 4.

7. "Le novità vedutesi ne' giorni passati per il cielo svegliano gli huomini anco non curiosi, e sforzano a levarsi di letto i più sonnachiosi e pigri della città di Roma." Galilei, *Le opere*, 12:422. See also Alessandro Tassoni's letter of the same date (December 1, 1618) in Tassoni, *Lettere*, 1:422.

8. "Viae, ac vitae suae institutis rationibus, omnium in se dum vixit conuertit oculos." Grassi, *De tribus cometis*, 6.

9. Grassi, *De tribus cometis*, 6.

10. "À dicendi magistris minimè discedendum mihi existimo"; "illustrem . . . clarissimae vitae circulum"; "eiusdem mortis." Grassi, *De tribus cometis*, 6.

11. Drake and O'Malley, *Controversy*, 9. "Par enim fuerat, ad lautiorem, ac splendidiorem coenam tantis hospitibus parandam, facem etiam lucidiorem accendi." Grassi, *De tribus cometis*, 7.

12. The idea that the ancients had populated the sky with countless animals and beasts appeared in a lecture delivered at the Collegio Romano near the beginning of the 1611–1612 academic year. Excerpta ex oratione habita in Instauratione studiorum Colleg. Rom., Galileo Collection (Gal.), Biblioteca Nazionale Centrale di Firenze, Ms. 84, fol. 71r–v. For the

etymology of "planets" as "wanderers" in early modern sources, see Copernicus, *De revolutionibus*, fol. 9r; Muñoz, *Libro*, fol. 5v.

13. On tropes and law in ancient rhetoric, see Eden, *Poetic*. On the porous and unstable boundaries between literal and figurative language, see Camillo Delminio, *Due trattati*, fol. 32r; González de Salas, *Nueva idea*, 2:576. See also Cummings, "Metalepsis."

14. Gracián, *Arte*, 133, 180.

15. Drake and O'Malley, *Controversy*, 10. "Ex ipsius foetus vultu, & colore . . . Sic enim oculos, sic ille manus, sic ora ferebat." Grassi, *De tribus cometis*, 7. In a work published around the same time as Grassi's lecture, astronomer Manuel Bocarro Francês suggested that comets are "portraits" (*retratos*) that resemble the planet from which they have been generated. Bocarro Francês, *Tratado*, fol. 16v.

16. Drake and O'Malley, *Controversy*, 10. "Foetum bonae obstetrices excipiant." Grassi, *De tribus cometis*, 8.

17. Heidarzadeh, *History*, 1–19. For Aristotle's view on comets and its exposition by Jesuit professor Manuel de Góis in the manual intended for use in the society's schools, see Aristotle, *Meteorologica*, 38–57; Góis, *Commentarij*, 28–31, §1.6–7, 342b25–345a10. On seventeenth-century cometary theory, see Gualandi, *Teorie*.

18. Tycho had claimed that the comet of 1577 orbited the sun. Brahe, *Tychonis Brahe*, 203.

19. There is consensus that the cause of Galileo's belligerent reaction was not Grassi's conclusion about the comet's location but rather Grassi's tacit adoption of Tycho's geoheliocentric system instead of the heliocentric model advanced by Nicolaus Copernicus. Tycho placed the earth at the center of the universe, with the stars, the moon, and the sun orbiting the earth and the planets orbiting the sun. Conversely, in *De revolutionibus orbium coelestium* (1543), Copernicus claimed that the earth orbited the sun, as did the other planets. Some of the works that were part of the dispute between Grassi and Galileo have been translated in Drake and O'Malley, *Controversy*. See Biagioli, *Galileo, Courtier*, 267–311; Galluzzi, *Lynx*, 246–299. For a different interpretation of Galileo's motivations vis-à-vis Grassi, focused on their different understanding of the telescope, see Gal and Chen-Morris, *Baroque Science*, 79–113.

20. Moss, *Novelties*, 220–227.

21. Biagioli, *Galileo, Courtier*, 270–273.

22. Eamon, "Science as a Hunt," 406. See also Daston, "What Can Be," 46–47.

23. "Le comete della poesia." Marino, *Epistolario*, 153. Both Campeggi and Achillini were members of the Accademia dei Gelati in Bologna. The date of Achillini's letter to Campeggi, November 30, 1609, is given in Fulco, "Marino," 303.

24. "Legitimi lumi . . . stelle naturali." Marino, *Epistolario*, 154.

25. "Stelle addottive e lumi adulterini." Marino, *Epistolario*, 153.

26. On conceits as artificial, see Manzini, *Delle meteore rettoriche*, 115–117, and also chapter 4.

27. Recent scholarship has nuanced the notion of a stark contrast between medieval wariness toward novelty and early modern fascination with it. See Ingham, *Medieval New*.

28. Hanafi, *Monster*, 187–217.

29. Grassi, *De tribus cometis*, 2. On the engraving, see Guiducci and Galilei, *Discorso*, 442–447. Grassi may have used the engraving as a teaching aid as he lectured. See Moss, *Novelties*, 220.

30. On comets as signs of upcoming political, social, or religious upheaval, see Genuth, *Comets*, 11–88. See also Bailey, Clube, and Napier, *Origin*, passim.

31. On the artificial and the natural, see Bensaude-Vincent and Newman, *Artificial*.

32. Drake and O'Malley, *Controversy*, 5. "Nouarum vsque adeo rerum appetens est humanus animus; N. N. vt vel ipsam, aliquando, bonorum diuturnitatem fastidiens, eandem mali alicuius vicissitudine efficere, scilicet, cupiat meliorem . . . Optabamus, ò iniqua hominum vota, ex hoc portentorum genere oriri aliquid, quod oculos . . . sua saltem nouitate oblectaret ac pasceret." Grassi, *De tribus cometis*, 3. It is possible but highly unlikely that Grassi added the passage on novelty, absent from the manuscript preserved in Varia de Cometis, Jesuit Collection (Ges.), Biblioteca Nazionale Centrale di Roma, Ms. 458, fols. 17r–23r, when he sent the oration for printing.

33. "Ita enim compositi sumus ut nos cotidiana, etiamsi admiratione digna sunt, transeant, contra minimarum quoque rerum, si insolitae prodierunt, spectaculum dulce fiat. Hic itaque coetus astrorum, quibus immensi corporis pulchritudo distinguitur, populum non convocat; at, cum aliquid ex more mutatum est, omnium vultus in caelo est." Seneca, *Natural Questions*, 226–227, §7.1. Here and in the following, I quote Thomas H. Corcoran's translation.

34. "Quamdiu ordo servatur. Si quid turbatum est aut praeter consuetudinem emicuit, spectamus interrogamus ostendimus; adeo naturale est magis nova quam magna mirari." Seneca, *Natural Questions*, 228–229, §7.1. See also Glorioso, *De cometis*, 3–4.

35. Plutarch, *Moralia*, 11:210–211, §29, 919a–b.

36. "La nostra natura è amica della nouità." Peregrini, *Delle acutezze*, 57.

37. Dooley, *Social History*, 35.

38. Dooley, *Social History*, 1–2. On early modern skepticism, see Popkin, *History*. On astronomical and astrological news, see Reeves, *Evening News*. On certainty in early modern Europe, see Fuchs and García Arenal, *Quest*; Tutino, *Shadows*; Tutino, *Uncertainty*.

39. On Rome's *banchi*, Madrid's *mentideros*, and London's news market at "Old Saint Paul's," see, respectively, Hunt, "Conclave"; Castro Ibaseta, "Mentidero"; Rolfe, "St Paul's."

40. Such claims were found in orations delivered by Francesco Maria Fattori and Francesco Sacchini in 1596 and 1597. Francesco Maria Fattori, De moderandis ingeniis, APUG Collection, Archivio della Pontificia Università Gregoriana, Ms. 117, fol. 40r; Francesco Sacchini, De ingenij praestantia. Oratio Habita . . . in renouatione studiorum III Non. Nouembris 1597, APUG Collection, Ms. 117, fol. 50v. See also Pico della Mirandola, *Oration*, 116–123.

41. That was Fattori's view in 1596. For early modern Jesuit perspectives on student advising, see Casalini, "Discerning Skills"; Patiño Loira, "Policing Talent."

42. Astrain, *Historia*, 5:111. In 1625, Vitelleschi refused to give permission to Emanuele Tesauro, then still a member of the Society of Jesus, to skip censorship when he intended to dedicate the manuscript of a tract on conceits to a patron. Zanardi, "Sulla genesi (I)," 456–457.

43. Augustine of Hippo, *Sancti Augustini Confessionum*, 182–186, §10.34–35. Augustine's remark related to Aristotle's suggestion that we enjoy the imitation of things that would disgust us if experienced directly. Aristotle, *Poetics*, 168–169, §4, 1448b4–19. On Augustine and curiosity, see Walsh, "Rights."

44. On the renewal of curiosity in the early modern period, see Blumenberg, *Legitimacy*; Daston, "Curiosity"; Ginzburg, "High"; Kenny, *Uses of Curiosity*.

45. Nieremberg's family had German origins. They had arrived in Madrid with the retinue of King Philip II's sister, Maria of Habsburg. Hendrickson, *Jesuit Polymath*, 9–12. On Nieremberg and novelty, see 92–106. The classic studies on novelty in early modern science are Thorndike, "Newness"; Hall, "Spirit." On the same topic, see also Garrod, *Cosmographical Novelties*, 2.

46. Nieremberg, *Historia*, 1. On Nieremberg's attitude toward knowledge and nature, see Marcaida, *Arte*; Marcaida, "Juan Eusebio Nieremberg."

47. Aristotle, *Metaphysics*, 1:2–3, §A.1, 980a22. On the travailed and multifaceted history of Aristotle's *dictum* from antiquity to the early modern period, see Daston and Park, *Wonders*, 109–120.

48. Nieremberg, *Curiosa filosofia*, fol. 2r. In addition to Nieremberg's *Prolusion* (1629), delivered at Madrid's Reales Estudios, he also published in the vernacular *Curiosa filosofia* (1630) and *Oculta filosofia* (1633), united in a single volume in 1643 and, again, in 1649. Additionally, he published *Historia naturae, maxime peregrinae* (1635) in Latin. On Nieremberg's *Prolusion*, delivered on February 19, 1629, see Marcaida, "Juan Eusebio Nieremberg," 21–23.

49. Nieremberg, *Prolusion*, fols. 6v, 38v–39r. In 1637, preacher Diego Niseno compared the intellect to a falcon, or *neblí*, which grows in audacity as the heron flies higher during the chase. Niseno, *El politico*, fol. 241v.

50. Compare, for instance, Nieremberg's book and chapter titles promising "plura & noua & curiosa" (Book 2) or "iucunda & curiosa multa" (Book 6) in *Historia naturae* (1635) to *Don Quijote*'s allusions to "la nueva y agradable aventura" ("the new and pleasant adventure," §1.28), "la rara aventura" ("the rare adventure," §1.52), or "el discreto, nuevo y suave coloquio" ("the wise, new, and pleasant conversation," §2.13). Nieremberg, *Historia*; Cervantes Saavedra, *Don Quijote*.

51. "Otras estrellas adulterinas." Nieremberg, *Curiosa filosofia*, fol. 180r.

52. Nieremberg, *Curiosa filosofia*, fols. 1v–2r.

53. Scholars such as Nieremberg and Grassi were particularly conscious of the intellect's fondness for novelty, but they were not the first to consider it in detail. In 1550, physician Girolamo Cardano claimed that any topic so out of reach that the impression is given that it is being shared privately with us tends to be particularly pleasant. Cardano, *Hieronymi Cardani*, 290–291. The feeling that we know something that no one else does might explain the cult of the secrets of nature and art that for centuries fed streams of popular literature, and then informed scientific practice in European courts. See Eamon, *Science and the Secrets*; Eamon, "Science as a Hunt," 397.

54. On music's relation to mathematics and science, see Gouk, "Music," 9; Brown, "Evolution"; Gozza, *Number*; Pesic, *Music*. Interestingly, the 1599 Jesuit plan of studies or *Ratio studiorum* made little mention of music. Romano, *La contre-réforme mathématique*, 616–617.

55. Drake and O'Malley, *Controversy*, 5. "Ita in cantu, ac sono minus modulos illos amamus, quos perpetua, ac stabilis uocum concordia comitatur, multoque iucundior musica est, si dissonis, consonisque uocibus componatur, si collisis durius sonis, concors illa uocum uis eneruetur subinde, atque frangatur." Grassi, *De tribus cometis*, 3.

56. Galileo's reaction to *De tribus cometis* rightly reminds us that the state of the field following the discoveries was less congruous than what Grassi aimed to suggest. In fact, the community of astronomers accepted the discoveries long before it agreed on the interpretation that should be ascribed to them.

57. Approximately a century later, the first dictionary of the Real Academia Española (Spanish Royal Academy), published between 1726 and 1739, argued that mixing dissonance and consonance was the source of music's greatest pleasure, suggesting that this was by then a widespread view. *Diccionario*, s.v. dissonancia. It also argued that dissonance, when used with artifice, "entertains the senses, and causes one to long for consonance, which thus becomes more pleasurable" ("entretiene el sentido, haciéndole desear a la consonancia, que después percibe con más gusto"). *Diccionario*, s.v. ligadura. However, only a century before Grassi, the role of dissonance seems to have been not only more limited but also motivated by different reasons. A character in Baldassare Castiglione's popular *Il Libro del Cortegiano* (*Book of the Courtier*, 1528) suggested that the use of slight touches of dissonance conveyed *sprezzatura* or naturalness. Dissonance countered the impression, easily suggested by the use of uninterrupted consonance, that the musician had paid excessive attention to avoiding a mistake. Castiglione, *Il Libro*, 62–63. For the increasing acceptance of dissonance in seventeenth-century writings on music, see Cohen, "'La Supposition,'" 63.

58. Rothman, "Johannes Kepler's Pursuit"; Cohen, *Quantifying Music*, 13–33. The first half of the seventeenth century saw the publication of other treatises on universal harmony by Robert Fludd (1617), Marin Mersenne (1636), and Athanasius Kircher (1650). See Gouk, "Music," 9; Fabbri, *Cosmologia*.

59. Artusi, *L'arte*, 11. Artusi's definition resembled the one advanced in 1574 by Jesuit professor Francisco de Toledo and the one provided by Kircher in 1650. Toledo, *D. Francisci Toleti*, fol. 91v, §2c8tx87; Kircher, *Musurgia universalis*, 1:81. The fact that the invention of polyphony dates only to the medieval period means that, unbeknownst to early modern scholars, the ancients had understood consonance as a relationship among sounds in a sequence, rather than among simultaneous sounds. Tenney, *History*, 9. Kepler cited Artusi in Kepler, *Harmony*, 249–250; Kepler, *Ioannis Keppleri Harmonices Mundi*, 82.

60. "Insieme l'vno e l'altro vnire, ouero mescolare non si possono." Artusi, *L'arte*, 40. Kircher's definition was virtually the same. Kircher, *Musurgia universalis*, 1:81.

61. Artusi, *L'arte*, 36.

62. Artusi, *L'arte*, 36.

63. "Perche la natura non ha fatto cosa alcuna indarno; l'arte le và dimodo accommodando, che l'udito infinitamente se ne compiace." Artusi, *L'arte*, 25.

64. Hadot, *Veil*, 25. On *Natura nihil agit frustra*, see also Scholar, *Je-Ne-Sais-Quoi*, 82.

65. Augustine of Hippo, *City of God*, 980, quoted in Hanafi, *Monster*, 13. The original reads, "Portentum ergo fit non contra naturam, sed contra quam est nota natura." Augustine of Hippo, *Corpus*, 40:5:2:531, §21.8.

66. Augustine made his point about insects in his reply to the Manicheans' claim that God's creation was faulty. See Augustine of Hippo, *On Genesis*, 72–74; Gronewoller, *Rhetorical Economy*, 65–75.

67. Fernández de Córdoba, *Examen*.

68. Kepler, *Harmony*, 249. "Primarium illud & praecipuum Cantus Melodici condimentum, Dissonantia." Kepler, *Ioannis Keppleri Harmonices Mundi*, 82. Like Artusi,

Kepler contrasted "major dissonances," which are unpleasant but useful for underscoring strong emotions and highlight *pathos*, with "minor dissonances," which enhance harmony when they are blended with consonances. Kepler, *Harmony*, 97, 249–250; Kepler, *Ioannis Keppleri Harmonices Mundi*, 82. Kepler's distinction was a recent development. For Antonfrancesco Doni in the mid-sixteenth century there was only what Kepler called "major dissonances." Antonfrancesco Doni, Libri de Tuoni o Modi della Musica, Biblioteca Marucelliana, Ms. A.287.II, fol. 97r.

69. Kepler, *Harmony*, 247. "Ut concordantias aut meras, aut brevissimo Concinnarum dissonantiarum interpunctu fermentatas faciant; non eas tamen perpetuo tenore identicas, nec easdem deinceps; sed ipsâ successionum alternatione ad delectationem variatas." Kepler, *Ioannis Keppleri Harmonices Mundi*, 80.

70. Kepler, *Harmony*, 250–251. "Sale aut aceto in Opsonijs." Kepler, *Ioannis Keppleri Harmonices Mundi*, 83. See Rothman, "Johannes Kepler's Pursuit," 40.

71. "El sabor del pensamiento, y la sal del entendimiento." Nieremberg, *Curiosa filosofia*, fol. 2r.

72. Sforza Pallavicino, *Trattato*, 118.

73. As noted by Matteo Peregrini also in 1639, *sales* was a name used by Cicero to designate what, mutatis mutandis, seventeenth-century individuals called conceits. Peregrini, *Delle acutezze*, 12. On conceits and cured meats, see Minozzi, *Sfogamenti*, 221–222. See also chapter 4.

74. Minozzi, *Sfogamenti*, 269; Gracián, *Arte*, 254.

75. On conceits as spicy (*piccanti*), see Manzini, *Lettera apologetica*, 19; Minozzi, *Sfogamenti*, 192–193.

76. Eraldo Bellini successfully refuted the attribution of the poem to Achillini, the poet who called conceits "poetry's comets." Bellini, *Agostino Mascardi*, 29–32.

77. On the date of Mascardi's discourse and the authorship of the poem, see Bellini, *Agostino Mascardi*, 16–32. As I argue in chapter 3, Mascardi followed in the footsteps of a previous assessment of conceits by the Jesuit Famiano Strada, his former teacher at the Collegio Romano.

78. "Dilettano al pari d'una studiosa dissonanza, in vn perfetto concetto . . . perché in fatti sono con maggior merauiglia riceuute le cose, che insperatamente, e fuori del pensamento humano succedono." Mascardi, *Orationi*, 324. On conceits and dissonance, see also Blanco, "Humanismo rezagado," 134.

79. Biagioli, *Galileo, Courtier*, 289–290.

80. Sydor, "La concordia discors," 586. For a different chronology of Sarbiewski's stay in Rome, see Langbein, *Commentatio*, xliiii–xlv.

81. Nedzinskaitė, "M. K. Sarbievijaus," 44.

82. Sarbiewski, *Wykłady*, 2. Two manuscripts with student transcriptions of Sarbiewski's views are now in Maciej K. Sarbiewski, *De acuto et arguto . . .*, Czartoryski Library, Ms. 1446, fols. 273r–295v and Ms. 1858, fols. 325r–347v. I exclusively quote from the 1958 edition except where otherwise noted.

83. "Viris in Polonia, Germania, Gallia Italiaque doctissimis." Sarbiewski, *Wykłady*, 10, and also 2.

84. Sarbiewski, *Wykłady*, 18. See also Langbein, *Commentatio*, lvii–lviii.

85. Sarbiewski, *Wykłady*, 2.

86. *De acuto et arguto* bore the subtitle *Seneca et Martialis*, referring to the primacy of Martial and Seneca in the invention of conceits. See the introduction.

87. In the wake of Julius Caesar Scaliger's *Poetices libri septem* (1561), early modern scholars identified two parts in every epigram: a narrative or descriptive segment, followed by a surprising and ingenious remark derived from that narration or description. Scaliger, *Ivlii Caesaris Scaligeri . . . Poetices libri septem*, 170. See Blanco, *Les rhétoriques*, 164–165; Nowicki, *Die Epigrammtheorie*, 54–60; Mantero, "Énigmes," paras. 4–9.

88. I quote the translation by D. R. Shackleton Bailey in Martial, *Epigrams*, 1:51. "Delicias, Caesar, lususque iocosque leonum / Vidimus—hoc etiam praestat harena tibi— / Cum prensus blando totiens a dente rediret / Et per aperta vagus curreret ora lepus. / Unde potest avidus captae leo parcere praedae? / Sed tamen esse tuus dicitur: ergo potest." Sarbiewski, *Wykłady*, 12. Sarbiewski so appreciated the poem that he rewrote it with a martyr replacing the hare. Nedzinskaitė, "'Finis Epigrammatis,'" 287–288.

89. Published in Vilnius in 1640, Radau's *Orator extemporaneus* started to circulate widely only in its 1650 edition, which appeared under the name of Georg Beckher in Amsterdam. I exclusively quote from the 1650 edition. Radau's familiarity with Sarbiewski is evidence of the circulation of his views among northeastern European Jesuits. See Nedzinskaitė, "M. K. Sarbievijaus."

90. "Quod dictum includat in se rationem ex una parte dissentanei, et ex altera parte rationem consentanei, hoc est—ut clarius dicam—quod illud dictum ad rem ipsam ex una parte uideatur non pertinere, ex altera autem uideatur ualde pertinere." Sarbiewski, *Wykłady*, 6. See Blanco, *Les rhétoriques*, 169–176; Snyder, *L'estetica*, 39–40; Lachmann, "'Problematic Similarity'"; Woods, *Gracián*, 18–20. According to Radau, Sarbiewski meant that, "on the one hand, the sentence's predicate and subject are suitable to one another; on the other hand, they are not" ("praedicatum et Subiectum orationis ex una parte secum conueniunt, ex alia uero parte secum disconueniunt"). Beckher and Radau, *Orator extemporaneus*, 35. Yet the reference to subject and predicate was Radau's innovation. Woods, *Gracián*, 19–20.

91. Mentions of "dissonance" in intellectual rather than sensory contexts were common beyond Mascardi's example. In Miguel de Cervantes's romance *Los trabajos de Persiles y Sigismunda* (*The Travails of Persiles and Sigismunda*, 1617), we read that lies make "dissonance in the intellect" ("disonancia en el entendimiento"). Cervantes Saavedra, *Los trabajos*, 303. Cervantes intended dissonance from the reader's perspective. On dissonance from the speaker's perspective, as a discrepancy between thoughts and words, see Martin, "Inventing."

92. Petau had claimed that a conceit should contain *dissentaneum* and *consentaneum*, so as to appear simultaneously "against the common use and expectation" ("παρὰ προσδοκίαν") and "greatly united with the matter of which it speaks" ("cum re, qua de agitur maximè coniunctum"). Petau, *Epistolarum*, 238–239.

93. Sarbiewski, *Wykłady*, 10.

94. Horace, *Satires*, 328–329, §1.12, l. 19.

95. Speaking at the Collegio Romano in 1597, Sacchini praised "that disagreeing agreement of music" ("discors illa Musicae concordia"). Francesco Sacchini, De ingenij praestantia. Oratio Habita . . . in renouatione studiorum III Non. Nouembris 1597, APUG Collection, Archivio della Pontificia Università Gregoriana, Ms. 117, fol. 48r. Sarbiewski's

suggestion that conceits were also expressions of *concordia discors* was highly successful in Jesuit educational settings, as attested by a manuscript, likely from the second half of the seventeenth century, which ties *concordia discors* to Peregrini's views on conceits, dating its contents to after 1639. See Quod optandum certe fuit, Jesuit Collection (Ges.), Biblioteca Nazionale Centrale di Roma, Ms. 1000, fol. 99r–v. The codex records several orations initially delivered in Italy that a speaker seems to have repurposed for a new audience as he traveled to Vilnius, precisely the university in which Sarbiewski had studied about half a century earlier. Conceits traveled widely throughout Jesuit Europe, and they persisted over time.

96. Nieremberg, *Oculta filosofia*, 1–3 et passim; Bartoli, *Dell'huomo*, 94–103. See also Tasso, *Discorsi*, fol. 22r; Gracián, *El Criticón*, 1:36–38; Clark, *Thinking*, 43–105. *Concordia discors* appeared in the title of pamphlets addressing contemporary religious and political turmoil, such as Willem van der Lindt's *Concordia discors, sive Querimonia Catholicae Christi Iesu Ecclesiae* (Cologne, 1583). Kepler also connected music and politics. For him, universal harmony was a blueprint for a more harmonious society in which religious difference, a form of dissonance, could be resolved into consonance. Kepler, *Ioannis Keppleri Harmonices Mundi*, fols. *2r–*4r; Kepler, *Harmony*, 2–5; Rothman, *Pursuit*. On a different vein, Henry F. Fullenwider has proposed a direct link between Sarbiewski's mention of *concordia discors* and Platonism, identifying Johannes Baptista Serranus, listed in *De acuto* as one of Sarbiewski's interlocutors, as Jean de Serres, the commentator on Plato in the 1578 edition published by Henri Estienne. Fullenwider's thesis is appealing but inaccurate. It relies on the 1958 edition of Sarbiewski's tract, which reads "Johannes Baptista Serranus" instead of "Johannes Baptista Ferrarius," a reference to Jesuit professor Giovanni Battista Ferrari. Ms. 1446 reads "Ferrarius," and Ms. 1858, "Serrarius." Maciej K. Sarbiewski, *De acuto et arguto . . .*, Czartoryski Library, Ms. 1448, 283; Ms. 1858, 334; Sarbiewski, *Wykłady*, 18. Sarbiewski describes his interlocutor as someone who "professed to be sharp in his style" ("se acutum iam oratione profitetur"), a feature that, as I discuss in chapter 3, aligns with Ferrari's advocacy of conceits. See Fullenwider, "Concors Discordia."

97. Busi, "Giovanni Pico," 295–297. See also chapter 4.

98. Likewise, speaking of well-formed plots, Aristotle wrote that in any whole, the end is that which follows "necessarily or for the most part" ("ἢ ἐξ ἀνάγκης ἢ ὡς ἐπὶ τὸ πολύ") from the previous event. Aristotle, *Poetics*, 177, §7, 1450b30; Heath, "Universality."

99. Aquinas, *Summa*, 3.2:78, §3.99.9. I am quoting Vernon J. Bourke's translation. The original reads, "ordo enim inditus rebus a Deo, secundum id est quod in rebus frequenter accidere solet, non autem ubique, vel secundum id quod est semper." Aquinas, *Summae*, 421.

100. Daston and Park, *Wonders*, 121.

101. On Bacon, see Daston, "What Can Be," 40–41; Weeks, "Francis Bacon," 132–133. On *praeter naturam* and "preternatural philosophy," see Daston, "What Can Be"; Daston, "Nature of Nature."

102. "Naturales ac consuetas mutationes quae in aëre sunt." Cardano, *De rerum varietate*, 1.

103. Kirwitzer, *Observationes cometarum*, 3; Nieremberg, *Prolusion*, fol. 9v. Surekha Davies distinguishes among three traditions that intersect in early modern notions of monstrosity: monsters could be nature's occasional error in the process of creating like

from like, as formulated mainly by Aristotle; portents or omens signifying the future, as in Cicero's *De diuinatione*; and/or wonders embodying nature's boundless and unrivaled ability to bring forth variety. Davies, "Unlucky." See also Hanafi, *Monster*, 1–52; Céard, *La nature*, 3–21; Park and Daston, "Unnatural Conceptions," 23.

104. Cardano, *De rerum varietate*, 1; Tacke, *Coeli Anomalon*.

105. In 1705, English astronomer Edmund Halley claimed that the comet he had tracked in the summer of 1682 had to be the one observed by Kepler in 1607. In 1619, however, "all comets were mysterious interlopers, moving unpredictably through the solar system." Broughton, "First Predicted Return," 123. Even after Halley's discovery, new comets would continue to appear unannounced, and while their periodicity discredited the claim that they were signs of upcoming political and social turmoil, they remained ominous in popular imagination and the danger that one might collide with earth became increasingly pressing. See Barnett, "Between Pleasure"; Stewart, "Science."

106. "Aequalem rationem dissentanei et consentanei." Sarbiewski, *Wykłady*, 40. In 1548, humanist Francesco Robortello also distinguished poetic imitation *secundum* (that is, *iuxta*) *naturam* and *praeter naturam*. Robortello, *Francisci Robortelli*, 87–88. Late sixteenth-century scholars of poetics such as Lodovico Castelvetro, Giason De Nores, Torquato Tasso, Jacopo Mazzoni, and Francesco Patrizi increasingly underscored that poetic plots should imitate events that occurred *praeter naturam* rather than *secundum* or *iuxta naturam*, a view that became widespread, not coincidentally, when the craze for conceits reached its early stage.

107. Echoing Sarbiewski's course of 1618–1619, Radau claimed that *contra naturam* conceits were the sharpest of all. Beckher and Radau, *Orator extemporaneus*, 35–36.

108. "Non est quidem contra naturam rei nec tamen etiam est usitatum & naturale." Beckher and Radau, *Orator extemporaneus*, 36.

109. "Admiratio vero nascitur ex inopinato, quod nimirum audiamus aliquid, quod non exspectabamus cuiusque causa nos latebat, ac proinde accidere illud non putabamus. Nam ut Aristoteles docet, admiratio nascitur ex ignoratione causarum. Si rursus haec admiratio coniunctam habet delectationem (ut semper oportet accidere in acuto a nobis percepto, admiramur enim illud cum delectatione), debet habere ex altera parte apparentiam aliquam causae illius, quae latebat, propter quam accidere illud videamus, quod accidere posse non sperabamus." Sarbiewski, *Wykłady*, 14. Petau's letter to Sarbiewski conveyed a similar idea. Petau, *Epistolarum*, 239.

110. Wonder was universally recognized as the aim of conceits. See, for instance, Peregrini, *Delle acutezze*, 45–46, 58; Tesauro, *Il cannocchiale aristotelico*, 535–536. On the centrality of wonder to the early modern period's scientific and artistic sensibilities, see Daston and Park, *Wonders*; Céard, *La nature*; Marr, "Introduction"; Campbell, *Wonder*. On wonder and poetics, see Esteve, "Les poètiques"; Minsaas, "Poetic Marvels"; Patiño Loira, "Acuity."

111. The mentions of wonder in Nieremberg's work are simply too numerous to list. See, for instance, Nieremberg, *Prolusion*, fols. 21r–23r; Nieremberg, *Curiosa filosofia*, fol. 4v; Nieremberg, *Oculta filosofia*, fols. 2r, 113r–114r; Nieremberg, *Historia*, 5.

112. Sarbiewski, *Wykłady*, 14. Peregrini would likewise underscore that a conceit connected objects in merely apparent rather than true ways. Peregrini, *Delle acutezze*, 33–46.

113. Daston and Park, *Wonders*, 118. See also 110–112.

114. "Parto d'Ignoranza," "De l'ignoranza effetto." Talentoni, *Discorso*, fols. *5v, **2v. Talentoni's book originated in a lecture he delivered in Milan's Accademia degli Inquieti (Academy of the Restless) in November 1596.

115. Talentoni, *Discorso*, 49; Plato, *Theaetetus*, 155–156, 155D; Aristotle, *Metaphysics*, 1:12–13, §1.2, 982b12–24. For a related passage, see Aristotle, *Art of Rhetoric*, 124–127, §1.11, 1371a20–b25. See also Bowe, "Thaumata."

116. Talentoni, *Discorso*, 29–30, 43.

117. Descartes, *Oeuvres*, 11:373, 380–382; Daston, "Curiosity," 397.

118. Wonder was a reaction to the object's sheer novelty, where "new" could mean (depending on the case) incredible, too beautiful, too large, or unheard-of. See Talentoni, *Discorso*, 41. In 1565, humanist Lionardo Salviati pictured the intellect as it falls prey to wonder "stumbling" ("uno intoppo") into something that prevents it from going forward on its way to knowledge. Salviati, *Cinqve lezzioni*, fol. 13r–v. A similar reluctance to pinpoint the object of wonderment in more precise terms may be found in a work on poetics composed by philosopher Francesco Patrizi da Cherso in 1587 and preserved in manuscript form in Parma's Biblioteca Palatina. See Patrizi, *Della poetica*, 2:268, 306–307.

119. Talentoni rejected the identification of wonder with desire proposed by Aquinas's fifteenth-century commentator, Cardinal Cajetan. Talentoni, *Discorso*, 12–13, 17–18. This did not prevent Jesuit professor Alessandro Donati, a fellow of Grassi and Sarbiewski, from linking wonder with fear in 1631. Gillet, "Note," 235.

120. Sarbiewski, *Wykłady*, 16.

121. "Statim incipit praeterea delectari, dum quodammodo utrique parti favet et in utramque distrahitur, et rursus in utriusque unione mirifica unitur, seu consentit, etiamsi antea putaverit aliud dissentaneum tantum fuisse rei, iam tamen esse consentaneum, vel e contra." Sarbiewski, *Wykłady*, 16–18.

122. "Unione et affinitate dissentanei et consentanei pullulantis ex ipsa materia, de qua est oratio." Sarbiewski, *Wykłady*, 10–12.

123. Sarbiewski, *Wykłady*, 14–16.

124. Park, "Organic Soul," 469. See Bono, "Medical Spirits"; Hirai, "Spirit."

125. Stolberg, "Emotions," 113–114, et passim. The expression "the soul's instrument" was proverbial. For instance, Talentoni, *Discorso*, 63; Kircher, *Musurgia universalis*, 2:204. On the movement of the spirits, see also Talentoni, *Discorso*, 63; Cardano, *Hieronymi Cardani*, fol. 245v.

126. "Nel dolore, e nel timore fuggon subito del tutto al cuore, e non ardiscono scuoprirsi; la doue nella marauiglia non si può dir propriamente, che [i.e., gli spiriti] fuggano, ma si ben, che stieno sospesi, ambigui, e sopra di se." Talentoni, *Discorso*, 66–67. For a precedent, see Fracastoro, *De sympathia*, fol. 23r. Lionardo Salviati's claim in 1565 that wonder provoked paleness had forced Talentoni to explain why that was true without conceding that wonder contained fear. Salviati, *Cinqve lezzioni*, fol. 13r. Descartes did not find any external or physical mark associated with wonder, concluding that it was a phenomenon occurring in the intellect alone—unlike passions or emotions, which take place once the object has been identified as desirable or not. Descartes, *Oeuvres*, 11:381. Talentoni located wonder not in the intellective soul, which was believed to be entirely disembodied and immortal, but in the sensitive soul, which was the site of the emotions and existed "organi-

cally" with the body, with the spirits bridging the material and the immaterial as well as explaining how it was possible for the emotions to turn one's face pale or red. Talentoni, *Discorso*, 10–11. See also Park, "Organic Soul"; Stolberg, "Emotions," 114.

127. I quote from Bacon, *Historie*, 175–176. For the original in Latin, see Bacon, *Historia*, 252–253. In *Sylva Sylvarum* (1626–1627), Bacon also distinguished between wonder and fear. Shirilan, "Francis Bacon," 79.

128. Cardano claimed that if too many spirits abandoned the heart in a burst of joy, they might kill the individual. Cardano, *Hieronymi Cardani*, fol. 245v. Excessive wonder could leave the spirits entirely immobilized. Astonishment (*stupore*), for instance, transformed the subject into a statue, thereby becoming a hindrance rather than a stimulus to intellectual and even vital activity. Talentoni, *Discorso*, fols. †1r–2r. On astonishment, see also Scaliger, *Exotericarum Exercitationum*, fol. 424v; Suárez, *Doctoris Francisci Suarez*, 213; Descartes, *Oeuvres*, 11:382–383; Daston, "What Can Be," 46. Early modern dictionaries etymologically connected astonishment and stone. White, "Language," 34.

129. "Ut uno cognito aliud ex natura sua latere postulet et consequenter aptum sit non cognosci. Itaque non est necessaria ipsa actualis non cognitio, seu ipsum formale inopinatum." Sarbiewski, *De acuto*, 18. Not everyone agreed that conceits afforded repeated, let alone endless, pleasure. See Strada, *Prolusiones Academicae*, 174.

130. "L'attention portée à l'un nous dissimule nécessairement l'autre." Blanco, *Les rhétoriques*, 176.

131. Snyder, *L'estetica*, 39–40. In 1591, scholar Gregorio Comanini described Arcimboldo's *Vertumnus* as a union of contraries. Connelly, *Grotesque*, 46; Ossa-Richardson, *History*, 14.

132. "Caput cum oculis, naso, barba." Kircher, *Ars magna*, 807.

133. Kircher, *Ars magna*, 810.

134. Werrett, *Fireworks*, 53; Werrett, "Watching," 169–170.

135. Partenio, *Della imitatione poetica*, 39.

136. Mazzoni, *Della difesa*, 410. On Galileo and Mazzoni, see Shea, "Galileo's Copernicanism," 211.

137. On subtlety, see the introduction.

138. Cardano, *Hieronymi Cardani*, 1–3; Scaliger, *Exotericarum Exercitationum*, 1–3. On Scaliger, see Sakamoto, *Julius Caesar Scaliger*.

139. Scaliger, *Exotericarum Exercitationum*, 1.

140. "Quoties admiramur coniunctionem causae cum effectu. Túncque est iudicium subtilis cuiuspiam intellectionis." Scaliger, *Exotericarum Exercitationum*, fol. 424r.

141. Aubert, "'Subtile'"; Mori, "Mathematical Subtleties," 13–14.

142. Talentoni, *Discorso*, 50–54. The episode also appeared in Bartoli, *Dell'huomo*, 19–20. Its source was Aelian's *Various History* (third century CE). See Aelian, *Historical Miscellany*, 488–489.

143. "Quien viesse las pieças de que constaua la estatua de Minerua, que labró Phidias cada vna de por sí, no haría caso dellas, pero encajadas, y trabadas todas assombró al mundo; y mucho más si se reparaua el arte con que todas ellas venían a engaçarse [*sic*], y trabarse en el escudo de la Diosa en que estaua el rostro del Artifice." Nieremberg, *Oculta filosofia*, fols. 123v–124r. It was the second time that Nieremberg invoked the anecdote about the shield. See also Nieremberg, *Prolusion*, fol. 2r–v.

144. Variants of the story appeared in Tesauro's early work on heraldic devices or *imprese* and in Bernardino Gómez Miedes's *Comentariorum de sale* (1572). Tesauro, *Idea*, 35; Gómez Miedes, *Comentarios*, 1:47–48. See also Harrison, "Composition," 107–108.

145. "Està esmaltado vn bulto de Dios, vn rostro de su Artifice." Nieremberg, *Oculta filosofia*, fol. 124r. Pseudo-Aristotle's *De mundo* connected nature's artifice with that of Phidias's statue. Harrison, "Composition," 108.

146. The difference between events being perceived as products of chance or as resulting from a superior form of agency, such as fate or divine providence, played a role in sixteenth-century debates on Aristotle's *Poetics*. See Vega Ramos, "El azar."

147. "Ubi opus eximij alicuius artificis aspeximus, atque admirati sumus, si possumus etiam uultum hominis libenter intuemur, nempe quod uidere causam, ex qua profectum est praeclarum aliquod opus satisfacere admirationi aliqua ex parte uideatur. Quare cum ingenium oratoris causa translationis sit, in illud mens auditoris fertur eiusque aspectu uoluptatem capit." Famiano Strada, De elocutione, APUG Collection, Archivio della Pontificia Università Gregoriana, Ms. 1283, fol. 227r. There is another version of the passage in APUG Collection, Ms. 1163, fol. 160v.

148. "Se truecan las manos, que lo más admirable de la naturaleza, parece que es lo que imita al arte; esto es su artificio, y traça." Nieremberg, *Oculta filosofia*, fols. 113v–114r.

149. "Su ingenio, su traça, su armaçon, su orden, sus correspondencias; al fin su arte." Nieremberg, *Oculta filosofia*, fol. 114r.

150. Nieremberg, *Oculta filosofia*, fols. 121v–122r.

151. Nieremberg, *Oculta filosofia*, fols. 113r–114r. On Nieremberg's views of sympathy and antipathy, see fols. 2v–3r and passim. On Descartes and nature as a machine, see Des Chene, *Spirits*.

152. Anna Grasskamp has noted that early modern German sources used the adjective "artificial" or "artful" (*künstlich*) indistinctly for both natural and crafted objects. Grasskamp, "Unpacking," 205–207, also 305n16. It is worth mentioning that, in addition to *praeter naturam* phenomena, Nieremberg also spoke of nature's artifice in connection with insects and similar creatures, which denoted God's ability to work at small scale. Nieremberg, *Prolusion*, fol. 15v. See chapters 3 and 4 for more on the topic.

153. Bredekamp, *Lure*, 31–36; Daston, "Curiosity," 398. Bacon also saw the preternatural and the artificial as interconnected. The preternatural was evidence that nature can stray from its usual course, a process that art and experiment could replicate and carry even further. Weeks, "Francis Bacon," 132–133.

154. Nieremberg, *Oculta filosofia*, fols. 113r–114r.

155. Kircher, *Mundus subterraneus*, vol. 2, fol. (*)4r.

156. "Naturae id industria elaboratum, quod tantum monstraret in artificii structura ingenium." Athanasius Kircher to Duke August, January 3, 1660, Herzog August Bibliothek, Ms. BA II,5, 355.

157. On early modern art and its materials, see Vélez-Posada, "Handsteine"; Oosterhoff, "Ingenious Materials"; Smith, *Body*; Cole, "Cult."

158. Castiglione, *Il Libro*, 59–61. In 1570, Lodovico Castelvetro claimed that effort and diligence (*fatica* and *industria*) made artworks pleasant. Castelvetro, *Poetica*, fol. 40r. On effort and its ambivalent relation to ingenuity, see Zagoury, "Minerva"; Suthor, *Bravura*, 3.

159. The piece was a fake. In 1678, Kircher's former assistant, Giorgio De Sepi, reported that a young prince inadvertently broke the piece while he was visiting the museum, releasing a fetid smell. De Sepi, *Romani collegii*, col. 44A. Kircher would not have been surprised by a kind of counterfeit that permeated the market of curiosities. One could say that art was always trying to imitate nature at its most artificial. On "fake" or counterfeit *naturalia*, see Findlen, "Inventing."

160. Peregrini, *Delle acutezze*, 42.

161. "Non si può formar cosa alcuna, di raro, e mancando affatto l'artificio altro di pregio non può sperarsi, che vna buona, e chiara connessione sillogistica: e cosi all'intelletto molto soddisfare, ma non già punto all'ingegno. Egli è dunque mestiere, che 'l mezo, o gli estremi, o tutti, sieno artificiosi, e come direbbe il Retore, figurati." Peregrini, *Delle acutezze*, 42–43.

162. Peregrini, *Delle acutezze*, 41–42.

163. "Molto rara, e campeggiante." Peregrini, *Delle acutezze*, 42.

164. "Non dall'imparar, o veder molte cose in poche parole, ancorché questa conditione non vi manchi; ma principalmente dalla particular virtù dell'ingegno del poeta, che ammirabilmente risplende in esso chiaramente dipende." Peregrini, *Delle acutezze*, 66. It was a pleasure born "from the special sagacity of the wit that shines in it" ("dall'accortezza speciale d'ingegno, che sfolgoreggia in esso"). Peregrini, *Delle acutezze*, 113. Peregrini might have been inspired by the debate on Aristotle's *Poetics* during the second half of the sixteenth century. In 1570, Lodovico Castelvetro had suggested that a spectator who wonders at the novelty or unexpectedness of a certain reversal or recognition in the plot also wonders at the acuity of wit displayed by the poet who put the events together. Castelvetro, *Poetica*, fols. 39v–40r, 117v. Castelvetro's younger contemporary, Giason De Nores, voiced a similar idea. De Nores, *Poetica*, fol. 2v; De Nores, "Discorso," 392. In 1642, Gracián called the author's ingenuity one of the causes of a conceit. Gracián, *Arte*, 424. Tesauro repeated Peregrini's claim that in conceits we are struck by novelty and the author's ingenuity. Tesauro, *Il cannocchiale aristotelico*, 337.

165. On comets as conceits devised by nature's ingenuity, see Tesauro, *Il cannocchiale aristotelico*, 108–110.

166. Peregrini, *Delle acutezze*, 8. See also Aristotle, *Art of Rhetoric*, 408–415, §3.11, 1412a4–b10. On Aristotle's ἀστεῖα, see Schenkeveld, "Ta Asteia."

167. Peregrini, *Delle acutezze*, 41–42. In 1650, Peregrini resorted one more time to the distinction between intellect (*l'intelletto*) and ingenuity (*l'ingegno*), associated with imagination. Peregrini, *I fonti*, 39–40. Strada's course on rhetoric in approximately 1606–1607 linked the pleasure of tropes to the author's ingenuity without renouncing the Aristotelian link between learning and pleasure. According to Strada, the question lay with *what*, exactly, one learns from tropes. As he saw it, Aristotle did not simply mean that the audience discovers how the terms paired by metaphor fit with one another; they also learn about "the author's ingenuity, which is metaphor's cause and brings us pleasure as we learn it, for Aristotle taught that to learn the cause of a thing is to know it perfectly" ("Ingenium authoris, tamquam causa translationis, quae causa dum deprehenditur, uoluptatem cumulat, dum, rei causam intellexit, hoc enim esse perfecte scire, nempe rem per causam cognoscere Aristoteles docuit"). Famiano Strada, De elocutione, APUG Collection, Archivio della Pontificia Università Gregoriana, Ms. 1283, fol. 227r. For another version, see

APUG Collection, Ms. 1163, fol. 160v. Strada conciliated his views with those of Aristotle by resorting to the latter's fourfold division of causality. To notice the author of a metaphor thus became a form of learning—specifically, of learning "its efficient cause."

168. On the relation between labor and contemplation of nature, see Picciotto, *Labors*, 10.

169. Long, *Openness*, 4.

170. Piccolomini, *Piena, et larga parafrase . . . nel terzo libro*, 80–82; Sforza Pallavicino, *Considerazioni*, 146–184. Pallavicino's method was reminiscent of the one developed by Giulio Camillo Delminio in the first half of the sixteenth century. See Camillo Delminio, *Due trattati*, fol. 32v. On Martial and plagiarism, see Seo, "Plagiarism."

171. "Capital commune . . . monete della cassa del publico." Manzini, *Delle meteore rettoriche*, 124.

172. Tesauro, *Il cannocchiale aristotelico*, 583. For the comparison between Euclid and Martial that inspired Tesauro, see Peregrini, *Delle acutezze*, 42.

173. On the distinction between penetrating and shaping reality, see chapter 2.

174. See the introduction.

175. In one of the few passages in which Gracián seemed to echo Peregrini, he associated a conceit's artificial—as opposed to natural—character with the fact that it is perceived as beautiful rather than true. Gracián, *Arte*, 140.

176. "L'ingegno, occhio dell'animo, non possa incontrarsi in raggio di bellezza altretanto saporoso al suo gusto; quando quello, che in altro ingegno gratiosamente fiammeggi." Peregrini, *Delle acutezze*, 48–49.

177. "Più simpatico, & amabile." Manzini, *Delle meteore rettoriche*, 102–103.

178. Plutarch, *Moralia*, 8:378–379, §5.1.

179. Peregrini suggested as much by underscoring the link between "figurative language" and the notion of figure as "shape." Peregrini, *Delle acutezze*, 42–43.

180. "La luz que derechamente termina atormenta la potencia . . . Que os hagáis política." Gracián, *Arte*, 395–396.

181. "Abrió los ojos la verdad, dio en andar con artificio, usa desde entonces las invenciones, introdúzese por rodeos, vence con estratagemas, pinta lexos lo que está muy cerca, propone en estraño sujeto lo que quiere condenar en el propio, apunta a uno para dar en otro, deslumbra las passiones, desmiente los afectos, y por ingenioso circunloquio viene siempre a parar en el punto de su intención." Gracián, *Arte*, 396. Gracián believed in human sensory and intellectual faculties' preference for artifice over nature. See Gracián, *Arte*, 139–140.

182. On Adam as a model of perfect intellectual ability, see Huarte de San Juan, *Examen*, 170, 584–585; Picciotto, *Labors*.

183. Peregrini, *Delle acutezze*, 4, 33–34, 41–42, 48–49, 57–58, 158; Gracián, *Arte*, 139–140, 143, 156.

184. Lipsius, *Politicorum*, 201–202; Snyder, *Dissimulation*, 125–126.

185. Gracián, *Oráculo manual*, 101. See the introduction.

186. On Gracián's advice for life at court, see chapters 2 and 4. Lipsius took the expression "in Romulus's dregs" from Cicero, *Letters to Atticus*, 1_132–133, §2.1.8. Gary Remer has claimed that Lipsius's pragmatic views on political prudence were based on Cicero's rhetorical principles even more than on Tacitus's histories. Remer, "Justus Lipsius." Conversely, on Lipsius's Tacitism, see Waszink, "Your Tacitism."

187. "Lo que cabe en el poder de la Naturaleza"; "contingente, i no monstruoso." Lopes da Veiga, *Heraclito*, 176, 189.

188. Nieremberg, *Practica*, fol. ¶4r.

189. "Dopò lunga pellegrinatione, alcuna spoglia di Oltramarino Monstro arrecando, frà le attonite radunanze la spiega, e prende dalle attonite labra i muti applausi." Tesauro, *Panegirici*, 1659, 2:263. On conceits and monstrosity, see Hanafi, *Monster*, 187–217. For a new approach to animals as spectacle, see Beusterien, *Transoceanic Animals*.

190. Tesauro, *Il cannocchiale aristotelico*, 538. Tesauro called the kind of "preachable conceits" that had become fashionable in sermons a product offered to the faithful as "exotic merchandise" (*pellegrina merce*). Tesauro, *Il cannocchiale aristotelico*, 99. He would have approved the claim made in 1592 by poet and natural philosopher Giulio Cortese that inventors of conceits were like merchants who reveal unseen commodities as they return from remote lands. Cortese, *Prose*, 23.

191. Preti and Ercolino, "Grassi"; Feingold, "Grounds," 137. Plutarch had listed rainbows and comets as comparable examples of rare and new phenomena that attracted attention to the detriment of nature's regular events. Plutarch, *Moralia*, 11:210–211, §29, 919a–b. The rainbow followed comets in Cardano's *De subtilitate* and, consequently, in Scaliger's reply to Cardano. Cardano, *Hieronymi Cardani*, fols. 81v–93v; Scaliger, *Exotericarum Exercitationum*, fols. 122v–132v. As a topic of natural philosophy, the rainbow boasted a history dating back to at least Aristotle's *Meteorology*. See Smith, *From Sight*, 26–28.

192. However, the relationship between topic and language in Grassi's oration was far from simple. His lectures of 1617 and 1619 were highly technical and intended for audiences with knowledge of mathematics. He used conceits only in the preamble, where they are countless and boost the feeling of wonder at the topic. It is telling that, in the lecture on the three comets, he marked the end of the preamble with a warning that, from there, he would talk "plainly" (*diserte*), that is, without conceits. Grassi, *De tribus cometis*, 8.

193. Baldi, *De gli automati*, fol. 10r. Talentoni likewise recalled Plato calling Iris, the messenger of the gods, "the child of Thaumas, that is, wonder." Talentoni added that Iris was the rainbow, which causes wonder with its figure and variety of colors. Plato, *Theaetetus*, 155–156, 155D; Talentoni, *Discorso*, 49. The link between wonder and the rainbow in Greek myth was proverbial. See Francesco Sacchini, De ingenij praestantia. Oratio Habita . . . in renouatione studiorum III Non. Nouembris 1597, APUG Collection, Archivio della Pontificia Università Gregoriana, Ms. 117, fol. 49r, and Halley, "De Iride," 714. See also Fisher, *Wonder*, 11–14.

194. Aristotle, *Minor Works*, 330–331; Monantheuil, *Aristotelis Mechanica Graeca*, 6, 847a12–14. Engineers used their knowledge of nature's behavior to obtain effects that nature alone could not accomplish. Aristotle, *Physics*, 1:172–173, §2.8, 199a15–17; Newman, *Promethean Ambitions*, 17. Thus, mechanics did not do violence to nature, but rather fit with Aristotle's claim that art *imitates* or *perfects* nature. Schiefsky, "Art," 85–87.

195. Baldi, *De gli automati*, fol. 10r.

196. Preti and Ercolino, "Grassi."

197. Tesauro, *Il cannocchiale aristotelico*, 126–128.

198. "Con misterioso artifitio questa spiritosa poetessa." Tesauro, *Il cannocchiale aristotelico*, 108–109.

199. Gracián, *Agudeza*, 1:320–324; Alemán, *Guzmán de Alfarache*, 1:432–434; Blanco, "Guzmán." See Varo Zafra, "Grupos tacitistas españoles." For the link between courtly and picaresque forms of deceit, see Ruan, *Pícaro*.

200. Blumenberg, *Die Lesbarkeit*, 108–120.

201. "Sustenta . . . su Magestad." Nieremberg, *Prolusion*, fol. 18r–v. Nieremberg was surely aware that, for Aquinas, any phenomena that occur "beyond the order that God has placed within natural things" (*praeter ordinem naturalibus inditum rebus à deo*) served as God's means of displaying power, not out of frivolity but because that is the most suitable way for humans to know him, which is the reason why they have understanding in the first place. Aquinas, *Summa*, 3.2:79; Aquinas, *Summae*, 421, §3.99.10.

202. Gracián, *Obras completas*, 7. See chapter 2.

Chapter 2 Gossip Made of Glass

1. "Mírame con piedad, y arda el cometa, / Filis, que agora pálido nos mira." Leonardo de Argensola, *Rimas*, 36–37.

2. Mexía, *Discurso*. See also Parker, *Global Crisis*, xxvi–xxvii, 11–12.

3. "Os barbaros de Africa & os hereges do norte . . . com pouco custo." Nájera, *Discursos astrologicos*, fol. D2r.

4. Bocarro Francês, *Tratado*; Leitão, "Manuel Bocarro Francês"; Carolino, "Manuel Bocarro Francês."

5. Francesco Ingoli also subscribed to Tycho's views on the location of comets in the treatise he finished and signed on February 21, 1619. Francesco Ingoli, *De Cometa Anni 1618 Tractatus . . .*, Biblioteca Vallicelliana, R. 56, ins. 61, fols. 776r–811r.

6. Drake and O'Malley, *Controversy*, 6; Grassi, *De tribus cometis*, 4.

7. "Nulla iam caeli pars nostram effugit aciem, neque tanti; vt antea, apud nos est Lunae pulchritudo; Veneris, ac Mercurij choreis licuit interesse; quin & solem puduit foedum se aliquando à nobis spectatum: Martis terrae appropiquantis insidias deteximus; Iouis, ac Saturni stipatores, nequicquam hactenus sese abdentes, in apertum duximus. Soli igitur Cometae supererant, lynceis hisce oculis spectandi." Grassi, *De tribus cometis*, 3–4.

8. Drake and O'Malley, *Controversy*, 5; Grassi, *De tribus cometis*, 3. The contrast between "lippientes" and "Lyncei oculi" may have reminded the audience of Horace's Epistle 1.1, ll. 28–29, which, as quoted by Erasmus, likewise opposed "Lynceus" and "lippus." Erasmus of Rotterdam, *Erasmi Roterodami Adagiorum*, fol. 119v. See also García Romero, "To Have." For the invention of the telescope, see Van Helden, "Invention"; Reeves, *Galileo's Glassworks*; Ruestow, *Microscope*.

9. Peregrini, *Delle acutezze*, 41–43.

10. Shea and Bascelli, *Galileo's Sidereus Nuncius*, 55. On the opposition to Galileo's claim that the moon was rugged, see Miller, "Seeing."

11. The accusation against Venus and Mercury was connected only in part with the telescope. As early as the fifth century CE, Martianus Capella suggested that the two planets orbited the sun. As mentioned in chapter 1, Tycho Brahe added Mars, Jupiter, and Saturn to the list, giving rise to a new geo-heliocentric model of the universe.

12. Tycho's observation that the distance between Mars and the earth was variable put his disciple, Johannes Kepler, on the path to realizing that the planet's orbit was not circular but elliptical, despite the opinions of Aristotle and Ptolemy to the contrary. Later vari-

ants of Grassi's joke about "the snares of Mars" attest to its success. In a dialogue published in Milan in 1619, Mars maimed Ptolemy, that is, compromised Ptolemy's system of the universe. Besomi and Camerota, *Galileo*, 206. For a similar reference dating also to 1619, see Malapert, *Caroli Malaperti*, 35. For a later allusion to the threat that Mars's orbit posed to the Ptolemaic system, see Carolino, "Cometas," 355–356.

13. *Stipator* is Latin for "attendant," a member of a retinue. It is essentially synonymous with *satelles*, which, under Kepler's influence, became the technical term for bodies orbiting around planets. Galileo called Jupiter's moons "planets," a term that designated moving as opposed to fixed celestial bodies. Others called them *astites* and *asseclae*. See Snellius, *De maculis*, 4.

14. Marino, *Epistolario*, 153. See chapter 1.

15. Aristotle, *Art of Rhetoric*, 352–353, §3.2, 1404b4–5. On artifice and deceit, see Covarrubias Orozco, *Tesoro*, s.v. artimaña. On the ambivalence toward artifice and its reflection in law, see Cook, "Before the Fetish," 10–33.

16. Webster, *Curious Eye*, 15–24. On "the mind's eye," see also Frede, "Plato"; Aristotle, *Metaphysics*, 1:2–3, §A.1, 980a22.

17. Shea and Bascelli, *Galileo's Sidereus Nuncius*, 3. On Galileo and *avvisi*, see Reeves, *Evening News*, 101–134; Rosen, "Title"; Wilding, "Galileian Angels," 69–71. On *nuncius* as the Latin word for "newsletter" or "gazette," see Infelise, "History," 246.

18. Dooley, *Social History*, 4. On Pope Pius V's decree against *avvisi*, see Dooley, *Social History*, 26–27.

19. On traditional cosmological and cosmographical beliefs, see Garrod, *Cosmographical Novelties*, 20–30.

20. Galluzzi, *Lynx*, 21–24.

21. Maelcote, "Nuntius Sidereus." Maelcote and his colleagues rejected the most controversial consequences of the discoveries, especially Galileo's support of Copernicanism.

22. Galilei, *Le opere*, 11:229.

23. For Grassi's activities around 1611, see Preti and Ercolino, "Grassi." Discussions of astronomy and optics well into the late seventeenth century often started with up-to-date lists of discoveries expressed through strings of conceits. The Jesuit Flaminio Figliucci did that as he described Galileo's discoveries around 1611. See Figliucci, *Stanze*; Guerrini, "Le Stanze." In 1623, Johannes Faber did the same in a poem, as did Giovanni Battista Hodierna decades later. Galilei, *Il saggiatore*, fol. A1r–v; Hodierna, *Opvscoli*, fol. +3v. See also De Armas, "Maculate Moon," 60.

24. Gal and Chen-Morris, *Baroque Science*, 1–6. On early modern instrumentality, see Cypess, *Curious*.

25. Reeves, *Evening News*; García Santo-Tomás, *Refracted Muse*.

26. "Solium Iovis fidenter inspeximus; Lunae faciem omnem, omnesque maculas accuratissime perspeximus, & stellas penè omnes . . . quasi numeratò recensuimus." Snellius, *De maculis*, 3. For the attribution of the work, see Vermij, *Calvinist Copernicans*, 44–45.

27. On Lucian's *Icaromenippus*, see Boadas, "Lucien"; Maus de Rolley, *Élévations*. On Maestlin, see Maestlin, *De Astronomiae hypothesibus*, fol. A2r; Barker and Goldstein, "Realism," 249.

28. Lucian, *Downward Journey*, 300–303. Snellius quoted Lucian's passage. Snellius, *De maculis*, 18.

29. Snellius, *De maculis*, 3–4. On the invention of the telescope as an act of "stupefying arrogance," see Borel, *De vero telescopii inventore*, 2; Malet, "Early Conceptualizations," 242. See also Arnaudo, "Il microscopio aristotelico."

30. "Tyrus inquam illa . . . superatis terrarum angustijs, per mare Erythraeum elapsa, Asiae & Africae littora lustrare, & externa maria tentare, Natura propemodum reluctante & invita, ausa est." Snellius, *Tiphys Batavvs*, fol. *2r.

31. Snellius, *Tiphys Batavvs*, fols. *2r–*3r.

32. Galilei, *Le opere*, 11:18; Galluzzi, *Lynx*, 28–29. In 1623, Johannes Faber mentioned Galileo alongside Columbus and Vespucci. Galilei, *Il saggiatore*, fol. A1r. In the same year, poet Giambattista Marino likened Galileo to Columbus and Tiphys in *Adone* (1623). Marino, *Adone*, 1:977, §10.45.

33. See, for instance, Góngora, *Soledades*, 279.

34. Reeves, *Galileo's Glassworks*, 15–46.

35. See the introduction.

36. Íñiguez de Lequerica, *Sermones funerales*, fol. 24v; Gracián, *Arte*, 138. On the lynx and the eagle as emblems of acuity, see Gracián, *Arte*, 396. On the poet Luis de Góngora as "the eagle of conceits," see Gracián, *Arte*, 157.

37. Sarbiewski, *Wykłady*, 10. See chapter 1.

38. Ferrari, *Orationes*, 1625, 227.

39. "D'hauer felicemente penetrato assai, conditione propria delle cose Acute." Peregrini, *Delle acutezze*, 28. For Tesauro, sharpness rendered conceits "better apt to pierce through living flesh" ("pungono meglio la carne viua"). Tesauro, *Il cannocchiale aristotelico*, 340. He implied that sharpness enabled conceits to leave a deeper impression, echoing remarks from early modern physicians who ascribed to those statements accompanied by wonder greater odds of being remembered. See Tesauro, *Il cannocchiale aristotelico*, 276; Gutiérrez de Godoy, *Disputationes phylosophicæ*, fol. 140v. In 1648, the Jesuit José de Ormaza likened a conceit to the sting of a horsefly. Ormaza, *Censura*, 12.

40. "Ille profecto acribus, ac bstrus est oculis." López, *Oratio*, fol. 2v; "quod penitiora rimatur, intima permeat, & sese totum in rerum medullas insinuat." López, *Oratio*, fol. 1r. See also Rodríguez Beltrán, *La agudeza*.

41. López, *Oratio*, fol. 1v.

42. Period dictionaries registered a duality between material and intellectual penetration. "Penetrate" could mean "to move inside, to the inner parts" and, secondly, to grasp some truth. Crusca, *Vocabolario* (1612), s.v. penetrare. Someone is "penetrating" (*penetrante*) when he or she grasps that which is most hidden about a subject, "and thus we call penetrating the one who is exceedingly sharp" ("y assi llamamos penetrante al demasiado de agudo"). Covarrubias Orozco, *Tesoro*, s.v. penetrar. On acuity and penetration, see especially Ripa, *Iconologia*, 10; Solervicens, *La poètica*, 171–172.

43. "Abdita, et intima naturae introspexisse." Ingoli, De Cometa Anni 1618 Tractatus . . . , Biblioteca Vallicelliana, Ms. R. 56, ins. 61, fol. 778r.

44. Lüthy, "Atomism"; Aubert, "'Subtile.'"

45. Jesuit mathematician Cristoph Clavius called Euclid and his demonstrations "sharp" or "acute" (*acutum*). Specifically, he praised Euclid's "acutissimas, subtilissimasque . . . demonstrationes" and described him as "geometra acutissimus." *Euclidis Elementorum*, fols. A4v, b4r. On acuity and ingenuity, see Marr et al., *Logodaedalus*.

46. Van Helden, "Invention"; Reeves, *Galileo's Glassworks*.

47. The lynx appeared in the title pages of Della Porta's *Phytognomonica* (1588) and the second, augmented edition of *Magia naturalis* (1589). When Galileo, a member of the Accademia dei Lincei since 1611, reacted to Grassi's lecture of 1619, Cesi and his colleagues sided with him. The alliance between Galileo and Cesi's group has led some to take Grassi's reference to the telescope as "these lynx eyes" as a jab at Galileo. Drake and O'Malley, *Controversy*, 361; Moss, *Novelties*, 222. While there is no doubt that Grassi's conceit about the sky emptied of secrets because of the telescope aimed directly at Galileo, the phrase "Lyncei oculi" was ubiquitous around 1619. For its use in reference to the telescope prior to Grassi's lecture, see Snellius, *De maculis*, 17. For more on the Accademia dei Lincei, see Freedberg, *Eye*; Galluzzi, *Lynx*; Brevaglieri, *Natural desiderio*. Grassi's reference to "these lynx eyes" may have been purely conventional. Conversely, the volume of *Disquisitiones mathematicae* published in the name of Scheiner's pupil Johannes Georg Locher pointed a finger directly at the Accademia dei Lincei in a poem signed by the author's brother Heinrich. Heinrich likened Johannes Georg to a perspicacious and lofty eagle that should be preferred to "the lynxes" of the Accademia. Locher and Scheiner, *Disquisitiones mathematicae*, 5. Locher's *Disquisitiones mathematicae* appeared in 1614, soon after the Accademia published Galileo's letters on sunspots in direct response to Scheiner's work on the topic.

48. "Hauendo questa eletta, acciò ne sia vno stimulo, e sprone continuo di ricordarci dell'acutezza della vista, non de gli occhi corporali, ma della mente, necessaria per le naturali contemplazioni, che professiamo; e tanto più douendosi in queste procurare di penetrar l'interno delle cose, per conoscere le loro cause, & operazioni della natura, ch'interiormente lauora, come con bella similitudine dicesi che la Lince faccia col suo sguardo, vedendo non solo quel ch'è di fuori, ma anche ciò che dentro s'asconde." Stelluti, *Persio tradotto*, 37. See also Galluzzi, *Lynx*, 32–37.

49. Pliny the Elder, *Natural History*, 8:86–87, §28.32. Other than Pliny, there is only one source from antiquity—specifically, an appendix of proverbs—associating acuity to the lynx. García Romero, "To Have," 79.

50. The name Lynceus originated in a toponym and is unrelated to "lynx." García Romero, "To Have." In seventeenth-century England, Joseph Hall maintained that Adam was like Lynceus until he was expelled from Paradise. Picciotto, *Labors*, 200.

51. Snellius, *De maculis*, 4; Drake and O'Malley, *Controversy*, 5; Grassi, *De tribus cometis*, 4.

52. "Con lo sguardo penetraua la grossezza de' muri . . . dalla Sicilia numeraua le naui ch'vsciuano dal porto di Cartagine." Stelluti, *Persio tradotto*, 37–38. On Lynceus, see also Kircher, *Mundus subterraneus*, vol. 2, fol. (*)2v.

53. "Onde Talpa il Ceruiero / Appò te GALILEO / Fora, & argo senz'occhi, / orbo Lincèo. . . . Onde ciò, ch'altrui cela / Natura entro nel seno, / Aperto si riuela / A l'vno, e l'altro tuo sguardo sereno." Galilei, *Il saggiatore*, fol. a3r. See also Daston, "Curiosity," 391–392.

54. "Primum reperisse fodinas metallorum." Erasmus of Rotterdam, *Erasmi Roterodami Adagiorum*, fol. 119v. See García Romero, "To Have," 83–84.

55. The accusation that mining defaced the earth dated to at least the late fifteenth century, and probably much earlier. Asmussen, "Spirited Metals."

56. Merchant, *Death*, 1–41; Hadot, *Veil*, 95–96. For a response to criticism against Merchant's book, see Merchant, "Scientific Revolution." On feminism and science, see Keller, *Reflections*.

57. Nieremberg, *Oculta filosofia*, fols. 94v–95r.

58. "Si toda la comtemplacion de la naturaleza es apacible, y gustosa aun con su primera vista, y considerada, solo por la corteça . . . mucho mas amena y agradable serà quando se penetran sus secretos, y se entra en lo hondo de sus misterios . . . violaremos su mas guardado retiro, llegaremos a lo arduo, a lo dificultoso, a lo innacesible della, a su mayor sacramento, que es la simpatia y antipatia, como hablan los Griegos, esto es vna secreta conformidad, y auersion, que parece, o ay en las cosas con que se executan efectos admirables por lo extraordinario, y anomalo, que tienen a la vista, y lo inuisible y oculto de sus causas." Nieremberg, *Oculta filosofia*, fol. 1r–v. See also Marcaida, "Juan Eusebio Nieremberg," 102. For Merchant's criticism, see Merchant, "Scientific Revolution," 515. On early modern epistemology and surfaces, see Savoia, "Knowing."

59. "Se atiende a la correspondencia que hazen los estremos cognoscibles entre sí; y esta Agudeza contiene aquel Epigrama de Marcial, que ha merecido más aplausos que sílabas. Careó a Diaulo, antes Médico y después sepulturero, y dixo: *Nuper erat Medicus; nunc est Vespillo Diaulus: Quod Vespillo facit fecerat et Medicus.*" Emphasis original. Gracián, *Arte*, 148.

60. Gracián, *Arte*, 146.

61. To cast lines or "echar líneas" was a reference to the activity of the mathematician. See Dávila y Heredia, *Tienda*, 12.

62. "Cárealas con el sujeto, y unas con otras entre sí; y en descubriendo alguna conformidad, o proporción, que digan unas con otras, exprímela con sutileza." Gracián, *Arte*, 146–147. As has been mentioned, Gracián was the only seventeenth-century scholar to attribute the invention of conceits also to the intellect, rather than to imagination alone.

63. "Poner una persona cara à cara con otra." *Diccionario*, s.v. carear.

64. Bacon, *Works*, 4:263; Hadot, *Veil*, 35–36, 93, 120. While scholars such as Carolyn Merchant and Evelyn Fox Keller interpret Bacon's references to experimentation with nature as violent, Peter Pesic has claimed that Bacon envisioned "a heroic mutual struggle, not the torture of a slavish victim." Pesic, "Wrestling," 81. See also Weeks, "Francis Bacon." Nieremberg, who cherished the prospect of natural philosophy forcing its way into nature's private or secret rooms, saw it necessary to point out that nature also has agency as object of knowledge, and takes it so far as to "torture and torment" those who think they can conquer them through ingenuity. Nieremberg, *Curiosa filosofia*, fol. 117r.

65. In 1665, Athanasius Kircher spoke of solutions, sublimations, "and the rest of the torments of fire" (*caeteraque ignis tormenta*). Kircher, *Mundus subterraneus*, 2:331.

66. Gracián, *Arte*, 143.

67. Peregrini, *Delle acutezze*, 41–42. See chapter 1.

68. "Penetra le più lontane & minute *Circonstanze* di ogni suggetto, . . . lequali cose giacciono in qualunque suggetto aggomitolate & ascose." Tesauro, *Il cannocchiale aristotelico*, 119; "come i Metalli dentro delle miniere." Tesauro, *Il cannocchiale aristotelico*, 608, emphasis original.

69. Tesauro, *Il cannocchiale aristotelico*, 608.

70. Asmussen and Long, "Introduction," 14.

71. Tesauro, *Il cannocchiale aristotelico*, 119. On juggling and conceits, see chapter 4.

72. See Frare, *Per istraforo*, 85–89.

73. "Si fermano nella superficie del Vocabulo"; "riflessiuamente penetra & inuestiga le più astruse notioni per acoppiarle." Tesauro, *Il cannocchiale aristotelico*, 336–337, 585–586.

74. Gracián, *Arte*, 133.

75. "Interna verò ipsius animi sensa, Astrologus licet Lynceus, ne optico quidem suo Telescopio introspexerit." Grassi, *Libra astronomica*, 5.

76. Guiducci and Galilei, *Discorso*, 140–141. Only with Giuseppe Biancani's 1620 *Sphaera mundi* did the Society of Jesus start its more or less official endorsement of the Tychonic system, with caveats. Carolino, "Making," 313; Gambaro, "Geo-Heliocentric Models."

77. "Iouis asseclas, / Patrisue Saturni . . . Nigras / in sole . . . retectas / labes." Gattei, *On the Life*, 304–305, ll. 10–11, 34–36. Barberini's letter of June 8, 1624, to the Medici Grand-Duke of Tuscany Ferdinando II praised Galileo in similar terms. Galilei, *Le opere*, 13:183–184.

78. "Turba satellitum / Hinc inde praecedit, colentes / Officijs comites sequuntur." Gattei, *On the Life*, 303–306.

79. Snyder, *Dissimulation*, 12.

80. Gattei, *On the Life*, 306–307, l. 70. Barberini echoed Cicero's question beginning with "Quis tam lynceus," found in a letter to Marcus Varro dated April, 46 BCE. Cicero, *Letters to Friends* 2:150–151, §9.2.3; see also García Romero, "To Have," 86.

81. "Con la seditiosa materia de' suoi Annali, e delle sue Historie, fabbricaua certi occhiali, che pernitiosissimi effetti operauano per li Prencipi, perche posti al naso delle persone semplici di modo assottigliauano loro la vista, che fino dentro le budelle faceuano veder gl'intimi, e più reconditi pensieri altrui." Boccalini, *De' ragguagli*, 329–330. See Reeves, *Evening News*, 105–120; García Santo-Tomás, *Refracted Muse*, 91–107.

82. "Principum aulas, principum interiorem vitam, consilia, iussa, facta." Lipsius, *C. Cornelii Taciti*, 5.

83. Boccalini, *De' ragguagli*, 330.

84. Anonymous, Oratio in instauratione studiorum, APUG Collection, Archivio della Pontificia Università Gregoriana, Ms. 1086, fol. 39r. The oration is in fols. 29r–48v,and has also been preserved in Veterum ac nostrorum temporum in disciplinarum acquisitione comparatio per singulas aetates. Barberiniani Latini (Barb. Lat.), Biblioteca Apostolica Vaticana, Ms. 1749, fols. 120r–137v. Neither manuscript mentions who delivered the oration or when.

85. The term "reason of state," first used by Guicciardini, became popular with the publication of Giovanni Botero's *Della ragion di stato* (*On the Reason of State*). See Botero, *Della ragion*, fol. *2r–v; Ammirato, *Discorsi*, 228–231. See also Tuck, *Philosophy*; Burke, "Tacitism, Scepticism"; Höpfl, *Jesuit Political Thought*. On Boccalini and reason of state, see Boccalini, *De' ragguagli*, 102; Bradford, "Stuart Absolutism," 136.

86. Boccalini, *De' ragguagli*, 331.

87. Oratio in instauratione studiorum, APUG Collection, Archivio della Pontificia Università Gregoriana, Ms. 1086, fols. 39v–42r. See also Boccalini, *De' ragguagli*, 102, 404–405. On secrecy and politics, see Stolleis, *Arcana Imperii*.

88. Gracián built a pun between the two meanings of the term *gafas*, meaning both "the bender of a crossbow" and "eyeglasses." Gracián, *El Criticón*, 1:185–194; Ruan, *Pícaro*, 117–120, 137.

89. Gracián followed the version of the story of Momus featured in Lucian's *Hermotimus*. Lucian, *How*, 298–299.

90. Gracián, *Oráculo manual*, 101. See the introduction.

91. Gracián, *Oráculo manual*, 155. On the cuttlefish, see Pliny the Elder, *Natural History*, 3:218–219, §9.45.

92. Gracián, *El Discreto*, 330; Gracián, *Oráculo manual*, 131. "La advertencia . . . la contracifra de intenciones." Gracián, *Oráculo manual*, 208. On Gracián's world, conceived in terms of attack and defense, see Chartier, "El proceso civilizatorio," 14.

93. For Gracián's mention of "that experimental science" ("aquella ciencia experimental"), see Gracián, *El Discreto*, 364.

94. "Haze concepto el sabio de todo, aunque con distinción caba donde ai fondo y reparo; y piensa tal vez que ai más de lo que piensa, de suerte que llega la reflexión adonde no llegó la aprehensión." Gracián, *Oráculo manual*, 122. On *reparo*, see also Gracián, *El Discreto*, 364.

95. See also chapter 4.

96. The debate over the term "crisis" in Gracián's *Arte de Ingenio* remains open. See Green, "Sobre"; Vaíllo, "Crisis"; Koselleck and Richter, "Crisis."

97. Vega Ramos, "El azar," 153.

98. Gracián, *Arte*, 240. For Martial's poem, see Martial, *Epigrams*, 2:222–223, §8.79.

99. "No se contentó con la vulgar sencilla narración de la Historia . . . no parava en la corteza de los sucessos, sino que transcendía a los más reservados retretes de la intención." Gracián, *Arte*, 239. See also Gracián, *Agudeza*, 1:287–288.

100. Gracián, *Arte*, 170; Gracián, *Agudeza*, 1:82.

101. "Assí como el obrar con artificio y con reflexo nace de ventaja de ingenio, assí el descubrir esse artificio y el notarlo es sutileza dos vezes primorosa." Gracián, *Arte*, 239–340. See also Gracián, *Agudeza*, 1:288.

102. "Conoscere dalla lunga le astutezze degli spagnoli." Tesauro, *Il cannocchiale aristotelico*, 128.

103. Hadot, *Veil*.

104. Bacon, *Works*, 4:42; Nieremberg, *Oculta filosofia*, fol. 1r–v. The Latin word *penetralia*, used by Bacon for nature's inner chambers, was the same term that in Barberini's letter to Grand-Duke Ferdinando II Medici referred to the reality of the planets.

105. Giglioni, "Philosophy"; Daston, "Curiosity," 391. On Bacon as a tutor, see Bradford, "Stuart Absolutism," 138.

106. Covarrubias's dictionary of 1611 explained that a mystery was "anything locked under a veil, made of facts, words, or any other kind of signs" ("qualquiera cosa que está encerrada debaxo de velo, o de hecho, o de palabras, o de otras señales"). Covarrubias Orozco, *Tesoro*, s.v. misterio.

107. Tamayo de Vargas, *Restauracion*, fol. ¶¶1r; Páez de Castro, *Memoria*, 75–76. Campanella claimed that Galileo's *Sidereus Nuncius* was a messenger who had had a glimpse of God's secrets (*arcana Dei*). Galilei, *Le opere*, 11:16.

108. Carlo Ginzburg argues that, at least until the seventeenth century, "the warning against knowledge of 'high' things has been referred to various (but intertwined) levels of reality," including nature, religion, and politics. Ginzburg, "High," 32.

109. "Ya en conceptuosa panegiri, ya en ingeniosa crysi." Gracián, *Arte,* 146–147. In 1648, Gracián clarified what he meant: "Digo alabando o vituperando." Gracián, *Agudeza,* 1:40. See also Gracián, *Agudeza,* 1:315.

110. "Todo gran ingenio es ambidextro: gran destreza es discurrir a dos vertientes." Gracián, *Arte,* 211. See also Gracián, *Agudeza,* 1:181.

111. "Todo superior gusto la estima porque lastima." Gracián, *Arte,* 238.

112. Strada, *Prolusiones Academicae,* 32–34. See also Mascardi, *Dell'arte historica,* 209. Royal chronicler of Castile Antonio de Herrera y Tordesillas wrote that Tacitus resembled more a prosecutor than a historian. Herrera y Tordesillas, *Discursos morales,* 19.

113. "Ipsi inter se mortales auersi, & suspiciosi, sibique ab aliorum perfidia metuentes." Strada, *Prolusiones Academicae,* 31–32. The virtue of trust (*fides*), which Botero opposed to cunning, was a foundation of society for Jesuit historians such as Juan de Mariana and Pedro de Ribadeneira. Höpfl, *Jesuit Political Thought,* 147. Drawing on Augustine, Herrera y Tordesillas claimed that history ought to promote trust among individuals, and between them and the government. Herrera y Tordesillas, *Discursos morales,* 1.

114. "Instrumento que halla mancha en el sol, y averigua mentiras en la luna, y descubre lo que el cielo esconde, es instrumento revoltoso, es chisme de vidrio, y no puede ser bienquisto del cielo." Quevedo, *La hora,* 311. Quoted also in Reeves, *Galileo's Glassworks,* 11. On curiosity as fault finding, see Picciotto, *Labors,* 196–197. Grassi's list of discoveries of 1619 commented humorously on the telescope's artifice. However, for a different interpretation, holding that Grassi lamented that nature appeared less perfect through the lens, see Gal and Chen-Morris, *Baroque Science,* 111–112.

115. "Saeculum christallinvm . . . Lentibus muniti cum Caelo pugnant, & stellas, victores, capiunt!" Hodierna, *Protei Caelestis,* 4.

116. "Natura istessa si stupisce, e mostra / Mestitia, e duol, d'hauer prodotto vn figlio / Ch'ogni segreto suo suela, e dimostra." Hodierna, *Opvscoli,* fol. +3r. See also Tesauro, *Il cannocchiale aristotelico,* 128.

117. "Verdadero enigma de los maldicientes." Pérez García, "Un texto académico," 367.

118. Leonardo de Argensola, *Obras sueltas,* 2:271. See also Delage, "Tacitisme," 327–328.

119. Von Staden, "Physis," 29–33; Hadot, *Veil,* 92–93. The auscultation of inner causes by the physician became a model for the writers attacked by Argensola. They fashioned themselves as readers of minds inspired by Tacitus, who later synthesized the rules of political behavior that they conjectured from the rulers' thoughts and intentions in aphorisms after the manner of Hippocrates. That was the case of Baltasar Álamos de Barrientos, who, in *Tácito español* (1614), taught his reader how to "grasp" the inner evils and desires as though from pulses" ("como por los pulsos conocer los vicios y codicias interiores"). Álamos de Barrientos, *Tácito español,* fol. †8r–v. Uroscopy also served as a proxy for the conjecture of intentions from people's behavior. See Matteo Renzi, Tratado del privado perfecto, Biblioteca Nacional de España, Ms. 10633, fol. 4v. On uroscopy, see Stolberg, *Uroscopy.*

120. On the continuity between reading bodies and minds, see Clark, *La Rochefoucauld,* 39–54; Patiño Loira, "Policing Talent."

121. "Nos *Lynceos* possumus nuncupare . . . Ferunt hosce videre quae abdita in penitis terrae visceribus, venas aquarum & metallorum, thesauros, & sub sarcophagis sita cadauera." Del Río, *Disquisitionum Magicarum,* 1:35, emphasis original. On dowsers, see Dym,

Divining Science; Dillinger, "Divining Rod." In 1628, Quevedo paired lynx and *zahorí* in the title of *El lince de Italia u zahorí español*. See García Santo-Tomás, *Refracted Muse*, 183–186.

122. "Nonnulli vim hanc humori melancholico . . . adscribunt." Del Río, *Disquisitionum Magicarum*, 1:35; Stelluti, *Persio tradotto*, 38.

123. "Los colores, por las vmedades, y yerua que encima nace: como lo saben los que buscan minas." Rocamora y Torrano, *Sphera*, fol. 5r–v. On sixteenth-century claims that it was the science of mirrors, a division of mathematics, that taught how to find underground streams, see Reeves, *Galileo's Glassworks*, 36.

124. "Ciertas yervas, que los signifiquen." Nieremberg, *Oculta filosofia*, fols. 94v–95r.

125. "La pratica de segnali portici da la benignita de la natura." Biringuccio, *De la pirotechnia*, fol. iiiiv.

126. "Anatomía de los espíritus, de los naturales, de las inclinaciones." Gracián, *Obras completas*, 124.

127. "Ni aun de resquicios para penetrar el más reservado interior." Gracián, *El Discreto*, 310.

128. "Hay zahoríes de entendimiento que miran por dentro las cosas, no paran en la superficie vulgar, no se satisfacen de la exterioridad." Gracián, *El Discreto*, 312. On metaphor, see Tesauro, *Il cannocchiale aristotelico*, 376–377, 706, 728.

129. "Torciendo, y aun fingiendo, las intenciones . . . No se contenta tal vez con descubrir el artificio en la intención, sino que se arroja a fingirle." Gracián, *Arte*, 239–240.

130. Galluzzi, *Lynx*, 46.

131. On *Los anteojos de mejor vista* (*Los antoios de meior vista* in its early modern edition), see García Santo-Tomás, *Refracted Muse*, 140–148.

132. Fernández de Ribera, *Los anteojos*, 47–48.

133. "En el mismo ser que son." Fernández de Ribera, *Los anteojos*, 47, 67. Works of fiction imagined Disabuse (*Desengaño*) in different guises. *Desengaño* was the name of the ragged old man who showed the narrator the truth about passersby in *El mundo por de dentro* (*The World from Within*), an allegory published by Quevedo in 1612. In 1659, Gracián introduced the character of "El Descifrador" ("The Decipherer") in the third part of *El Criticón* to teach the protagonists how to decrypt a world that existed in cipher. Ribera placed optics in the role that Quevedo assigned to elucidation and Gracián to cryptography.

134. García Santo-Tomás, *Refracted Muse*, 143.

135. "Tan estraño, nuevo y prodigioso." Fernández de Ribera, *Los anteojos*, 47.

136. "Un altro Creatore . . . cose che finhora non si sapeva che fossero state create" (May 11, 1624). Gabrieli, *Atti*, 1038–1039, 1047–1048. See Freedberg, *Eye*, 152–154; Galluzzi, *Lynx*, 386.

137. Seventeenth-century utopias often opposed discovering worlds to inventing them out of "pure wit." Cavendish, *Blazing World*, 121. In *The New Atlantis*, however, Bacon listed "Columbus, that discovered the West Indies" together with technological innovations such as ships and glass. Bacon, *Works*, 3:165–166. The continuity between discovery and invention is evidenced by Girolamo Cardano's claim in 1550 that mechanical clocks were one of the inventions that "remained concealed" (*latebant*) until recently, a turn of phrase that suggested, even if only metaphorically, that technologies have some form of existence as they wait to be invented. Cardano, *Hieronymi Cardani*, fol. 267v.

138. Quoted in Freedberg, *Eye*, 154.

139. As late as 1776, Enlightenment-era critic José de Cadalso still mocked pseudo-scholars who fashioned themselves as magicians just because they could explain "that a louse resembles an elephant when looked at through the microscope" ("cómo un piojo parece Elephante en el microscopio"). Cadalso, *Los eruditos*, 31. Yet at times, the insight that one thing's inner truth resided in its similarity to another could be meaningful in natural-philosophical terms, as in the claim that the microscopes collected by Gracián's patron Vincencio Juan de Lastanosa displayed a world in which "from a dead flea a horrible monster emerges, and from the smallest flower seeds truly grown-up fruits appear, such as citrons, lemons, oranges, melons" ("De vna pulga muerta se muestra vn horrible monstruo, y de menudissimas semillas de Flores crecidissimas frutas como son Cidras Limones, Naranjas, Melones"). Narración de lo que le pasó a don Vincencio Lastanosa a 15 de octubre del año 1662 con un religioso docto y grave, Hispanic Society of America, Ms. B-2424, fols. 55v–56r. See also Reula Baquero, *El camarín*, 93. Here, the notion that the seed was the fruit in miniature could point to preformation theories in vogue. See Pinto-Correia, *Ovary*, 20–23 and passim.

140. Cavendish, *Blazing World*, 142–144. See also Webster, *Curious Eye*, 68–102; Walters, "Optics." Descartes likewise saw lenses and mirrors as deceitful. Malet, "Early Conceptualizations," 255. On mirrors and deception, see Garzoni, *La piazza universale*, 896–903; López Pinciano, *Philosophía antigua poética*, 276; Suárez de Figueroa, *Plaza Universal*, fols. 346r–348v; López, *Oratio*, fol. 10r.

141. "Omnia à nobis visa multò reuera, ac videntur, alia esse." Kircher, *Ars magna*, 834. See Waddell, *Jesuit Science*, 141.

142. Lüthy, "Atomism," 18.

143. "Están ya en el objeto las agudezas . . . llega y levanta la caça el ingenio." Gracián, *Arte*, 426. For natural philosophy as a hunt without sweat, see Nieremberg, *Prolusion*, fol. 39r; Eamon, "Science as a Hunt"; Lewis, "Francis Bacon," 135–136.

144. On imagination, see the introduction.

145. "Perché faccia oggetto di prospettiua ingegnosa." Peregrini, *Delle acutezze*, 42.

146. "Faccia a faccia . . . senza interpretamento di lingua mentitrice." Tesauro, *Il cannocchiale aristotelico*, 40.

147. "Del diletto di tante belle Arti sermonali." Tesauro, *Il cannocchiale aristotelico*, 40–41. See also Torre, "Rimirandolo," 46. See Tesauro, *La filosofia morale*, 198. In 1647, the account of the Duke of Parma's funeral service in Piacenza cathedral described a statue of Silenus, complete with a reference to Plato's *Symposium*, with a window open in his chest and inside the inscription *Pretiosa latent* ("That which is precious hides inside"). Raulino, *Pompa*, 25.

148. "Il corto, & natural linguaggio degli Angeli . . . essendo del human genio . . . ammirar maggiormente la verità vestita, che ignuda." Tesauro, *Il cannocchiale aristotelico*, 40–41.

149. "Ogni obietto schifoso e laido . . . vna Tarantola, vna Botta, vn Aspido, vno Scorpione, veduto vicino cagiona afrezza, ma se tu lo miri da lungi col Cannocchiale; con diletto lo miri: per la nouità dell'ingegnoso ordigno che'l rappresenta." Tesauro, *Il cannocchiale aristotelico*, 168.

150. Aristotle, *Poetics*, 169, §4, 1448b15–19. As I already noted, Augustine had ascribed to curiosity the ability to render ugly objects pleasing, precisely what the lens does, according to Tesauro. Augustine of Hippo, *Sancti Augustini Confessionum*, 182–186. See chapter 1.

151. Aristotle mentioned elsewhere that taking a portrait for its sitter counted as equivocation. Aristotle, *Categories*, 12–13 §1, 1a1–8.

152. Like Peregrini, Tesauro assumed that once the intellect noticed novelty in the way conceits represented reality, it also experienced beauty as a result. In 1646, however, Pietro Sforza Pallavicino criticized Peregrini for mixing up the two notions of novelty and beauty. Sforza Pallavicino, *Considerazioni*, 142–146. Yet Pallavicino's distinction was the exception, and seventeenth-century sources tended to conflate novelty, pleasure, and beauty. A character in Juan Enríquez de Zúñiga's novella *Amor con vista* (1625) wakes up from a dream in which, carried upward by an eagle, he sees the world transformed into "an anthill or a written paper" ("vn hormiguero, ò vn papel escrito"). As distance becomes excessive, it acts as a form of mediation that defamiliarizes the object like the microscope that transforms fleas into elephants, making earth look like "a Flemish painting, where I saw things as though they were painted" ("vn lienço de Flandes, donde miraba las cosas como pintadas"). Enríquez de Zúñiga, *Amor*, fol. 50v. The narrator's first reaction as he sees the earth from a new and unfamiliar perspective is to aestheticize it.

153. Tesauro, *Il cannocchiale aristotelico*, 376–377.

154. Gilman, *Curious Perspective*, 67–87; Snyder, "Art."

155. Strehlau, "Fernrohr," 130. Tesauro's term for the aperture resounds with the mention of *gli strafóri* in reference to Christ's wounds in one of his discourses. Tesauro, *Panegirici, et ragionamenti . . . Volume terzo*, 329. See also Torre, "Rimirandolo," 47.

156. Tesauro, *Il cannocchiale aristotelico*, 376.

157. Del Monte, *Guidubaldi . . . Perspectivae*.

158. Niceron, *La perspective curieuse*, fol. êiiijv; Fratini and Moriconi, "Datazione." Prior to Niceron, in 1625, Pietro Accolti had published a chapter on anamorphosis. Accolti, *Lo inganno*, 48–50. After Niceron, the topic reemerged in numerous publications, including the Jesuit Mario Bettini's *Apiaria Universae Philosophiae Mathematicae*, published starting in 1642.

159. Niceron's anamorphic device may be found in Museo Galileo, inv. 3196.

160. "La reflexión . . . filosofando artificiosamente . . . vuelven atrás." Gracián, *El Criticón*, 1:21–22. On reflection and the invention of conceits, see Güntert, "Filosofar."

161. Quoted in Daston, "Curiosity," 400–401. Gracián and Hooke took exception with Nieremberg's advice, discussed in chapter 1, that one should focus on rare and unusual phenomena. They believed that, rather than focus on aspects in which nature has used artifice, humans should apply intellectual or optical artifice to make familiar objects appear new.

162. Picciotto, *Labors*, 188–211; Wilding, "Galileian Angels"; Schaffer, "Regeneration."

163. "T'insegna . . . che vna cosa sia l'altra . . . acennandoti vna Propietà; lascia che tu comprendi l'altra col tuo intelletto." Tesauro, *Il cannocchiale aristotelico*, 710.

164. Gracián, *Arte*, 138.

165. Gracián, *Obras completas*, 1479–1480.

166. On self-fashioning, see Martin, "Inventing." On dissimulation and simulation, see Snyder, *Dissimulation*. On mental reservation, see Tutino, *Shadows*.

167. Caussin, *Eloquentiae sacrae*, 73–78.

168. "Aquel que significa otro tanto más de lo que dize." Gracián, *Arte*, 414. See also Gracián, *Oráculo manual*, 129.

169. "Bizarros y de cerbelo gallardo." Leonardo de Argensola, *Obras sueltas*, 2:336. Similarly, English clergyman George Hakewill claimed that Tacitus interpreted motives just to display ingenuity. Bradford, "Stuart Absolutism," 131.

170. "[I] matti . . . meglio che i sani . . . sono conditionati à fabricar nella lor fantasia metafore facete, & simboli arguti: anzi la *Pazzia* altro non è che Metafora, laqual prende vna cosa per altra." Tesauro, *Il cannocchiale aristotelico*, 133, emphasis original. See also Zanardi, "Metafora," 55–60. Tesauro may have taken inspiration from a claim in Aristotle's *Poetics* that, in the interpretation of most early modern commentators, suggested that poets should be either gifted (εὐφυεῖς) or maniac (μανικοί). Tesauro, *Il cannocchiale aristotelico*, 118–119; Sifakis, "Aristotle."

171. "Leggere il rimanente nel petto di chi le scrisse." Tesauro, *Il cannocchiale aristotelico*, 44.

172. "Imagine abozzata più tosto che finita, doue l'ingegno intende, più che la lingua non parla." Tesauro, *Il cannocchiale aristotelico*, 41.

173. "Il doppio godimento . . . l'vn gode di dar vita nell'intelletto altrui, à vn nobil parto del suo: & l'altro si rallegra d'inuolar col proprio ingegno ciò che l'ingegno altrui furtiuamente nasconde." Tesauro, *Il cannocchiale aristotelico*, 41.

174. "Non richiedendosi minor sagacità nell'esporre, che nel comporre vna Impresa arguta & ingegnosa." Tesauro, *Il cannocchiale aristotelico*, 41.

175. "Auditoribus etiam nonnullis grata sunt haec, quae cum intellexerunt acumine suo delectantur, et gaudent non quasi audierint sed quasi invenerint." Quintilian, *Orator's Education*, 3:338–339, §8.2.21.

176. "At ego vitiosum sermonem dixerim quem auditor suo ingenio intellegit." Quintilian, *Orator's Education*, 3:336–337, §8.2.19. In 1493, Raffaele Regio emended the statement by adding a negative particle, *non*, before the verb *intelligit*. Regio thought that Quintilian was condemning a statement that the audience cannot understand, rather than the fact that they need to use their own wit as they do it. When, in 1516, Josse Bade removed Regio's emendation, it was too late, and the damage had been done. Regio's version coexisted for centuries with the one restored by Bade. The editorial history of the passage embodies the fraught and difficult reaction to the idea of audience cooperation. Regio, *Quintilianus*, n.p.; Bade, *M. Fabij Quintiliani*, fol. 160v.

177. "La riflessione sopra la parte, ch'egli hà nell'intelligenza . . . che l'ascoltatore in vn certo modo particolare cooperi con la propria sagacità. Cosa diletteuole, a segno, che molti si godono d'vdire, o leggere, cose oscure, perche, dice Quintiliano, *cum intellexerint, suo acumine delectantur, non quasi audierint, sed quasi inuenerint*. Perche in somma tutto quello, per lo quale *Fit in nobis aliqua cognitio, & existimatio, proprij boni*, come diceua S. Tomaso, ci e sempre dilettoso." Peregrini, *Delle acutezze*, 126. Peregrini knew perhaps that for Pseudo-Longinus (first century CE) the sublime was marked by a sense of "proud exaltation" (μεγαλαυχία) that made the audience feel as if they had created what they had heard. Aristotle, Longinus, and Demetrius, *Aristotle*, 178–179, §7.2. Interestingly, the first edition of Ignatius Loyola's *Spiritual Exercises* (1548), a foundational text of the Society of Jesus, opened with a eulogy of cooperative reading. Loyola, *Exercitia Spiritualia*, fols. A8v–B1r.

178. "No tanto nos contenta el ingenio que descubrimos en el Autor, como el que nos parece mostramos en alcançarle, y gustamos de leerle, porque en cada clausula nos vamos secretamente lisonjeando." Ormaza, *Censura*, 63. In 1572, Alessandro Piccolomini disputed

Aristotle's assertion that the pleasure of metaphor derives from learning. Piccolomini suggested that since it is natural for us to love ourselves more than we love others, when we learn something in a way that seems to demand effort from us, we take it to be partly ours and feel delighted due to self-love. Piccolomini, *Piena, et larga parafrase . . . nel terzo libro,* 73. In fact, Aristotle said something similar in *Art of Rhetoric,* 126–127, §11.1, 1371b25–27. Piccolomini also drew inspiration from a passage in Aristotle's *Rhetoric* about a kind of syllogism that seduces the audience by helping them arrive at the conclusion by themselves. Piccolomini, *Piena, et larga parafrase . . . nel secondo libro,* 436–437. In 1622, Virgilio Malvezzi argued that in the laconic style used by Tacitus we take what we read to be "the offspring of our own intellect" ("il parto del propio intelletto"). Malvezzi, *Discorsi,* fol. *3v.

179. "Han . . . de descubrir lo mismo que ocultan, y dezir aun lo que callan artificiosas. Como las pinturas de primor, que aunque no descubren el reuerso, de tal suerte se entorna el ombro, y lados, que parece nos dan a ver las espaldas que ocultan." Ormaza, *Censura,* 62.

180. See, for instance, López, *Oratio,* fols. 5v–8v; Hanafi, *Monster,* 204–206. See also chapter 4.

181. That was the case of the Jesuit superior general Muzio Vitelleschi in a letter already mentioned, dated January 20, 1631. Astrain, *Historia,* 5:111.

182. Tesauro called *imprese* and preachable conceits "popular." Tesauro, *Il cannocchiale aristotelico,* 98, 704, 748–749. On the power of conceits to level social hierarchies, see Hanafi, *Monster,* 211. However, Tesauro clarified that by "popular audiences," he meant average individuals, not learned and refined, but not plebeians either. Tesauro, *Il cannocchiale aristotelico,* 746–747.

183. Tesauro, *Il cannocchiale aristotelico,* 99.

Chapter 3 At Small Scale

1. "Morendo appena nate, par che portino seco il funerale, e le facelle." Minozzi, *Sfogamenti,* 115.

2. On early modern natural-philosophical explanations of the light of fireflies, see Wolfe, "Illumination." On bioluminescence more generally, see Anctil, *Luminous Creatures,* 3–16; Harvey, *History.*

3. Peregrini, *Delle acutezze,* 42–46. See chapter 1.

4. Little is known about Minozzi's life. See Ghilini, *Teatro,* 1:190–191; Fumaroli, *L'âge,* 220–221. For Minozzi's publications, see Minozzi, *La Bibliotheca Medicea,* 33–42.

5. On Minozzi and Peregrini, see Croce, *I trattatisti italiani,* 18–19. Minozzi and Peregrini shared acquaintances and patrons, such as theologian Paganino Gaudenzi. For Minozzi's letters to Gaudenzi, see "Lettere scritte da Discepoli a Paganino Gaudenzio dall'anno 1635 fin all'anno 1641," Urbinates Latini (Urb. Lat.), Biblioteca Apostolica Vaticana, Ms. 1625, fols. 25r–26r, 34r–35v, 65r. For a poem that Peregrini dedicated to Gaudenzi, see Molina, "Matteo Peregrini poeta," 586.

6. "Diciamo in proverbio. Mostrar lucciole per lanterne, che è il dare a vedere, e ad intendere una cosa per un'altra." Crusca, *Vocabolario* (1612), s.v. lucciola. For other seventeenth-century occurrences of the proverb and its variants, see Besomi and Camerota, *Galileo,* 203; Tesauro, *Il cannocchiale aristotelico,* 531.

7. Fireflies have a history of standing for other things. In a brief piece published on February 1, 1975, mere months before his assassination, Pier Paolo Pasolini used fireflies

as code for individual and cultural difference, threatened by the homogenizing spotlight of consumerism. In Pasolini's wake, several political and environmental scholars have projected a variety of nostalgic implications onto the insects' dreamlike quality. See Pasolini, *Scritti corsari*, 160–168; Didi-Huberman, *Survival*.

8. Marvell, *Complete Poems*, 109, l. 5. I thank Robert Watson for the reference to Marvell's pairing of comets and fireflies. See also Watson, *Back to Nature*, 122–123. In a similar vein, in ninth-century Basra, scholar Al-Jahiz wrote of light-bearing insects that at night appear as shooting stars. Harvey, *History*, 48.

9. See, for instance, the chapter that Francesco Ingoli dedicated to the brevity of the third comet of 1618. Ingoli, De Cometa Anni 1618 Tractatus . . . , Biblioteca Vallicelliana, Ms. R. 56, ins 61, fol. 804r.

10. "I concetti eran lampi, che dalla ruota del Sol diuelti, discendeuano ad illustrar gli intelletti; hora son lucciole, che in vn solo dibattimento d'ali, partoriscono, e sepelliscono il lumicino innocente; hor son baleni, che prima si dileguano, che sien veduti, hor son fauille, che suaniscono à vn tratto; senza sodezza, senza maturità." Mascardi, *Orationi*, 330. On Mascardi, see Bellini, *Agostino Mascardi*; Tutino, *Shadows*, 40–73.

11. Strada, *Prolusiones Academicae*, 167–173.

12. Strada, *Prolusiones Academicae*, 173–174.

13. "Rotto e sciolto e discatenato." Peregrini, *Che al savio*, fol. b3r. Plutarch's *Life of Lycurgus* was the main ancient source on Spartan or "Laconic" brevity. Plutarch, *Lives*, 1:264–269, §19–20.

14. On Jesuit Ciceronianism, see Fumaroli, "Cicero"; Fullenwider, "Die Kritik."

15. See Puteanus, *De Laconismo Syntagma*; Ferro, *Federico Borromeo*.

16. Mascardi, *Orationi*, 110. Mascardi's remark imitated a similar caveat placed at the front of Nicholas Caussin's treatment of "the sharp brevity of style," published in 1619. Caussin, *Eloquentiae sacrae*, 73.

17. Mascardi, *Dell'arte historica*, 661–662; Jerome, *Select Letters*, 196–199. On Jerome's Letter 52, see Cain, *Jerome*.

18. Mascardi, *Dell'arte historica*, 661–662. See also Mascardi, *Orationi*, 330.

19. Yet in 1687 Francesco Fulvio Frugoni referred to Mascardi's discourse of 1622. Bellini, *Agostino Mascardi*, 31.

20. Mascardi's reputation as the scourge of conceits led the writer Giovanni Battista Manzini to assume that he was the author of a piece criticizing Manzini's style. Manzini replied in 1637 with "A letter in self-defense," but Mascardi later disavowed any responsibility in the affair and denied that his attacks against conceits and laconism had been intended for Manzini. See Bellini, *Agostino Mascardi*, 229.

21. Minozzi, *Sfogamenti*, 117, 264.

22. Minozzi, *Sfogamenti*, 258.

23. In 1624, Peregrini had identified Marino, the author most directly associated with conceits, as the summit of Italian vernacular poetry. Peregrini, *Che al savio*, fols. 2v–3r. Years later Peregrini wrote a poem on the death of Marino. Molina, "Matteo Peregrini poeta," 586.

24. "Qualunque volta per mia disgratia inciampo in così fatte compositioni, e veggio, che posto da parte ogni studio della più soda eloquenza, al solo brillar di questi spiriti sono riuolti, non mi risoluo se ingegni spiritosi li chiami, o spiritati; e parmi a punto che e' loro Autori habbiano sortita la semplicità di que' fanciulli, che tutto che'l Cielo arda di così vaghi

lumi nel tempo della sera, non alzano ad ogni modo pur vno sguardo ver quella parte, intenti solo a correr forsennatamente dietro allo sfauillar delle lucciole." Peregrini, *Delle acutezze*, fol. a6r–v. For Minozzi's criticism of Renieri's passage, Minozzi, *Sfogamenti*, 262–263.

25. See Graziosi, "Cesura," 64–74. In 1956, Franco Croce branded Peregrini's approach that of "a moderate baroque." With a grayish prudence, he was "a remora" that countered the excesses of Marino's unrestrained use of conceits. Croce, *Tre momenti*, 138 et passim. Mercedes Blanco believed that Peregrini's mistrust of conceits stood in stark contrast with his meticulous, and at times enraptured, analysis. Blanco, *Les rhétoriques*, 227–229; Croce, *I trattatisti italiani*, 13–15. Graziosi's revisionist position has received some attention in Marini, "Elisabetta Graziosi"; Bellini, *Agostino Mascardi*, 228. For Peregrini's objections to Mascardi, see Bisello, *Medicina*, 151–152. For Peregrini's praise of Marino, see Molina, "Matteo Peregrini poeta," 597 et passim.

26. For Mascardi's advocacy of moderation in the use of conceits, see Mascardi, *Dell'arte historica*, 659–660.

27. "A far l'amor con le Stelle, che correspondono assai bene." The letter is dated Genoa, June 1, 1640, and was sent to Galileo in Arcetri. Galilei, *Le opere*, 18:200.

28. Mascardi, *Orationi*, 330; Pliny the Elder, *Natural History*, 3:492–493, §11.34.

29. "Cur caelum intuearis, agricola? cur sidera quaeras, rustice? . . . ecce tibi inter herbas tuas spargo peculiares stellas." Pliny the Elder, *Natural History*, 5:346–349, §18.66–67. Pliny's injunction also appears in Muffet, *Insectorum*, 111. I quote from H. Rackham's translation.

30. Beagon, "Curious Eye," 75.

31. As was mentioned in chapter 1, for Giovanni Battista Manzini, the greater immediacy with which conceits were able to impact the reader was motivated by a preference, natural in humans, for the artificial over the natural. In chapter 2, I argued that, for Tesauro, tropes rendered "preachable conceits" capable of touching the audience in ways that were entirely out of reach for loftier and truer theological arguments. Manzini, *Delle meteore rettoriche*, 102–103; Tesauro, *Il cannocchiale aristotelico*, 98–99.

32. "Confesso io . . . la fugacità di tai concetti; tuttavia, a me dilettano in estremo quelle Veneri d'ingegno, che veramente son Veneri, tanto più dolci, quanto più brevi. Le viuacità, e le acutezze sono Circi ingegnose, che m'affascinano l'intelletto." Minozzi, *Sfogamenti*, 117.

33. "Mi trattengono, mi lusingano, mi rapiscono." Minozzi, *Sfogamenti*, 117–118.

34. Minozzi, *Sfogamenti*, 262. On spirits and life, see Hirai, "Spirit." See chapter 4.

35. Jessica L. Wolfe has pointed out that in "Upon a Glow-Worm That He Kept Included in a Crystal Viol" (*Occasional Reflections*, 1665), Robert Boyle wavered between utter fascination and contempt toward fireflies. She argues that fireflies "came to embody the paradoxes inherent in nature at large, illustrating the superior grandeur of the little while also turning topsy-turvy the order of things." Wolfe, "Illumination," 190–191. The fact that Boyle compared the striking and ostentatious light of fireflies to those who flaunt "flashes of Wit" or ingenuity places Wolfe's remarks—published after I finished writing *The Age of Subtlety*—in close proximity to the argument about conceits advanced in chapters 3 and 4. On human-animal interactions and emotional relationships, see Wagschal, *Minding Animals*.

36. Garrod, *Cosmographical Novelties*, 21–22.

37. On the notion of a ladder of nature or *scala naturae*, see Lehoux, *Creatures*, 19, 88; Harrison, "Natural History," 137; Leonhard, *Fertile Ground*, 22–27.

38. The distinction between "perfect," sexually reproducing animals and "imperfect," spontaneously generated ones was not Aristotelian but the product of thirteenth-century attempts at systematizing Aristotle's thought. Lehoux, *Creatures*, 6–7. It is worth noting that Aldrovandi, who believed that insects were born by spontaneous generation, countered the suggestion that they were in any way imperfect. Aldrovandi, *De animalibus insectis*, 3–4.

39. On spontaneous generation, see Lehoux, *Creatures*. See also Kress, "Aristotle"; Ogilvie, "Order," 224–229; Newman, *Promethean Ambitions*, 166–171; Leonhard, *Fertile Ground*, 28–31. On the notion of insect metamorphosis, see Ogilvie, "Order"; Pinto-Correia, *Ovary*, 25–26. I cover spontaneous generation in detail in chapter 4.

40. Scaliger was reacting to Cardano's 1550 *De subtilitate*. Scaliger, *Exotericarum Exercitationum*, 260–261; Cardano, *Hieronymi Cardani*, fol. 189r. For Ulisse Aldrovandi's approving mention of Scaliger's discovery, see Aldrovandi, *De animalibus insectis*, 495.

41. Mandabach, "Matter," 7–8. For the seventeenth-century fascination with the seemingly protean, indistinct order in which spontaneous generation was believed to thrive, see Leonhard, *Fertile Ground*. See also chapter 4.

42. Ogilvie, "Order," 222; Jorink, "Insects"; Neri, *Insect*. On the link between the fascination with insects and other small creatures and developments in the genre of still-life painting, see Barker, "Universe," 21.

43. "Segretario della Natura, Interprete dell'Vniuerso . . . Ne' suoi quasi immensi Volumi parlanti de gli Animali terrestri, volatili, acquatili, ed insetti hà quasi publicato al Mondo vn nuouo Mondo." Minozzi, *Sfogamenti*, 216.

44. Augustine of Hippo, *On Genesis*, 72–74. See Lehoux, *Creatures*, 67. On Augustine's justification of, and interest in, small animals, see Miller, *In the Eye*, 156–165; Gronewoller, *Rhetorical Economy*, 65–75. See also chapter 1.

Fernández de Córdoba, *Examen*, n.p.

45. Aldrovandi, *De animalibus insectis*, 1–2; Pliny the Elder, *Natural History*, 3:432–435, §11.1.

46. On subtlety, see the introduction and chapter 1.

47. Cardano, *Hieronymi Cardani*, 1–3; Scaliger, *Exotericarum Exercitationum*, 1–3.

48. "Longè maiora collocasse videantur miracula, quàm in enormibus alijs animalibus, vt vel sic videantur ostendisse non molis quantitate, sed virtutis amplitudine res ipsas dimetiri oportere." Aldrovandi, *De animalibus insectis*, 1.

49. Aldrovandi, *De animalibus insectis*, 2; Augustine of Hippo, *Corpus*, 25:6:1:55. The notion that a fly should be preferred to the sun because it is animated became commonplace among scholars advocating the right to study nature's smallest creatures. In connection with seashells, for instance, see Buonanni, *Ricreatione*, 36.

50. On the vegetative and the sensitive souls, see Des Chene, *Life's Form*. On the vegetative soul specifically, see Baldassarri and Blank, *Vegetative Powers*.

51. "No auia menester mirar tan alto, por mas contento se podria dar con la vista de vn animalejo, vna aueja, vn gusanillo, vn mosquito, en losquales conoceria mas el ingenio del artifice." Nieremberg, *Prolusion*, fols. 15v–16r.

52. "No le mandò que contemplasse los saraos de las estrellas en esse salon del firmamento, ni las danças de los Planetas, ni las mudanças de la Luna, y demás luzes del mundo, solo los animales, y las mas minimas auezillas se las recogio, y puso ante los ojos." Nieremberg, *Prolusion*, fol. 31v.

53. In 1635, Nieremberg claimed that speculation on "the animated substance" was the most excellent application of the intellect. Nieremberg, *Historia*, 1. Writing in 1642, Baltasar Gracián contended that the rays of the sun were merely material, and therefore inferior, to the animated, living, and sentient, but also intellective, rays that ingenuity had infused in a conceit. Gracián, *Arte*, 137.

54. For the reaction to Augustine's condemnation of curiosity, see chapter 1.

55. "Rerum natura nusquam magis quam in minimis tota sit." Aldrovandi, *De animalibus insectis*, 2; Pliny the Elder, *Natural History*, 3:434–435, §11.4.

56. Minozzi, *Sfogamenti*, 117. On smallness, see Stewart, *On Longing*, 37–69.

57. "Que' luminosi animaletti," "stellette volatili," "animate Fiammelle." The list also included "Arghi piccioli," "spiritelli volanti," and "fugacissime scintillette." Minozzi recalled that Mascardi compared conceits "to the small lights of fireflies" ("a' lumiccini di Lucciole"). Minozzi, *Sfogamenti*, 115–117.

58. Minozzi, *Sfogamenti*, 251; Ferrari, *Orationes*, 1627, 221–222.

59. Sarbiewski, *Wykłady*, 18. On the identification of Ferrari as the individual mentioned in Sarbiewski's manuscript, see chapter 1.

60. "Hac pulcherrima humanae eloquentiae particula." Sarbiewski, *Wykłady*, 2.

61. Ferrari, *Orationes*, 1627. A previous edition of Ferrari's orations had appeared in 1625.

62. On Ferrari's orations, see Freedberg, "From Hebrew," 41–42; Freedberg, *Eye*, 420.

63. On stylistic horseflies, see Estienne, *Dictionarivm*, s.v. acvlevs; Ormaza, *Censura*, 12. Minozzi argued that one learns conceits ("le acutezze, e gli spiriti più piccanti") "among the thorns of the junipers, and of other sharp and prickly plants" ("frà le spine de' Ginepri, et d'altre piante acute, e pungenti"). Minozzi, *Sfogamenti*, 192–193.

64. Ferrari, *Nomenclator Syriacus*, fol. a3r–v; Ferrari, *Orationes*, 1625, fols. 2r–3v. Gerardo Saraceni's *imprimatur* to the 1627 edition of Ferrari's orations praised the stings of Ferrari's style, which he considered particularly suitable to one of the bees of Pope Urban VIII, who was himself a poet of epigrams, well-versed in conceits. Ferrari, *Orationes*, 1627, fol. a3v. More than two decades later, the approval for the 1644 edition of Ferrari's *Orationes* still referred to the author as "aculeato." Ferrari, *Io. Bapt. Ferrarii Senensis . . . Orationes*, fol. *3v.

65. Anonymous, *Nevv Testament*, 404. The motto is from Romans 9:28.

66. The motto *Verbum breviatum fecit Dominus* was, in fact, a modified version of the translation provided in Jerome's *Vulgate Bible* ("Verbum breviatum faciet Dominus super terram").

67. "Hebraicae literaturae securitas" and "Stylus Hebraicus." Ferrari, *Orationes*, 1627, 216–247.

68. Jerome, *Select Letters*, 124–125. See Barr, "St Jerome's Appreciation," 286–287.

69. Barr, "St Jerome's Appreciation," 293–295.

70. Ferrari, *Orationes*, 1627, 216.

71. Ferrari, *Orationes*, 1627, 243.

72. "Diuinum Verbum ad puerilis corporis breuissimam periodum sese contraxit." Ferrari, *Orationes*, 1627, 233; also 213–214, 243.

73. In 1642, Baltasar Gracián resorted to the notion of mystery, religious in origin, to describe a kind of conceit that enticed readers with a deep and elusive meaning. Gracián, *Arte*, 166–171.

74. Ferrari, *Orationes*, 1627, 233–234.

75. Ferrari, *Orationes*, 1627, 245. In 1652, Manzini invoked Pliny's maxim in reference to the importance of polishing one's style. Manzini claimed that nature had used the same know-how, care, and finesse "to fabricate a firefly and a star." Manzini, *Delle meteore rettoriche*, 173. Ferrari compared conceits with gems, which, as Aldrovandi had argued a few decades earlier while advocating the study of insects, a discerning eye prefers to rocks. Ferrari, *Orationes*, 1627, 246–247. Those who are learned and busy, Ferrari also claimed, have little time to lose with loquacious explanations. Ferrari, *Orationes*, 1627, 237–238.

76. Ferrari, *De florum cultura*, 499; Freedberg, *Eye*, 38–46; Freedberg, "From Hebrew," 45–47.

77. "Homeri longissima Ilias, in breuissima membrana, quae nuce includeretur, ita scripta . . . pulcherrima contractio, quae Dantis Aligerij profusam comoediam humanum uix unguem superante codice tam feliciter clausit." Ferrari, *Orationes*, 1627, 245–246. On the idea, originating in Cicero, of Homer's *Iliad* transcribed in a nutshell, see Keller, "Charlatan's Ingenuity," 181.

78. Quintilian, *Orator's Education*, 1:402–403, §2.20. Cardano argued that it is licit to scorn useless forms of subtlety at the same time that we admire the ingenuity that crafted them. Cardano, *Hieronymi Cardani*, 294–302.

79. Francesco Maria Fattori, De moderandis ingeniis, APUG Collection, Archivio della Pontificia Università Gregoriana, Ms. 117, fols. 41r–42r. The mention of Myrmecides's miniature artworks may be found in a multitude of ancient authors, from Varro (first century BCE) to Plutarch, Athenaeus, and Aelian. It was featured in a passage of Pliny the Elder's *Natural History* frequently quoted in early modern sources. Pliny the Elder, *Natural History*, 10:34–35, §36.4. For one instance of its reception in a book of the same years as Ferrari's two orations, see Gramigna, *Fantasie varie*, 130–131. On Myrmecides's fame, see Squire, *Iliad*, 287–288.

80. Aristotle, *Poetics*, 177, §7, 1450b–51a.

81. Terzago, *Museum septalianum*, 114; Scarabelli, *Museo*, 214–216; Lugli, "La ricerca," 276.

82. "Alcune altre opere minutissime, ma altretanto habili a dar testimonio di vn grande ingegno." Scarabelli, *Museo*, 214–215. Francisco de Los Santos's *Brief Description* (*Descripción breve*, 1657) of the monastery of El Escorial near Madrid mentions a cabinet containing "precious, but minute things" ("otras cosas preciosas, aunque menudas"). Quoted in Reula Baquero, *El camarín*, 69.

83. Schott, *P. Gasparis Schotti*, fols. 3r–4r.

84. Carducho, *Dialogos*, fols. 156v–157r.

85. Fattori, De moderandis ingeniis, APUG Collection, Archivio della Pontificia Università Gregoriana, Ms. 117, fols. 40r–42v.

86. Manzini quoted Ferrari's oration "De Stylo Hebraico" in Manzini, *Delle meteore rettoriche*, 21. For Manzini's criticism of miniatures, see *Delle meteore rettoriche*, 135–136.

87. On Peregrini's notion of *acconcezza*, see Peregrini, *Delle acutezze*, 33–36.

88. Peregrini, *Delle acutezze*, fol. a5r; Minozzi, *Sfogamenti*, 263–266. With conceits, Minozzi argued, "si fa pompa d'ingegno." *Sfogamenti*, 267.

89. Garrard, "Not-So-Still Lifes," 66; Grootenboer, "Thinking," 107.

90. López, *Oratio*, fol. 3r–v. See chapter 2 for a discussion of the passage. López also launched a harsh invective against "idle arts." He shared Ferrari's referents, but used them with opposing goals. The writers Polycrates, Lucian, and Psellus, he claimed, earned enormous praise from writing on minuscule or vile creatures, such as mice, flies, and moths, deserving that the lines of Vergil's *Georgics* used in Ferrari's title page ("a work on a small subject, but of not small glory") be applied to them, but only in a sarcastic way. In the exaltation of the miniscule and vile, López saw a form of sophistry comparable to the invention of conceits, in which the quest for novelty occurred at the expense of truth. López, *Oratio*, fol. 7r.

91. Pliny the Elder, *Natural History*, 3:432–433, §9.1.

92. Daston, "Curiosity," 392, 396; Findlen, "Inventing," 301. On early modern cabinets of curiosities, *studioli*, and *Wunderkammern*, see Impey and Macgregor, *Origins*; Pomian, *Collectors*; Findlen, *Possessing*; Bredekamp, *Lure*; Jardine, *Worldly Goods*; Ago, *Gusto for Things*, 125–138.

93. Meadow, "Quiccheberg."

94. "Vn pretioso musaico di mille ingegnosi pensieri . . . di diamanti, di rubbini, di zaffiri, di carbonchi, e di perle composta." Bartoli, *Dell'huomo*, 399–400. I always quote Bartoli's *Dell'huomo di lettere* from its second edition, published in 1645. Pliny had mentioned only pearls. Pliny the Elder, *Natural History*, 10:174–175, §37.6.

95. "Lo sminuzzamento de' periodi trinciati in piccolissimi concisi, effetto della moltitudine di tante coserelle minute, ciascuna delle quali finisce il senso, e muta pensiere." Bartoli, *Dell'huomo*, 405.

96. Bartoli, *Dell'huomo*, 413. Galileo also disliked intarsia and compositions of *pietre dure* that created figures by placing different pieces together—of wood and stone, respectively. Bredekamp, *Galileo's Thinking Hand*, 45.

97. For Aristotle on the unity of the plot, see Aristotle, *Poetics*, 178, §8, 1451a16–35.

98. "Algun Concetico hilado à lo Sofistico." Lopes da Veiga, *Heraclito*, 183.

99. "Aquel monstruo de los ingenios; aquel Fenix de las agudezas don Luis de Góngora . . . el moderno Marcial." Suárez de Figueroa, *Pusílipo*, 260.

100. "Todas sus obras consisten en variedad de fracmentos; que aunque son pedaços de oro, es lastima no se aya aplicado ingenio tan superior a vna larga y proseguida; que sin duda, fuera un cielo, todo grabado de joyas." Suárez de Figueroa, *Pusílipo*, 260.

101. "Come vn filo di perle, nelle quali caduna delle parti è vna gioia, e tutto il corpo insieme è vn tesoro." Manzini, *Delle meteore rettoriche*, 118.

102. Cole, *Ambitious Form*, 78–89; Fried, *Moment*, 83–84, 174. For a similar mode of autonomous and portable copresence of objects in the context of sugar sculptures, see Kociszewska, "Displays," 451–452.

103. For Giambattista Marino's *a tasselli* composition in his *Galleria* and *Sampogna*, see Russo, "Introduzione," 12. Early modern scholars often contended that it was necessary to give the reader some respite from the continued succession of conceits. With that in mind, the Jesuit José de Ormaza argued in 1648 that preachers should moderate their ingenuity because conceits become more noticeable when they are placed in conjunction with less demanding moments. See Ormaza, *Censura*, 61–62.

104. "La moral trabaçón de un discurso." Gracián, *Arte*, 144–145.

105. "¿La agudeza libre, o la ajustada a un discurso; la suelta, o la encadenada en una traça?" Gracián, *Arte*, 375.

106. "En España siempre huvo libertad de ingenio, o por gravedad, o por cólera de la nación, que no por falta de inventiva. Sus dos primipilos, Séneca en lo juizioso y Marcial en lo agudo, fundaron esta opinión, acreditaron este gusto. Prudente aquel, nunca pudo sujetarse a los rigores de un discurso, a la afectación de una traça . . . Tributó nuestra Bílbilis a la gran emperatriz del orbe, no monstros como el África, sino al que lo fue de la agudeza y cultura. Entró Marcial en Roma destinado a la oratoria; mas su estremada prontitud, no sufriendo pigüelas de encadenada eloqüencia, se remontó libre en tantas puntas de agudeza quantos se eternizan epigramas." Gracián, *Arte*, 375–376.

107. On conceits and "the Spanish temperament," see the introduction.

108. "Este feraz siglo de ingenios, discurriendo todos a lo libre." Gracián, *Arte*, 376, 133–134.

109. Ferro, *Federico Borromeo*, 227–228. For the notion that the laconic style of Virgilio Malvezzi was "Spanish," see Delage, "Las vidas particulares," 67.

110. On ingenuity and noble birth, see Marr et al., *Logodaedalus*, 21, 44–45, 71; Fuchs, "Crusoe's Absence," 38–39.

111. Mascardi, *Prose vulgari*, 10; Biagioli, *Galileo, Courtier*, 263–264. Biagioli has argued that natural philosophers such as Francis Bacon often saw adherence to a system as servile. Biagioli, *Galileo, Courtier*, 308.

112. Hodierna, *Opvscoli*, fol. +3v.

113. "Censúranse en los más las Agudezas, antes por unas que por únicas, y homogéneos los pensamientos." Gracián, *Arte*, 136.

114. Gracián, *Arte*, 135.

115. Sarbiewski, *Wykłady*, 28.

116. Peregrini, *Delle acutezze*, 160–161.

117. Beckher and Radau, *Orator extemporaneus*, 39.

118. For the opposition between *ingenium* as inventiveness and *genius* as natural predisposition, see Aphorism 2 in Gracián, *Oráculo manual*, 101. The authors of *Logodaedalus* show that the term for *genius* in different languages could be translated at times as a divine or angelic influence that predisposed the individual, implying external rather than internal justifications for his or her individual characteristics, strengths, and limitations. Marr et al., *Logodaedalus*, 110–111. For the word history of the Latin word *genius* and the Italian word *genio*, see Marr et al., *Logodaedalus*, 28–36, 64–68.

119. Ormaza, *Censura*, 64–65.

120. Aquinas, *Summa*, 3.2:66–68. Cardano's notion of subtlety likewise "equated power and order with diversity." Findlen, "Jokes," 302. Athanasius Kircher exalted nature's generative powers as the ability to introduce variety in matter. Kircher, *Mundus subterraneus*, 2:328. It has been suggested that the fascination with variety that informed early modern cabinets of curiosity had a counterpart in the Mannerist style in art. Tatarkiewicz et al., *History*, 3:158–159; Eamon, *Science and the Secrets*, 407. On variety in general, see Fitzgerald, *Variety*. On rhetorical and poetic copiousness, see Cave, *Cornucopian Text*.

121. "Buen ejemplo nos da naturaleza, / que por tal variedad tiene belleza." Vega, *Arte nuevo*, 313, ll. 179–180. Lope de Vega revived a controversy over the genre of "tragicomedy" that had raged in Italy beginning in 1587, pitting the advocates of mixing tragic and comic

elements, spearheaded by Giovanni Battista Guarini, against Giason De Nores. Albalá Pelegrín, "El Arte nuevo." William Fitzgerald has observed that Plotinus (third century CE) justified the existence of evil in the world on aesthetic grounds, and did so precisely through a comparison with how, in tragedy, one could find "among its characters not only heroes, but also slaves and peasants who speak incorrectly." Expelling variety for the sake of a world in which only the good is present, Plotinus argued in *Enneads* 3.2.11, "would be to spoil the beauty of the whole." Fitzgerald concludes that "to prioritize the variety of creation is to identify creativity as the essence of divinity, over other characteristics such as justice." Fitzgerald, *Variety*, 2.

122. "Aun la misma Naturaleza, por atender a ella para más abellecerse, produce a veces cosas contrarias a su particular intento, como son los monstruos." Fernández de Córdoba, *Examen*. I have invoked Fernández de Córdoba's passage on monsters in chapter 1.

123. "Haec atque talia ex hominum genere ludibria sibi, nobis miracula, ingeniosa fecit natura." Pliny the Elder, *Natural History*, 2:526–527, §7.2. Quoted in Findlen, "Jokes," 296.

124. Carvallo, *Cisne*, 183; Monforte y Herrera (pseud.), *Relación*, fol. 9r–v.

125. Dalle Pezze, *Diálogos*, 77.

126. Escardó, *Rhetorica christiana*, fol. 53r. In 1596, Alonso López Pinciano argued that a plot of fiction has to be unified but varied, for "nature takes joy in the variety of things" ("la naturaleza se goza con la variedad de las cosas"). López Pinciano, *Philosophía antigua poética*, 196.

127. Tarquinio Galluzzi, De multiplici doctrinarum studio contrahendo . . . , APUG Collection, Archivio della Pontificia Università Gregoriana, Ms. 1153, fols. 38r–43v. See also Patiño Loira, "Policing Talent," esp. 142–145.

128. "La gran madre de la belleza." Gracián, *Arte*, 137; Gracián, *Agudeza*, 1:17.

129. "La uniformidad limita, la variedad dilata, y tanto es mas sublime quanto mas nobles entidades multiplica. No brillan tantos Astros en el Firmamento, campean flores en el prado, quantas se alternan sutilezas y conceptos en una fecunda inteligencia." Gracián, *Arte*, 142.

130. "Quae varietate sola placent." Bartoli, *Dell'huomo*, 1645, 404. It was a quote from Pliny the Elder, *Natural History*, 6:196–197, §21.27.

131. "Nature . . . in her most sportive mood, playful in her great joy at her varied fertility" ("lascivienti praesertim et in magno gaudio fertilitatis tam variae ludenti"). Pliny the Elder, *Natural History*, 6:160–161, §21.1; Findlen, "Jokes," 296.

132. On *lusus naturae*, see Findlen, "Jokes"; Findlen, "Ludic Postscript." See also Kircher, *Mundus subterraneus*, vol. 2, fol. (*)4r.

133. Leonhard, *Fertile Ground*, 28–31.

134. "La varietà capricciosa di questi Animali, con cui scherzò la Natura." Buonanni, *Ricreatione*, fol. †4r. On the link between nature's creativity and smallness ("in qualsiasi minimo lavorio"), see Buonanni, *Ricreatione*, fol. †5r. For Pliny on seashells, see Pliny the Elder, *Natural History*, 3:230–233, §9.52; Findlen, "Jokes," 296. In the sixteenth century, physician Ambroise Paré similarly wondered at "the uneconomical variety of seashells, which to him indicated 'that Nature, chambermaid of great God, disports herself in the manufacture of them.'" Kragh, "Inventive Nature," 16.

135. Minozzi, *Sfogamenti*, 251.

136. Huarte de San Juan, *Examen*, 343–346.

137. The tag *Natura nihil agit frustra* ("Nature does nothing in vain") was one in a series of general principles popularized by Aristotle, which "seem to define a certain constant behavior of nature as an art and knowledge within things." Hadot, *Veil*, 25. I mentioned the tag *Natura nihil agit frustra* in chapter 1.

138. Accolti, *Lo inganno*, 144.

139. Crusca, *Vocabolario* (1612), s.v. capriccio. Anton Francesco Doni's 1597 edition of Burchiello, a poet active in the first half of the fifteenth century, suggests that *capriccio* and *ghiribizzo* had closely related but not identical meanings. Its title mentioned both *capricci* and *ghiribizzi* in the company of other terms that often meant "conceits," such as *argutie* and *sali*. See Burchiello, *Rime*.

140. "Altri senza capo, altri col piè maggior del capo." Tesauro, *Panegirici*, 1659, 2:261–262. For other aspects of Tesauro's sermon, delivered on May 3, 1626, see chapter 1.

141. "Co' suoi confusi, ed abbozzati Arabeschi." Tesauro, *Panegirici*, 1659, 2:262.

142. Connelly, *Grotesque*, 54.

143. Connelly, *Grotesque*, 25–52. See also Guest, *Understanding*, 494–591; Guest, "Plato's Stag Goats." For the invective against grotesques by Cardinal Paleotti and others, see Acciarino, *Lettere*.

144. Acciarino, *Lettere*, 83; Andrés, *Dell'origine*, 1:429.

145. "Ingegno. Acutezza d'inventare e ghiribizzare, che che sia, senza maestro o avvertitore." Crusca, *Vocabolario* (1612), s.v. Ingegno, quoted in Marr et al., *Logodaedalus*, 55.

146. "Son pur Figure eleganti, & viuaci Argutie dell'ingegnosa *Natura*. Peroche, sicome le Argutie de' Poeti si chiaman *Fiori*: così i Fiori della Natura, si chiamano *Argutie*." Tesauro, *Il cannocchiale aristotelico*, 108, emphasis original.

147. "Con mille arguti, & ingegnosi concetti." Tesauro, *Il cannocchiale aristotelico*, 108.

148. "Il poeta, & l'arguto fauellatore." Tesauro, *Il cannocchiale aristotelico*, 91.

149. "Inquanto si adatta alla Materia da lui fabricata à principio, e disposita alla manutenzione dell'Vniverso . . . l'Intelletto Creante, & la Materia creata." Tesauro, *Il cannocchiale aristotelico*, 114–115. "Oltre ogni credenza ingegnosissime." Tesauro, *Il cannocchiale aristotelico*, 107–108. Tesauro expressed similar ideas in Tesauro, *Panegirici*, 1659, 2:107.

150. "Per mera pompa d'ingegno, di mostrarsi arguta & faceta." Tesauro, *Il cannocchiale aristotelico*, 108.

151. "Parendo che il Sol nascente, per far della terra vn Cielo, scuota le Stelle di Cielo in terra." Tesauro, *Il cannocchiale aristotelico*, 108.

152. "Argutamente inuero scherzò Mauritio Principe di Orange"; "comincerai da capo à scherzar co' translati, & argutezze"; "così puoi tu per otio, & per ischerzo, trouar forme più argute"; "in due maniere puoi tu Laconicamente scherzare." Tesauro, *Il cannocchiale aristotelico*, 128, 163–164, 526, 626. *Scherzi* could also refer to grotesques in painting. See Connelly, *Grotesque*, 38.

153. Tesauro, *Il cannocchiale aristotelico*, 551, 589.

154. Beckher and Radau, *Orator extemporaneus*, 37. For Martial on his epigrams as *lusus*, Laurens, *L'abeille*, 297.

155. In a poem by the ancient poet Claudian (ca. 370–ca. 404 CE) that was translated into the vernacular in 1589 and enjoyed considerable success in the seventeenth century, the god Jupiter looks amused at the ingenuity of the ancient mathematician and engineer Archimedes of Syracuse (ca. 287–ca. 212), who joked or played (*luditur*) as he replicated the

universe, Jupiter's work, at small scale and in self-moving form. Bernardino Baldi, *De gli automati*, fol. 7r. Automatons were jokes (*lusus*) played at nature's expense in a tradition with roots in ancient Greek and Roman culture. They engaged with nature's behavior and made a toy of it, just as conceits did with natural—that is, logical—reasoning.

156. In a letter mentioning Terzago and dated Milan, March 12, 1675, Settala notified Francesco Redi that Tesauro, who had been his professor, had died of apoplexy. Manfredo Settala to Francesco Redi, March 12, 1675, Redi Collection, Biblioteca Medicea Laurenziana, Ms. 211, 410.

157. Cassiano dal Pozzo, a friend of Tesauro, asked Vincenzo Leonardi to draw the piece for inclusion in his so-called paper museum (now Windsor Castle, Royal Collection 25481). See Freedberg, *Eye*, 32, 419; Rolfe, "Materia Medica." Amber inclusions of Baltic origins containing relatively large animals were mostly fakes. This is true of Settala's two frogs, but also of the piece with a frog and a lizard to which Daniel Hermann dedicated a poem in 1583. See Hermann, *De rana*. On amber inclusions, see King, "Collecting."

158. On 4.32, see Martial, *Epigrams*, 1:284–285. On 4.59, see Martial, *Epigrams*, 1:304–305. On 6.15, see Martial, *Epigrams*, 2:10–11. For a study of Martial's compositions on amber inclusions, see Rimell, *Martial's Rome*, 51–93. See also Laurens, *L'abeille*.

159. Cardano, *Hieronymi Cardani*, 132–133; Bacon, *Historia*, 408. The three epigrams by Martial often appeared side by side with early modern variations on them, as was the case in a manuscript from the first half of the sixteenth century curated by physician Gregor Duncker, who oversaw amber mining on behalf of Duke Albrecht of Prussia. In addition to a few passages dealing with the properties of amber, Duncker created a section containing the three epigrams by Martial accompanied by Latin poems on a caterpillar and a fly encased in amber, the work of the humanists Johannes Poliander (Johann Graumann) and Johann Apel. Gregor Duncker, ed., Prutenorum chronica, Herzog August Bibliothek, Ms. Cod. Guelf. 14.11 Aug. 4°, fols. 92r–95r. See also Polyakova, "Collecting," 66.

160. "Duas coaxantes importune ranas, succino Neptunus inuoluit, unde succino non leuis gratia conciliata." Terzago, *Museum septalianum*, 27. On amber gathering in Baltic shoals, see Tacitus, *Agricola*, 208–210. On amber inclusions as decorative objects in antiquity, see Watson, "Martial's Snake," 940.

161. Gracián, *Arte*, 174.

162. "Pezzo che in sè due ranocchie rinchiude, nella fatale lor mutolezza più che mai strepitose; e vna gocciola di acqua assai grossa, perche a quegli animali non manchi la sua palude." Scarabelli, *Museo*, 59.

163. "En vna gota de agua . . . se representa vna espaciosa laguna." Yrissarri, *Compendio*, 51.

164. Findlen, *Possessing*, 36.

165. Scarabelli, *Museo*, 59–60.

166. Tesauro, *Il cannocchiale aristotelico*, 605–629, 633. Present, post-Romantic sensibilities might find contradiction between a fertile and generative ingenuity and Tesauro's methodical art of variating conceits. Yet as Hannah Murphy explains in connection with early modern calligraphy, the idea of ingenuity was "connected to rather than opposed to the practice of disciplining." Murphy, "Calligraphy," 82.

167. "Fingi che in vna conuersation familiare sopraggiunga questa Nouella: *Vn Ape è morta dentro l'Ambra.* Vn dirà con Martiale: *Ell'hà saputo elegersi vna morte condegna.*

Vn'altro: *Neanche la Reina Cleopatra hebbe vn sì nobil sepolcro.* Vn'altro: *Dunque ell'è più pretiosa morta che viua.* E così più altri pensieri si possono andar cambiando, standosi sempre salda l'istessa Tema." Tesauro, *Il cannocchiale aristotelico,* 605, emphasis original.

168. A theme "non è che l'Argomento, ilquale ò altri à noi, ò noi à noi proponiamo, per fabricarui sopra gli scherzi arguti & concettosi." Tesauro, *Il cannocchiale aristotelico,* 605. "L'Altro Teorema non è men piaceuole, & vario, & profitteuole per isuegliar l'ingegno, & renderlo copioso di Concetti." Tesauro, *Il cannocchiale aristotelico,* 632, 636.

169. "Vn altro spiritoso esercitio . . . desta duo Intelletti arguti, à continuar sopra di alcuna Tema vn corso di Proposte, Risposte, Repliche, Picchi, e Ripicchi, per modo di vn Duello d'Ingegno." Tesauro, *Il cannocchiale aristotelico,* 637.

170. "S'Iddio stesso, il quale doueua muouere tutta l'humana generazione alle ammirazioni di se medesimo, *Ludens in orbe Terrarum,* fabricó l'Vniverso scherzando." Minozzi, *Sfogamenti,* 265, emphasis original. The passage is from Proverbs 8:31.

171. Findlen, "Jokes," 293.

Chapter 4 The Life of Conceits

1. *"Al Signor Cavalier Lorenzo Bernini.* Cedano, o buon Lorenzo, al tuo scalpello / Di Prometeo la face, / E'l teschio di Medusa. / La face in auuiuar corpi insensati, / E'l teschio in impetrir membri animati. / Poscia ch'ogni tua statua è si viuace, / Ed io resto sì stupito in mirarla, / Ch'ella par l'animato, io l'impetrito, / Ella rassembra il viuo, io lo scolpito." Stigliani, *Il canzoniero,* 445. For an analysis of Stigliani's poem, see Van Gastel, "Bernini's Metamorphosis," 198–199. Bernini was a client of Cardinal Scipione Borghese, the dedicatee of the 1623 edition of Stigliani's *Canzoniere.*

2. In 1585, Giovanni Paolo Lomazzo described Prometheus as "the first to shape images of men from earth, making them move with his art, as if they had spirit and life." Quoted in Marr, "Pregnant Wit," 8.

3. Quoted in Des Chene, "Forms," 139. The terms are taken from the course on Aristotelian physics prepared by Manuel de Góis (1543–1597) and published by Coimbra's Jesuit College in 1592. This and similar works issued in late sixteenth- and early seventeenth-century Coimbra served as textbooks in Jesuit schools and beyond. See Carvalho, "Manuel de Góis"; Carvalho, "Cursus Conimbricensis"; Casalini, *Aristotle in Coimbra.*

4. Baldi, *De gli automati,* fol. 3v.

5. "Non dipinta, ma viva." Vasari, *Delle vite,* 1:506, quoted in Jacobs, *Living Image,* 26–27. Vasari added wryly that others attributed Francia's death to poison.

6. Squire, "Making," 603–608. On "spirited" statues and early modern poetry, see Christian, "Poetry."

7. Fernández, *To Embody,* 3. See also Kimmel, "Automatons"; Riskin, *Restless Clock,* 22. Just as the sculptor employed a repertory of tricks to create the impression of life, Stigliani resorted to hyperbole and the symmetry of opposites, which, according to Emanuele Tesauro, had a numbing effect on the reader's intellect and "made a firefly pass for a star." Tesauro, *Il cannocchiale aristotelico,* 531. See also Gracián, *Arte,* 152, 221. On the hesitation between wonder and disbelief characteristic of conceits, see my analysis of Maciej K. Sarbiewski's *De acuto et arguto* (1626) in chapter 1. See also chapter 1 for the notion of logic as a counterpart to nature in Peregrini's *Delle acutezze* (1639).

8. For instance, "vero, e spirante / sù le tele animate," "l'imagine spirante," "in tela spirante," "dotto, animato lino." Bruni, *Le Veneri*, 198, 253, 299, 302. On Bernini's *Apollo and Daphne*, see Leporeo, *Villa Borghese*, fol. A11r; Ferrari, "Bernini."

9. "Della soavita, del solazzo, del vezzo, della persuasione, della fraude, dell'incantesimo dipinto." Bruni, *Le Veneri*, fol. †8v.

10. Minozzi, *Sfogamenti*, 117.

11. "Innocentissimi vivi fuochi di feste ed allegrezze notturne . . . animate Fiammelle." Minozzi, *Sfogamenti*, 116. The mower in Andrew Marvell's poem "The Mower and the Glowworm," mentioned in chapter 3, likewise called fireflies "living lamps." Marvell, *Complete Poems*, 109.

12. Minozzi, *Sfogamenti*, 254–255, 269. For conceits as "spirits," see Minozzi, *Sfogamenti*, 266.

13. "Vn non sò, che di vitale." Peregrini, *Delle acutezze*, 48. On "a certain je-ne-sais-quoi" more generally, see Scholar, *Je-Ne-Sais-Quoi*.

14. "Cuerpos vivos . . . con alma conceptuosa; . . . cadáveres que yazen en sepulcros de polvo y comidos de polilla." Gracián, *Arte*, 136–137.

15. Croce, *I trattatisti italiani*, 3. Croce was reacting against a more ambivalent assessment of Gracián's treatise by Marcelino Menéndez y Pelayo, who appreciated Gracián's "extremely original errors" even though he saw his work as a code of poetic intellectualism fit only for a generation that had lost its sanity. Croce mistranslated Gracián's expression "writing with soul" as "writing with the soul" or "with one's soul" ("con l'anima"), adopting a solution with Romantic reminiscences that suggests sincere or heartfelt writing. Gracián's concern lay not with subjectivity or affect but rather with the belief that one's writing should be endowed with life. As Sander De Boer clarifies for medieval and early modern Aristotelians, "saying that something has a soul is . . . equivalent to saying that it is alive." De Boer, "Imagination," 12. For Menéndez y Pelayo's assessment, see Menéndez y Pelayo, *Historia*, 2.2:535–541. See also Rabaté, "Materia."

16. Jacobs, *Living Image*, 2. See also Van Gastel, *Il Marmo Spirante*.

17. Athanasius Kircher distinguished between two kinds of objects in his museum: animals and works that "art makes appear to be alive." The divide between nature and art was, for him, one between life and lifelikeness. See Grafton, "Renaissance Histories," 189.

18. On Aldrovandi's argument and its indebtedness to Augustine, see chapter 3.

19. Carvalho, "Cursus Conimbricensis." On the debate over life's alleged uniqueness, see Lehoux, *Creatures*, 99–104.

20. Minozzi, *Sfogamenti*, 273–274.

21. See Manzini, *Lettera apologetica*, 42. On Manzini's *Delle meteore rettoriche*, see Pieri and Varini, *Il buratto*.

22. Marr, "Pregnant Wit"; Lewis, "Francis Bacon"; García Santo-Tomás, *Signos vitales*. On the etymology of "ingenuity" and its link to generation, see the introduction.

23. Huarte de San Juan, *Examen*, 186–194. On *conceptus* as "embryo," see Harvey, *Exercitationes*, 202; Pomata, "Innate Heat," 206.

24. Manzini, *Delle meteore rettoriche*, 115; Parker, "Concept."

25. On the role of heat in Aristotle's hierarchy of creatures, see Lehoux, *Creatures*, 19. On the female's inability to "concoct" the seed, see Aristotle, *Generation*, 65, §1.20, 728a18–21; Gelber, "Females," 173–174; Freudenthal, *Aristotle's Theory*, 24–25, 47–73. The notion that

women were colder than men permeated early modern medicine. See Huarte de San Juan, *Examen*, 627 et passim. Even women who wished to react against the misogyny explicit in similar views often took the fact for granted, only then to subvert its implications. This was the case of novelist María de Zayas in 1637. Zayas, *Novelas amorosas*, 159–161. In the second century C.E., the physician Galen hypothesized the existence of both a male and a female seed, against Aristotle's views. Pomata, "Innate Heat," 196–198.

26. Aristotle, *Generation*, 80–83, §2.2, 736a24–737b6. On the transfer of likeness in Hippocratic and Aristotelian thinking, see Lehoux, *Creatures*, 16–18; O'Connor, "Subjects"; Henry, *Aristotle on Matter*.

27. "A guisa di madre." Manzini, *Delle meteore rettoriche*, 115–116.

28. Aristotle, *Generation*, 141, §4.3, 767a36–b14; Freudenthal, *Aristotle's Theory*, 24–25.

29. "Ingenium illud magnum censendum est, quod ex se se fructificat ~ suosque habet foetus, ac partus." Baltasar López, Oratio pro instauratione studiorum . . . Anno 1639, Biblioteca Nacional de España, Ms. 8317, fol. 260r.

30. Baltasar López, Oratio pro instauratione studiorum . . . Anno 1631, Biblioteca Nacional de España, Ms. 8317, fol. 257r–v. On the soil's ingenuity, see Oosterhoff, "Ingenious Materials"; Marr et al., *Logodaedalus*, 159, 169.

31. "Masculi enim . . . est prae se se absque mutuato forinsecus . . . foetus mentis inceptaculo gignere, partus edere, . . . Contra uero inertis, atque torpentis ingenii est, ac nescio quid molle redolentis, atque foemineum aliena semper uirtute grauidari, ac sobolescere, nihilque nouum, ac rescens [*sic*] edere, nisi quod alterius industria sit partum, alterius negotio, atque labore repertum." Baltasar López, Oratio pro instauratione studiorum . . . Anno 1639, Biblioteca Nacional de España, Ms. 8317, fol. 261r–v.

32. Nikulin, "Memory."

33. "Suos habet foetus ac partus, quod foetificat ex sese insita uirtute, quod ingenita feracitate fructificat." Baltasar López, Oratio pro instauratione studiorum . . . Anno 1639, Biblioteca Nacional de España, Ms. 8317, fol. 261r.

34. Baltasar López, Oratio pro instauratione studiorum . . . Anno 1639, Biblioteca Nacional de España, Ms. 8317, fol. 261r. The contrast between parent and nurse predated Aristotle; it had crystallized in Aeschylus's tragedy *Eumenides* (fifth century BCE), where the god Apollo claimed that the mother was "not the parent, just a nurse to the seed" who, "like a stranger for a stranger, keeps / the shoot alive." Aeschylus, *Oresteia*, ll. 666–671.

35. "Ἥφαιστον . . . / ἐκ πάντων τέχνῃσι κεκασμένον Οὐρανιώνων." Hesiod, *Theogony*, 76–77; Detienne and Vernant, *Les ruses*, 17.

36. "In tutto è opera dell'intelletto, il quale, anzi creatore, che generatore, opera senza concorso, od aiuto presente della specie. E questi sono i concetti rettorici, e poetici, nella fantastica procreazion de' quali altro non contribuendo di presente la materia, che l'occasione, nascono spontanei dalla feracità di vn suolo, in cui la natiua fecondità germina, anche senza seme, e senza cultore." Manzini, *Delle meteore rettoriche*, 116. Natural philosopher Margaret Cavendish similarly argued that conception takes place when the intellect acts "freely and by rote," without aiming to imitate external objects. Otherwise, it should be called perception. Quoted in Walters, "Optics," 390.

37. "L'imitazione . . . presso di me, è vno de' più infelici, & inutili arredi, di che vn ingegno grande possa auualersi per far la maniera." Manzini, *Delle meteore rettoriche*, 49.

Manzini quoted from the late fifteenth-century exchange between Paolo Cortesi and Angelo Poliziano on parent-child resemblance as a metaphor for imitation. DellaNeva, *Ciceronian Controversies*, 8–9. On parent-child resemblance in early modern art, see Jacobs, *Living Image*, 11.

38. Della Porta, *Magiae natvralis*, 60–61; Muffet, *Insectorum*, 110. In 1618, Fortunio Liceti published a work entirely dedicated to spontaneous generation. Liceti, *De spontaneo viventium ortu*; Hirai, "Earth's Soul." For spontaneous generation in general, see Lehoux, *Creatures*; Fazzari, "Redi"; Harris, *Things*; Hirai, *Le concept*.

39. By the thirteenth century, propagation by spontaneous generation was enough to classify animals as "imperfect," in contrast with those born from parents. Albertus Magnus and especially Thomas Aquinas saw "fundamental differences between sexually generated animals (which were larger, more complex) and spontaneously generated animals (small, simple)." Lehoux, *Creatures*, 89. On the generation of bees and the recipe for its re-creation by humans, see Newman, *Promethean Ambitions*, 166–171; Lehoux, *Creatures*, 54–56.

40. On the notion of a species generating another one, see Toledo, *D. Francisci Toleti*, fols. 72v–73v. §2c4q11. See also Des Chene, *Life's Form*, 25; McLaughlin, "Spontaneous."

41. Lehoux, *Creatures*, 62–70.

42. Kircher's notion of *panspermia rerum* developed Augustine's views in detail. Kircher, *Mundus subterraneus*, 2:327; Leonhard, *Fertile Ground*, 33–35. Liceti's treatise on spontaneous generation also harmonized the views of Aristotle and Augustine. See Lehoux, *Creatures*, 98.

43. Karin Leonhard argues for the conceptualization of creativity as spontaneous generation in *sottobosco* paintings from the mid-seventeenth century. Leonhard, *Fertile Ground*. See also Jorink, "Snakes." Marisa Mandabach contends that Peter Paul Rubens, by representing creatures born from spontaneous generation, construes the task of the painter as "someone who transacts in nature's hidden creative energies." Mandabach, "Matter," 31. For the medieval period, see Steel, "Creeping Things." Intriguingly, Pamela Smith has argued that lizards and toads—creatures associated with spontaneous generation—topped the list of creatures most often favored for life casting. Smith, "Between Nature," 45.

44. Newman, *Promethean Ambitions*, 166–171 et passim.

45. Manzini, *Delle meteore rettoriche*, 52.

46. Manzini, *Delle meteore rettoriche*, 55. Early modern literature about artillery was adamant that fire always finds a way out, a notion that José de Ormaza connected to laconism in 1648. See Collado, *Platica Manval*, fol. 6r; Ormaza, *Censura*, 63.

47. On imagination and heat, see Huarte de San Juan, *Examen*, 340. Fantasy and imagination were often synonymous. See, for instance, Crusca, *Vocabolario* (1623), s.v. fantasia. At times, however, they designated two related but different faculties. See the introduction and Park, "Organic Soul," 470–471; Clark, *Vanities*, 39–77; Kemp, "From Mimesis"; Swan, "Counterfeit Chimeras"; Bundy, *Theory*.

48. In excess, however, heat burned the phantasms into fanciful creations, resembling the grotesques that decorated palaces and gardens, or the visions created by the vapors that usually rise from the stomach of somnolent or drunken individuals. Tesauro, *Il cannocchiale aristotelico*, 133–135; Mascardi, *Orationi*, fol. †5r.

49. Carvallo, *Cisne*, 96–98.

50. "θερμὸς ἀήρ." Aristotle, *Generation*, 80–83, §2.2–3, 736a1–737a1; 130, §3.11, 762a. Athanasius Kircher opposed vital heat to fire in *Mundus subterraneus*, 2:329. For Aristotle's views on the seed, see Meeusen, "Aristotle's Second Breath," 69; Lehoux, *Creatures*, 22–23; Freudenthal, *Aristotle's Theory*, 78. On vital or innate heat, see Moreau, "Innate Heat"; Pomata, "Innate Heat"; Freudenthal, *Aristotle's Theory*; Durling, "Innate Heat." On pneuma and spontaneous generation, see Wilson, "Heat."

51. Fruton, *Fermentation*, 3.

52. Tesauro, *Panegirici*, 1659, 2:264. On "the sun as a source of growth for all living things" but also minerals, see Smith, "Between Nature," 49.

53. Kircher, *Mundus subterraneus*, 2:256; Baldwin, "Alchemy," 47–49. On the debate over the ability of alchemy to imitate nature's processes and creations, see Newman, *Promethean Ambitions*.

54. Consider the title of Johann Ernst Burggrav's 1610 *Lampas vitae et mortis* (*Lamp of Life and Death*), reissued as *Biolychnium* in 1611. Keller, "Charlatan's Ingenuity."

55. Similar to vital heat, early modern scholars often "naturalized" Plato's notion of a poetic fury as fire. See Marr, *Rubens's Spirit*, 23, 76, 80–81. On the theory of humors and creativity, see Wittkower and Wittkower, *Born*; Klibansky et al., *Saturn*. In 1572, Alessandro Piccolomini called failed metaphors "cold"; Alonso López Pinciano used the same adjective in 1596 to describe plots that resort to deus ex machina. Piccolomini, *Piena, et larga parafrase . . . nel terzo libro*, 57–59; López Pinciano, *Philosophía antigua poética*, 213. "Coldness" ("τὸ ψυχρόν") was a common and familiar notion in Greco-Roman rhetorical treatises such as Pseudo-Longinus's *On the Sublime* and Pseudo-Demetrius's *On Style*. It applied to any joke that does not land, any trope that sounds off, and any plot that seems hard to believe.

56. "Questo, per essere opus vitale, è nobilissimo sopra tutti gli effetti dell'intelletto, sormontato in certa guisa ad vn non sò che di diuinità, si è vsurpato in proprietà il termine di concetto, che per sua natura generico, competteva tanto all'vna, quanto all'altra maniera di concepire." Manzini, *Delle meteore rettoriche*, 116–117.

57. Manzini, *Delle meteore rettoriche*, 117.

58. As explained in chapter 1, Peregrini also claimed that a conceit's use of figurative language made it artificial, as opposed to a syllogism's natural—meaning literal and unequivocal—argument. Peregrini, *Delle acutezze*, 42–43.

59. Manzini, *Delle meteore rettoriche*, 102.

60. Des Chene, *Life's Form*, 55–66.

61. "Senza inspirargli l'anima, che è il concetto, nel corpo, che è l'elocuzione, manifestamente ciechi, non vedono, che questa scrittura, senz'anima, e senza spirito, non goderà pur della vita, non che di quella immortalità, che suol esser la sola obiettiva cagione per la quale si scriue." Manzini, *Delle meteore rettoriche*, 117.

62. See Des Chene, *Spirits*; on the mechanistic philosophy and automatons, see Riskin, *Restless Clock*; Werrett, "Wonders."

63. "Spirito vitale delle morte pagine . . . Per miracolo di lei [i.e., argutezza], le cose Mutole parlano: le insensate viuono, le morte risorgono: le Tombe, i Marmi, le Statue; da questa incantatrice degli animi, ricevendo voce, spirito, e movimento; con gli Huomini ingegnosi, ingegnosamente discorrono. Insomma, tanto solamente è morto, quanto dall'Argutezza non è auuiuato." Tesauro, *Il cannocchiale aristotelico*, 21–22.

64. Cassiodorus, *Selected Letters*, 246; Schiefsky, "Art," 87.

65. Tesauro, *Il cannocchiale aristotelico*, 119; Doglio, "Introduzione," 27.

66. Famiano Strada, De poetica. Dictabat P. Famianus Strada in classe rethoricae anno 1608, APUG Collection, Archivio della Pontificia Università Gregoriana, Ms. 1188, fol. 2r.

67. Des Chene, *Life's Form*, 139. In 1618, Liceti distinguished Creation (divine and ex nihilo) from generation. Liceti, *De spontaneo viventium ortu*, 154; Lehoux, *Creatures*, 94.

68. "Tutto il fattibile." Manzini, *Delle meteore rettoriche*, 52.

69. "Che nulla ardisca, o tenti." Manzini, *Delle meteore rettoriche*, 50. Manzini argued that we "insult the nobility of our spirit, created free, when we place it on a leash" ("facciamo torto alla generosità del nostro spirito, creato libero, col metterlo alla catena," 53–54). Other references to magnanimity (*generosità*), appear on 94–95, 99–101. On generosity in connection with melancholy and creativity ("melancholia generosa"), see Klibansky et al., *Saturn*, 241–274.

70. Manzini, *Delle meteore rettoriche*, 132.

71. Minozzi, *Sfogamenti*, 273–274. The fact that Peregrini was from Bologna is likely to have made his dubious and ambivalent attitude toward conceits even worthier of blame in Minozzi's eyes.

72. Minozzi, *Sfogamenti*, 211–218. On sericulture as "cross-species symbiosis" between silkworms and humans, see Lincoln, *Brilliant Discourse*, 165.

73. Minozzi, *Sfogamenti*, 218–219.

74. Marsh, *Unravelled Dreams*, 22–33.

75. Malpighi, *Dissertatio epistolica*, 1. For Pliny's claim, see chapter 3.

76. Park and Daston, "Introduction," 1–2; Marsh, *Unravelled Dreams*, 34. Medici Grand-Duke Francesco I, Stradanus's patron, cultivated silkworms himself. Eamon, "Science as a Hunt," 396–397. On *Nova reperta*, see Markey, *Renaissance Invention*.

77. Vergil, *On Discovery*; Piccolomini, *Piena, et larga parafrase . . . nel terzo libro*, 80–82. On Piccolomini and metaphors, see chapter 1.

78. "Per vn piede impiccato." Minozzi, *Sfogamenti*, 219. On Menzani, see Belfanti, "Guilds," 581. On painting as punishment, see Edgerton, *Pictures*; Ortalli, *La pittura infamante*. On secrecy and invention, see Long, *Openness*.

79. "Hò letto, che in Danzica Città della Prussia vedeasi vn'Artifizio ingegnosissimo, con cui quanti Telai altri hauesse voluto in vna stanza faceansi lauorare ogni tela, ed ogni drappo da loro stessi, accommodati, e temperati per hore ventiquattro, appunto come gli ordegni degli Horologi. Mà perciocchè tanti huomini bisognosi, che con l'arte del tessere si viueano, sarebbon morti di fame, dal Magistrato della Città fu prohibita quella inuenzione, e l'Autore segretamente fatto affogare." Minozzi, *Sfogamenti*, 219–220. According to Pliny the Elder, a similar end awaited the inventor of a form of foldable and unbreakable glass (*uitrum flexile*) who was beheaded on the orders of Roman emperor Tiberius. Pliny the Elder, *Natural History*, 10:154–155, §36. 66.

80. Lancellotti, *L'Hoggidì*, 457. On socially and politically disruptive technology, see Reeves, *Galileo's Glassworks*, 29.

81. New inventions and technologies altered the conditions also of intellectual and creative labor. The printing press reshaped what counted as valuable work from a preacher who composed his sermon. In 1648, the Jesuit José de Ormaza complained that his contemporaries did not understand that widespread availability of printed anthologies of

patristic sources and commentaries had rendered erudition redundant: why read through countless volumes when the harvest one would reap after years of study could be bought in the nearest shop? Ormaza, *Censura*, 97–98.

82. Jesuit professor Francesco Lana Terzi contended that "the worms prove they are expert weavers by forming their tombs from silk, that is, from their entrails" ("i vermi, formando di seta, cioè delle sue viscere, il suo sepolcro, si dichiarano periti nell'arte del tessere"). Lana Terzi, *Prodromo*, 60.

83. Fernández de Ribera, *Los anteojos*, 39.

84. Petrarca, *Letters*, sec. 1.8; Seneca, *Epistles*, 2:276–285, §84. For the posterity of Seneca's metaphor of the bee, see Pigman, "Versions."

85. Castelvetro, *Poetica*, fols. 36v–37r.

86. See Courcelle, *Connais-toi toi-même*.

87. Bacon, *Works*, 4:92–93. The distinction between "reasoners" and "empirics" (or "men of experiment") dates back to antiquity. It was developed at length, for instance, by physician C. Cornelius Celsus (fl. first century CE). See Celsus, *On Medicine*, 1:6–41, *Prooemium* 12–75.

88. On the quarrel of the ancients and the moderns, see Fumaroli, "Les abeilles"; Salvatore, *Antichi*. Jonathan Swift opposed moderns spiders to ancients bees in 1704. Swift, *Tale*, 249.

89. See Tarquinio Galluzzi, De multiplici doctrinarum studio contrahendo . . . , APUG Collection, Archivio della Pontificia Università Gregoriana, Ms. 1153, fols. 38r–43v. Philo of Alexandria (fl. first century CE) connected the spider with a remarkable but potentially dubious independence. Dickerman, "Some Stock Illustrations," 124.

90. On dependence, see Lopes da Veiga, *El perfeto señor*, 24–25.

91. Ovid, *Metamorphoses*, 1:288–289, §6.7–8 and §6.10–13.

92. Biagioli, *Galileo, Courtier*, 264–265. Clizia Carminati has shown that distilling quintessences was a metaphor for the laconic style with conceits. Carminati, "Alcune considerazioni," 93–94. In 1622 and from a less favorable stance Mascardi had criticized writers who spent day and night distilling the matter of their brains hoping for a subtle and delicate conceit. Mascardi, *Orationi*, 106.

93. *Inkstand with a Madman Distilling His Brains* (ca. 1600, Italy), Metropolitan Museum of Art, New York, accession number 04.9.15, Rogers Fund, 1904, https://www.metmuseum.org/art/collection/search/188899.

94. Gramigna, *Fantasie varie*, 117–147. Interestingly, Gramigna also advocated imitating silkworms rather than bees. Yet he apparently believed that silkworms were, like bees, agents of "composite imitation" (191–193). I am grateful to Michele Lodone for the reference.

95. "Distillate le quintessenze dell'intelletto." It was a hint to the laconic style cultivated by Virgilio Malvezzi. Here Minozzi linked the cult of smallness or brevity in style with the result of heat applied to distillation. Minozzi, *Sfogamenti*, 260.

96. "Altresì suda bene, chi compon bene, e distillandosi l'intelletto in ingegnosi ritrouamenti, s'indebolisce, si stanca, e si consuma." Minozzi, *Sfogamenti*, 252. Minozzi echoed a formulation found in Francesco Berni's *Orlando innamorato*, published posthumously in 1542: "Let everyone turn to the hoe [i.e. in reference to the tool's sharpness] of their ingenuity, / and study, and tire themselves, and wear themselves thin" ("Hor dell'ingegno ogn'un la zappa pigli, / Et studi, & s'affatichi, & s'assottigli"). Berni, *Orlando innamorato*, fol. 97r.

It is hardly accidental that, when Filippo Buonanni quoted Berni's lines of verse in 1681, he replaced "study" (*studi*) with "sweat" (*sudi*). Buonanni, *Ricreatione*, 19.

97. López, *Oratio*, fol. 3r.

98. Bacon, *Works*, 3:285–286.

99. Diogenes Laertius, *Lives*, 2:264–265, §7.2.

100. Bourgogne, *Mundi lapis lydius*, 2–5, translated in Garrod, "Subtilis," 141. Old Testament precedents for the same idea include the claim in Isaiah 59:6 that "their [i.e., those of spiders] webs shall not become garments." Michalsky and Michalsky, *Spider*, 52.

101. Garrod, "Subtilis," 143.

102. "Faticano in lauorare concetti, che il più delle volte riescono sconciature, o sconcerti; fatture di vetro lauorate alla punta d'vna lucerna, che solo toccate, per non dir vedute, si spezzano: e pure quanto più fragili tanto più belle, *imò quibus pretium faciat ipsa fragilitas*." Bartoli, *Dell'huomo*, 403–404, emphasis original.

103. "Murrina . . . , quibus pretium faceret ipsa fragilitas. hoc argumentum opum, haec vera luxuriae gloria existimata est, habere quod posset statim perire totum." Pliny the Elder, *Natural History*, 9:4–7, §33.1. See Tressaud and Vickers, "Ancient Murrhine Ware"; Beretta, *Alchemy*.

104. Just as the luxury of glass stood in contrast to its humble components ("gravel, sand, and a few plants, albeit sometimes exotic ones"), the value of a conceit was exclusively a function of its ability to connect terms, however high or low, in new and unexpected ways. See Maitte, "Façon," 210–211.

105. The adjectives "subtle," "small," and "fragile," found together as early as 1550 in Girolamo Cardano's *De subtilitate*, informed Bartoli's indictment of conceits. "Opera parua, subtilia, fragilia." Cardano, *Hieronymi Cardani*, fol. 268r–v.

106. On honor as "a fine glass" that once broken cannot be repaired, see Ângela de Azevedo's play. Azevedo, *El muerto disimulado*, 212–213, ll. 1973–1975.

107. Gracián did so in *El héroe* (1637), *El discreto* (1646), and *Oráculo manual* (1647). On Gracián's social capital or *caudal*, see Sánchez, "Symbolic Wealth."

108. "Huir los empeños." Gracián, *Oráculo manual*, 129, 224. Mario Biagioli famously argued that "courtly science" in the age of Galileo depended on a similar mechanism: engaging in scholarly disputes could result in earning or losing capital—either promotion or losing face. Biagioli, *Galileo, Courtier*.

109. Gracián, *Arte*, 354–355.

110. "Ánima de sutileza." Gracián, *Arte*, 180.

111. "Ipsa recedente, omnia diffluunt & corrumpuntur." Toledo, *D. Francisci Toleti*, col. 40rA, 2cIqI. For the meanings of soul in Aristotle's *De anima*, see Aristotle, *On the Soul*, 66–73, §2.1, 412a–413a11.

112. "All'hora maggiormente lusingan gli spettatori, che con vna sfuggita di piè minacciano di cadere." Mascardi, *Orationi*, 324.

113. "Lo star sempre col piè pendente, toglie il piacere, e partorisce l'horrore." Mascardi, *Orationi*, 324.

114. López Pinciano, *Philosophía antigua poética*, 518. As discussed in chapter 1, in 1641 António Lopes da Veiga criticized the poets of his time for preferring the possible over the verisimilar. See Lopes da Veiga, *Heráclito*, 176.

115. Wallace-Hadrill, "Pliny the Elder"; Lao, "Luxury."

116. See Jerome's first epistle to Paulinus of Nola. Jerome, *Select Library*, 6:99, §53.7.

117. "Ita enim ex ore quasi ex reserata apotheca . . . multos, ac varios lemniscos evomit, vt modo purpureum, modo viridem, deinde caeruleum, luteumque." López, *Oratio*, fol. 8r. For Seneca's passage, see Seneca, *Epistles*, 1:294–295, §45.8–9.

118. "La versabilità, velocemente raffronta tutte queste Circonstanze infra loro, ò col Suggetto: le annoda ò diuide; le cresce ò minuisce . . . acenna l'vna per l'altra . . . come i Giocolieri i lor calcoli. E questa è la Metafora, Madre delle Poesie, de' Simboli, & delle Imprese. Et quegli è più ingegnoso, che può conoscere & accopiar circonstanze più lontane." Tesauro, *Il cannocchiale aristotelico*, 119. On Tesauro's notion of versability, see chapter 2.

119. Marco Arnaudo has identified Tesauro's *giocoliere* with the ancient *calcularii* or *acetabularii* who played cup-and-ball routines at the fair. Arnaudo, "Sul significato," 8. See also Zanardi, "Metafora," 40.

120. Ingenuity-related language often designated both manual and intellectual skills in romance languages. See, for instance, Pons d'Icard, *Libro*, fols. 263r–266v. See also the introduction.

121. Keller, "Charlatan's Ingenuity," 181.

122. "Cum agilitate pedum vel manuum." Del Río, *Disquisitionum Magicarum*, 1:43. For Barwich's claim, see Barwich, "Sleight." On *praestigia*, see Clark, *Vanities*, 78–122; Arnaudo, "Sul significato," 7. On jugglers and charlatans, see Keller, "Charlatan's Ingenuity."

123. Bacon, *Works*, 4:270; Clark, *Vanities*, 79. See also Rossi, *Francis Bacon*.

124. "Naturas intimas rerum callebat." Del Río, *Disquisitionum Magicarum*, 1:18. For a distinction between a true and a false magic, intended only for the sake of spectacle ("verâ vel circulatoriâ"), see Del Río, *Disquisitionum Magicarum*, 1:112.

125. Del Río, *Disquisitionum Magicarum*, 1:18–19. Zoroaster was hailed as "the inventor of magic" in one of Petrarch's sonnets, quoted twice in Crusca, *Vocabolario* (1612), s.v. inventore, magico. On Zoroaster, see Copenhaver, *Magic*, 25–34; De Jong, "Zoroaster."

126. "Seminari d'incanti . . . un Zoroastro novello componitore di spiritose magie." Minozzi, *Sfogamenti*, 117–118. See Tesauro, *Il cannocchiale aristotelico*, 21–22.

127. Sumrall, "Natural Magic," 3.

128. "Exactior quaedam arcanorum naturae cognitio, qua caelorum ac siderum cursu & influxu, & sympathijs atque antipathijs rerum singularum obseruatis, suo tempore, loco ac modo res rebus applicantur, & mirifica quaedam hoc pacto perficiuntur, quae causarum ignaris praestiogosa vel miraculosa videntur." Del Río, *Disquisitionum Magicarum*, 1:20.

129. On magic in the history of early modern science, see Copenhaver, *Magic*, 3–24.

130. On occult causes or properties, see, for instance, Scholar, *Je-Ne-Sais-Quoi*, 82–128; Daston and Park, *Wonders*, 127; Waddell, *Jesuit Science*; Eamon, "Science as a Hunt," 399.

131. Sumrall, "Natural Magic," 2–3.

132. Copenhaver, *Magic*; Sumrall, "Natural Magic," 4–5.

133. "Pars practica scientiae naturali." Quoted in Sumrall, "Natural Magic," 4–5. For natural magic as the counterpart to natural philosophy, see Bruno, *Opere magiche*, 160–161.

134. "Plurimae compos potestatis, abditis scatens mysterijs, delitescentium rerum qualitates, proprietates, ac totius Naturae cognitionem elargiens, docensque ex rerum consensu, dissensuque ita dissociare, vel mutua, & opportuna applicatione coniungere, vt ea

opera efficiat, quae vulgus vocat miracula, omnem admirationem, humanumue captum excedentia." Della Porta, *Io. Bapt.*, 2. I quote from the 1589 edition, the earliest to contain twenty instead of four books.

135. Daston, "Nature of Nature," 157; Sumrall, "Natural Magic," 1.

136. Sumrall, "Natural Magic," 1. On Francis Bacon and natural magic, see Weeks, "Francis Bacon," 117. In 1646, Kircher distinguished magic that merely seeks the knowledge of causes from magic that resorts to the knowledge of causes to create new and unheard-of effects. Kircher, *Ars magna*, 769–770; Waddell, *Jesuit Science*, 100. On ink made of fireflies, see Della Porta, *Magiae natvralis*, 60–61; Muffet, *Insectorum*, 110.

137. Nieremberg claimed that the magician mixed a set of causes that obtained effects rarely seen, doing on purpose what nature did fortuitously in order to bring about wondrous results. Nieremberg, *Oculta filosofia*, fol. 202r. See also Marcaida, "Juan Eusebio Nieremberg," 73. Jesuit natural philosopher Francesco Lana Terzi described natural magic in similar terms as late as 1670. See Lana Terzi, *Prodromo*, 59–60. On the relations between natural magic, occult causes, and the study of nature, see Eamon, *Science and the Secrets*; Waddell, *Jesuit Science*.

138. In fact, Della Porta had devoted much attention to tricks that were mostly intended as a form of courtly spectacle. See Sumrall, "Natural Magic," 6.

139. Del Río, *Disquisitionum Magicarum*, 1:12. Also, "Illa [i.e., operatrix magia] miros effectos producit." Del Río, *Disquisitionum Magicarum*, 1:21–22. For Minozzi's claim, see Minozzi, *Sfogamenti*, 117.

140. "Vna molta vicendeuole acconcezza tra le parti." Peregrini, *Delle acutezze*, 33.

141. "Alguna conformidad, o proporción." Gracián, *Arte*, 146–148. See also Gracián, *Arte*, 143.

142. Peregrini, *Delle acutezze*, 28.

143. Minozzi called conceits "spirits" on a number of occasions. Minozzi, *Sfogamenti*, 116, 193, 262, 266, 269.

144. "Spirits" was also the second name in the title of Peregrini's *Delle acutezze, che spiriti, vivezze, e concetti, volgarmente si appellano*.

145. Minozzi was responding to a remark by Vincenzo Renieri, included in Peregrini's *Delle acutezze*, in which Renieri claimed that the inventors of conceits, who self-styled themselves as spirited (*spiritosi*), would be more aptly called *spiritati*, that is, frantic or possessed by the devil. A precedent for the pun was already in Mascardi's *Dell'arte historica* (1636). Minozzi, *Sfogamenti*, 262; Peregrini, *Delle acutezze*, fol. a6v; Mascardi, *Dell'arte historica*, 659. On "spirit," see Bono, "Medical Spirits"; Hirai, "Spirit."

146. "Gli Spiriti mi cauan l'anima, se però non debbo dire, che mi dan l'anima." Minozzi, *Sfogamenti*, 262.

147. Minozzi, *Sfogamenti*, 118. On *seminarium*, see Hirai, "Bodies," 405.

148. Alchemists considered matter as "spirited" or "ensouled." Rampling, "Spirited Matter," 20. Francis Bacon distinguished spirits, or pneumatic matter, from tangible matter. Rusu, "Deceiving," 26.

149. Park, "Organic Soul," 469. On different kinds of spirit with different functions, see Bono, "Medical Spirits."

150. Ovid, *Metamorphoses*, 2:80–85, §10.243–297. Vasari, *Delle vite*, 3:488. See also Weststeijn, "'Painting's Enchanting Poison,'" 141.

151. Vasari, *Delle vite*, 3:488.

152. In the mid-sixteenth century, poetic theorist Jacques Peletier du Mans argued that invention is like blood that becomes diffused through the poem and deserves to be called its life or soul. Peletier du Mans, *L'art poëtique*, 19. Considering that early modern physicians believed that spirits traveled with the blood, Peletier's metaphor is consonant with Vasari's description of Bronzino's *Pygmalion* and Pontormo's *Raising of Lazarus*.

153. Hirai, "Bodies," 404–405. See also Clericuzio, "Internal Laboratory," 53.

154. "Elemento terrae vero & genuino." Kircher, *Mundus subterraneus*, vol. 2, fol. (*)4r.

155. Kircher, *Mundus subterraneus*, 2:328. The experiments that Kircher directed at the Jesuit Collegio Romano demonstrate that he had both a theoretical and a practical interest in alchemy, or what he called "spagyrica ars." Kircher, *Mundus subterraneus*, 2:329. On the practice of alchemy in the Society of Jesus, see Baldwin, "Alchemy."

156. "Omnia foecundat, vivificat, nutrit, conservat." Kircher, *Mundus subterraneus*, 2:331.

157. "Semen sal animatum dici possit." Kircher, *Mundus subterraneus*, 2:332.

158. Kircher, *Mundus subterraneus*, 2:331. On salt, see Roos, *Salt*. On Kircher, matter, and life, see Chang, "Alchemy."

159. Kircher, *Mundus subterraneus*, 2:331.

160. Manzini, *Delle meteore rettoriche*, 55, 116–117.

161. "In fimetis, stabulis, columbarii, latrinis, oletisque." Kircher, *Mundus subterraneus*, 2:331–332.

162. Giovanni Maria Lancisi, Io. Ma. Lancisij Studia Iuuenilia Inchoata ab Anno 1672, Lancisi Collection, Biblioteca Lancisiana, Ms. 197 LXXVI.1.6, fols. 115r–155r.

163. Willis, *Diatribae Duae Medico-Philosophicae*, 1.

164. Kepler, *Harmony*, 247, 250–251; Kepler, *Ioannis Keppleri Harmonices Mundi*, 80, 83.

165. Huarte de San Juan, *Examen*, 345.

166. Rocamora y Torrano, *Sphera*, fol. 4r.

167. Sarbiewski, *Wykłady*, 10.

168. Freudenthal, *Aristotle's Theory*, 74–105.

169. Freudenthal, *Aristotle's Theory*, 12–19.

170. "Causa generationis & corruptionis rerum omnium." Kircher, *Mundus subterraneus*, 2:396; Baldwin, "Alchemy," 53; Chang, "Alchemy," 326.

171. "Communem quandam rerum caussam seu potius occasionem." Gómez Miedes, *Comentarios*, 1:26–27.

172. "Qua in re diuinam quandam uim ille exercet atque animi uice fungitur, dum ueluti spiritus quidam corpori immersus, quicquid occupauerit, cogit subsistere." Gómez Miedes, *Comentarios*, 1:138–139.

173. Gómez Miedes, *Comentarios*, 2:694–895.

174. Minozzi, *Sfogamenti*, 221–222. See chapter 1.

175. "Reliquie del Sole già tramontato; Occhi dell'Estate, che con belle vicende di luce, e d'ombre hor s'aprono, ed hor si chiudono; Arghi piccioli fiammeggianti, che fan Lincea la Talpa ombrosissima della Notte; Stellette volatili della Terra, Piropi alati, Baleni erranti, animate Fiammelle, Atomi sfauillanti, ed infine innocentissimi viui Fuochi di feste, ed allegrezze notturne." Minozzi, *Sfogamenti*, 115–116.

176. On nature and emblematic thought, see Ashworth, "Natural History."

177. "Il moto con nuoua Filosofia è cagione di luce, non di calore." Minozzi, *Sfogamenti*, 116.

178. Minozzi, *Sfogamenti*, 216. On Aldrovandi, see chapter 3.

179. Similar to Mascardi, Minozzi assumed that the light of fireflies flickered only as they flew, which follows a claim advanced by Pliny and later rejected by Aldrovandi that the insects were always lit up and that the wings simply obscured the light while they were not flying. Pliny the Elder, *Natural History*, 3:492–493, §11.34; Aldrovandi, *De animalibus insectis*, 4. For Mascardi's echo of it, see chapter 3 and also Mascardi, *Orationi*, 330.

180. See chapter 1 for Aristotle's understanding of the object of *scientia* as that which happens "always of for the most part."

181. "Mirasi in esse [i.e., le lucciole] con istrana generazione la luce nata dall'ombre, il moto partorito dalla quiete, ed il fuoco da gli humori notturni." Minozzi, *Sfogamenti*, 116.

182. On early modern reactions to ambiguity, see Ossa-Richardson, *History*, 4, et passim.

183. See note 38, and also chapter 3, for spontaneous generation and for Scaliger's argument that fireflies reproduced sexually, exceptional until the late seventeenth century.

184. Karin Leonhard has likened the darkness of the soil in which spontaneous generation takes place, which Minozzi called "shadow," to the painting's surface as a site of fertile creativity. Leonhard, *Fertile Ground*, 37–40.

185. Des Chene, *Life's Form*, 23. On experience, see Dear, "Meanings." See also Gregoriâc and Grgiâc, "Aristotle's Notion"; Lehoux, *Creatures*, 32–53.

186. Antonio Telesio was the father of the natural philosopher Bernardino Telesio. His poem "Tantula nocte uolans uolucris micat aere tantum" was quoted by Aldrovandi (1602), Muffet (posthumously, 1634), and Kircher (1646). See Aldrovandi, *De animalibus insectis*, 498; Muffet, *Insectorum*, 111; Kircher, *Ars magna*, 23.

187. Daston, "Cold Light."

188. Tesauro, *Il cannocchiale aristotelico*, 585.

189. For the attention to new and rare particulars in early modern natural philosophy, see chapter 1.

190. Daston and Park, *Wonders*, 109–134; Waddell, *Jesuit Science*.

191. See chapter 3.

192. Stigliani, *Il canzoniero*, 266–267.

193. Tassoni, *Varietà*, 10. The substances listed by Tassoni appeared with countless others in Thomas Bartholin's work on bioluminescence *De luce animalium* (*On the Light of Animals*, 1647). See Bartholin, *Thomae Bartolini*. On plant bioluminescence, see Gessner, *De raris*.

194. "Mi fà ridere lo Stigliani, quando le chiama à fargli lume, ed à supplire à gli occhi d'vna sua Gatta, Lucerna antica (com'egli dice) del suo Studio; ed appunto par, che voglia vender' altrui le Lucciole per Lanterne, mentre vuole, che gli seruano per Lumiere." Minozzi, *Sfogamenti*, 116–117. Stigliani borrowed the mention of cats' eyes as lamps from Torquato Tasso. See Tasso, *Poesie*, 881–882. I thank Eileen Reeves for the reference.

195. Stigliani must have composed the madrigals, which were absent from the 1605 edition of *Il Canzoniero*, around the years when he was writing *Nuovo Mondo*.

196. Fernández de Oviedo y Valdés, *La Historia General*, fol. CXVIr–v, §15.8.

197. Quoted in Thompson, "Monsters," 103–106.

198. Marvell, *Complete Poems*, 109, ll. 9–16. Cardano mentioned reading by the light of fireflies in suspicious proximity to a description of *cocuyos*. On *cocuyos*, see Cardano, *Hieronymi Cardani*, fol. 189r–v; Nieremberg, *Prolusion*, fols. 11v–12r; Muffet, *Insectorum*, 111–113.

199. Minozzi, *Sfogamenti*, 115.

200. Des Chene, *Physiologia*, 168–211.

201. Beckher and Radau, *Orator extemporaneus*, 53.

202. Aldrovandi, *De animalibus insectis*, 497.

203. Aldrovandi, *De animalibus insectis*, 494.

204. "Infirma est aetas cùm nescia rerum." Aldrovandi, *De animalibus insectis*, 498. See also Kircher, *Ars magna*, 23.

205. Kircher, *Ars magna*, 23.

206. Kircher, *Ars magna*, 22. See also Harvey, *History*, 103–104.

207. "Tum autem vel maximè, cum plures cincindelae, siue lampades simul ponerentur, quasi ambitiosa lucis gloria gestiens, superba lucentis humoris ornamenta vel maximè exercebat; diceres ipsa, vt aspiceretur ambire." Kircher, *Ars magna*, 22.

208. Edwards, "Digressing," 137.

209. On early modern accounts of the light of fireflies, see Wolfe, "Illumination."

210. Pimentel, *Rhinoceros*. On imagination in scientific discovery, see Daston, "Fear." On visualizing images and drawing them as knowledge making, see Bredekamp, *Galileo's Thinking Hand*; Lipking, *What Galileo Saw*; Rocke, *Image*. On invention and science, see Garrod, *Cosmographical Novelties*; Moss, *Novelties*.

211. "Es menester fingir primero mil rayas en el aire y componer muchos modelos y últimamente poner las manos." Huarte de San Juan, *Examen*, 191.

212. Gracián, *Arte*, 283–287.

213. Garrod, "Subtilis," 140.

Epilogue

1. "Diceres profecto quot tubos opticos, totidem belli tormenta in coelum esse directa, quae orbium illorum firmitatem, & inuicta hactenus mundi moenia non parum concusserint . . . At quid de cometis dicas? quorum vnum nuper ipsi vidimus magno impetu in solida coeli Peripatetici munimenta arietantem." Malapert, *Caroli Malaperti*, 35; Guerrini, "Tradizione astronomica," 231. For the same idea, see Nieremberg, *Curiosa filosofia*, fol. 182r.

2. On the solid spheres, see Rosen, "Dissolution."

Bibliography

Archives

APUG Collection. Archivio della Pontificia Università Gregoriana, Rome.
Barberiniani Latini (Barb. Lat.). Biblioteca Apostolica Vaticana, Rome.
Biblioteca Marucelliana, Florence.
Biblioteca Nacional de España, Madrid.
Biblioteca Vallicelliana, Rome.
Czartoryski Library, Krakow.
Galileo Collection (Gal.). Biblioteca Nazionale Centrale di Firenze, Florence.
Herzog August Bibliothek, Wolfenbüttel.
Hispanic Society of America, New York.
Jesuit Collection (Ges.). Biblioteca Nazionale Centrale di Roma, Rome.
Lancisi Collection. Biblioteca Lancisiana, Rome.
Metropolitan Museum of Art, New York.
Moreni Collection. Biblioteca Moreniana, Florence.
Museo Galileo, Florence.
Redi Collection. Biblioteca Medicea Laurenziana, Florence.
Urbinates Latini (Urb. Lat.). Biblioteca Apostolica Vaticana, Rome.

Primary Sources

Acciarino, Damiano, ed. *Lettere sulle grottesche (1580–1581)*. Rome: Aracne, 2018.

Accolti, Pietro. *Lo inganno de gl'occhi, prospettiva pratica*. Florence: Pietro Cecconcelli, 1625.

Aelian. *Historical Miscellany*. Translated by Nigel G. Wilson. Cambridge, MA: Harvard University Press, 1997.

Aeschylus. *The Oresteia*. Translated by Robert Fagles. New York: Viking Press, 1975.

Álamos de Barrientos, Baltasar. *Tácito español, ilustrado con aforismos*. Madrid: Luis Sánchez, 1614.

Aldrovandi, Ulisse. *De animalibus insectis libri septem, cum singulorum iconibus ad vivum expressis*. Bologna: Giovanni Battista Bellagamba, 1602.

Alemán, Mateo. *Guzmán de Alfarache*. Edited by José María Micó. 2 vols. Madrid: Cátedra, 1987.

Ammirato, Scipione. *Discorsi del signor Scipione Ammirato sopra Cornelio Tacito*. Florence: Filippo Giunti, 1594.

Andrés, Joan. *Carta del Abate D. Juan Andres al señor Comendador Frey Cayetano Valenti Gonzaga....* Madrid: Antonio de Sancha, 1780.

———. *Dell'origine, progressi e stato attuale d'ogni letteratura....* 7 vols. Parma: Stamperia Reale, 1782–1799.

———. *Lettera dell'abbate D. Giovanni Andres al Sig. Comendatore Fra Gaetano Valenti Gonzaga....* Cremona: Lorenzo Manini, 1776.

Anonymous. *The Nevv Testament of Iesus Christ, Translated Faithfully into English....* Rheims: John Fogny, 1582.

Aquinas, Thomas. *Summa Contra Gentiles*. Translated by Vernon J. Bourke. 4 vols. Garden City, NY: Doubleday, 1955–1957.

———. *Summae de Veritate Catholicae Fidei Contra Gentiles*. Edited by Pietro Antonio Uccelli. Rome: Typographia Polyglotta, 1878.

Aristotle. *Art of Rhetoric*. Translated by John Henry Freese. Cambridge, MA: Harvard University Press, 1926.

———. *Categories. On Interpretation. Prior Analytics*. Translated by Harold P. Cooke and Hugh Tredennick. Cambridge, MA: Harvard University Press, 1938.

———. *Generation of Animals and History of Animals I, Parts of Animals I*. Translated by C. D. C. Reeve. Indianapolis: Hackett, 2019.

———. *Metaphysics*. Translated by Hugh Tredennick. 2 vols. Cambridge, MA: Harvard University Press, 1933.

———. *Meteorologica*. Translated by H. D. P. Lee. Cambridge, MA: Harvard University Press, 1952.

———. *Minor Works*. Translated by W. S. Hett. Cambridge, MA: Harvard University Press, 1936.

———. *On the Soul. Parva Naturalia. On Breath*. Translated by W. S. Hett. Cambridge, MA: Harvard University Press, 1957.

———. *Physics*. Translated by Philip H. Wicksteed and Francis M. Cornford. 2 vols. Cambridge, MA: Harvard University Press, 1957.

———. *Poetics. Editio Maior of the Greek Text with Historical Introductions and Philological Commentaries*. Edited by Leonardo Taran and Dimitri Gutas. Leiden: Brill, 2012.

Aristotle, Longinus, and Demetrius. *Aristotle. The Poetics. "Longinus": On the Sublime. Demetrius. On Style*. Translated by Stephen Halliwell, W. F. Fyfe, and W. Rhys Roberts. Cambridge, MA: Harvard University Press, 1995.

Artusi, Giovanni Maria. *L'arte del contraponto....* Venice: Giacomo Vincenti, 1598.

Augustine of Hippo, Saint. *The City of God*. Translated by Henry Bettenson. London: Penguin, 1972.

———. *Corpus Scriptorum Ecclesiasticorum Latinorum*. Edited by Joseph Zycha. Vol. 25:6:1. Vienna: F. Tempsky, 1891.

———. *Corpus Scriptorum Ecclesiasticorum Latinorum*. Edited by Emanuel Hoffmann. Vol. 40:5:2. Vienna: F. Tempsky, 1900.

———. *On Genesis. Two Books on Genesis Against the Manichees and On the Literal Interpretation of Genesis: An Unfinished Book*. Translated by Roland J. Teske. Washington, DC: Catholic University of America, 1991.

———. *Sancti Augustini Confessionum libri XIII*. . . . Edited by Martin Skutella and Luc Verheijen. Turnhout: Brepols, 1981.

Azevedo, Ângela de. *El muerto disimulado / Presumed Dead*. Edited by Valerie Hegstrom. Translated by Catherine Larson. Liverpool: Liverpool University Press, 2018.

Bacon, Francis. *Francisci Baronis de Verulamio, Vice-Comitis Sancti-Albani, Historia vitae & mortis*. . . . London: John Haviland, 1623.

———. *The Historie of Life and Death: With Observations Naturall and Experimentall for the Prolonging of Life*. . . . London: I. Okes, 1638.

———. *The Works of Francis Bacon*. . . . Edited by James Spedding, Robert Leslie Ellis, and Douglas Denon Heath. 7 vols. London: Longman, 1857–1859.

Bade, Josse, ed. *M. Fabij Quintiliani oratoriarum institutionum Lib. XII. Una cum annotationibus Raphaelis Regii*. . . . Paris: Josse Bade and Jean Petit, 1516.

Baldi, Bernardino. *Di Herone Alessandrino, De gli automati, ouero, Machine se mouenti libri due*. . . . Venice: Girolamo Porro, 1589.

Bartholin, Thomas. *Thomae Bartolini Casp. Filii De luce animalium libri III*. Leiden: Frans Hacke, 1647.

Bartoli, Daniello. *Dell'huomo di lettere difeso et emendato: parti due*. Rome: Heirs of Francesco Corbelletti, 1645.

Beckher, Georg, and Michael Radau. *Georgi Beckheri Elbingensis Orator extemporaneus*. . . . Amsterdam: Louis Elzevir, 1650.

Berni, Francesco. *Orlando innamorato*. . . . Milan: Andrea Calvo, 1542.

Bettinelli, Saverio. *Dell'entusiasmo delle belle arti*. Milan: Giuseppe Galeazzi, 1769.

Biringuccio, Vannoccio. *De la pirotechnia: libri .x. dove ampiamente si tratta non solo di ogni sorte & diuersita di miniere*. . . . Venice: Venturino Roffinello, 1540.

Bocarro Francês, Manuel. *Tratado dos cometas que apareceram em Novembro passado de 1618*. Lisbon: Pedro Craesbeeck, 1619.

Boccalini, Traiano. *De' ragguagli di Parnaso di Traiano Boccalini Romano: Centuria prima*. Venice: Pietro Farri, 1612.

Borel, Pierre. *De vero telescopii inventore: cum brevi omnium conspiciliorum historia*. The Hague: Adrian Vlaq, 1655.

Botero, Giovanni. *Della ragion di stato libri dieci*. Venice: Giolito, 1589.

Bourgogne, Antoine de. *Mundi lapis lydius siue vanitas per veritatem falsi accusata et conuicta*. Antwerp: Widows of Joannes Cnobbarius, 1639.

Brahe, Tycho. *Tychonis Brahe Dani De mundi aetherei recentioribus phaenomenis: Liber secundus*. . . . Hven/Uraniborg: Christophorus Weida, 1588.

Bruni, Antonio. *Le Veneri*. . . . Rome: Giacomo Mascardi, 1633.

Bruno, Giordano. *Opere magiche*. Edited by Michele Ciliberto. Milan: Adelphi, 2000.

Buonanni, Filippo. *Ricreatione dell'occhio e della mente. Nell'osseruation' delle Chiocciole, proposta a' curiosi delle Opere della Natura*. Rome: Il Varese, 1681.

Burchiello. *Rime del Burchiello fiorentino comentate dal Doni. Et piene di capricci, fantasie, umori, strauaganze, grilli, frenesie, ghiribizzi, argutie, motti e sali*. . . . Edited by Anton Francesco Doni. Vicenza: Heirs of Perin, 1597.

Cadalso, José de. *Los eruditos a la violeta, ó curso completo de todas las ciencias*. . . . Madrid: Antonio de Sancha, 1772.

Camillo Delminio, Giulio. *Due trattati dell'eccellentissimo M. Iulio Camillo. . . .* Venice: Giovanni Farri, 1544.

Caravita, Giovanni. *Arte di ben concettare. . . .* 2 vols. Palermo: Heirs of Pietro dell'Isola, 1690.

Cardano, Girolamo. *De rerum varietate libri XVII.* Basel: Heinrich Petri, 1557.

———. *Hieronymi Cardani Medici Mediolanensis De Svbtilitate Libri XXI.* Nuremberg: Johannes Petreius, 1550.

Carducho, Vicente. *Dialogos de la pintura: Su defensa, origen, essencia, definicion, modos, y diferencias.* Madrid: Francisco Martínez, 1633.

Carvallo, Luis Alfonso de. *Cisne de Apolo.* Edited by Alberto Porqueras Mayo. Kassel: Edition Reichenberger, 1997.

Cassiodorus, Senator. *The Selected Letters of Cassiodorus: A Sixth-Century Sourcebook.* Oakland: University of California Press, 2020.

Castelvetro, Lodovico. *Poetica d'Aristotele, vulgarizzata et sposta.* Vienna: Gaspar Stainhofer, 1570.

Castiglione, Baldassare. *Il Libro del Cortegiano.* Edited by Amadeo Quondam and Nicola Longo. Milan: Garzanti, 1981.

Caussin, Nicolas. *Eloquentiae sacrae et humanae parallela libri XVI.* Paris: Sébastien Chappelet, 1619.

Cavendish, Margaret. *The Blazing World and Other Writings.* Edited by Kate Lilley. London: Penguin, 2004.

Celsus, Aulus Cornelius. *On Medicine.* Translated by Walter George Spencer. 2 vols. Cambridge, MA: Harvard University Press, 2014.

Cervantes Saavedra, Miguel de. *Don Quijote de la Mancha: edición del Instituto Cervantes (1605, 1615, 2015).* Edited by Francisco Rico et al. Madrid: Real Academia Española, 2015.

———. *Los trabajos de Persiles y Sigismunda.* Edited by Laura Fernández, Ignacio García Aguilar, Isabel Lozano-Renieblas, and Carlos Romero Muñoz. Madrid: Real Academia Española, 2017.

Cicero, Marcus Tullius. *Letters to Atticus.* Edited and translated by D. R. Shackleton Bailey. 4 vols. Cambridge, MA: Harvard University Press, 1999

———. *Letters to Friends.* Edited and translated by D. R. Shackleton Bailey. 3 vols. Cambridge, MA: Harvard University Press, 2001.

Claudian. *Claudian.* Translated by Maurice Platnauer. Cambridge, MA: Harvard University Press, 1922.

Clavius, Christoph. *Euclidis Elementorum libri XV. Accessit XVI. De solidorum regularium comparatione. . . .* Rome: Vincenzo Accolti, 1574.

Collado, Luis. *Platica Manval de Artilleria en la qual se tracta de la Excelencia de el Arte militar, y Origen de ella. . . .* Milan: Paolo Gottardo Ponzio, 1592.

Copernicus, Nicolaus. *De revolutionibus orbium coelestium.* Nuremberg: Johann Petreius, 1543.

Cortese, Giulio. *Prose.* Edited by Maurizio Slawinski. Turin: Res, 2000.

Covarrubias Orozco, Sebastián de. *Tesoro de la lengua castellana o española.* Madrid: Luis Sánchez, 1611.

Crusca, Accademia della. *Vocabolario degli Accademici della Crusca. . . .* Venice: Giovanni Alberti, 1612.

———. *Vocabolario degli Accademici della Crusca. . . .* Venice: Jacopo Sarzina, 1623.

Dalle Pezze, Francesca, ed. *Diálogos en que se contienen varias materias y se explican algunas obras de D. Luis de Góngora.* Verona: Fiorini, 2007.

Dávila y Heredia, Andrés. *Tienda de antojos politicos.* Valencia: Gerónimo de Villagrassa, 1673.

DellaNeva, Joan, ed. *Ciceronian Controversies.* Cambridge, MA: Harvard University Press, 2007.

Della Porta, Giambattista. *Io. Bapt. Portae Neapolitani Magiae Naturalis Libri XX. . . .* Naples: Orazio Salviani, 1589.

———. *Magiae natvralis, sive De miracvlis rervm libri 4. . . .* Naples: Mattia Cancer, 1558.

Del Monte, Guidobaldo. *Guidubaldi . . . Perspectivae libri sex.* Pesaro: Girolamo Concordia, 1600.

———. *Mechanicorum liber.* Pesaro: Girolamo Concordia, 1577.

Del Río, Martín. *Disquisitionum Magicarum libri sex, in tres tomos partiti.* 3 vols. Louvain: Gerardus Rivius, 1599–1600.

De Nores, Giason. "Discorso intorno a que' principii, cause et accrescimenti. . . ." In *Trattati di poetica e retorica,* edited by Bernard Weinberg, 3:373–419. Bari: Laterza, 1970–1974.

———. *Poetica. . . .* Padua: Paolo Meietti, 1588.

Descartes, René. *Oeuvres de René Descartes.* Edited by Charles Adam and Paul Tannéry. 12 vols. Paris: Léopold Cerf, 1897–1913.

De Sepi, Giorgio. *Romani collegii societatis Jesu museum celeberrimum. . . .* Amsterdam: Joannes Janssonius a Waesberge, 1678.

Diccionario de la lengua castellana. Madrid: Francisco del Hierro, 1726–1739.

Diogenes Laertius. *Lives of Eminent Philosophers.* Translated by R. D. Hicks. 2 vols. Cambridge, MA: Harvard University Press, 1925.

Drake, Stillman, and C. D. O'Malley, eds. *The Controversy on the Comets of 1618: Galileo Galilei, Horatio Grassi, Mario Guiducci, Johann Kepler.* Philadelphia: University of Pennsylvania Press, 1960.

Enríquez de Zúñiga, Juan. *Amor con vista: lleva una summaria descripcion del mundo, ansi de la parte Elemental, como de la Aetherea.* Madrid: Juan Delgado, 1625.

Erasmus of Rotterdam, Desiderius. *Erasmi Roterodami Adagiorum Chiliades Tres, Ac Centuriae fere totidem.* Venice: Aldo Manuzio, 1508.

Escardó, Juan Bautista. *Rhetorica christiana o idea de los que dessean predicar con espiritu, y fruto de las almas. . . .* Mallorca: Heirs of Gabriel Gualp, 1647.

Estienne, Robert. *Dictionarivm, sev Latinae linguae thesaurus.* Paris: Robert Estienne, 1543.

Fernández de Córdoba, Francisco. *Examen del Antídoto o Apología por las Soledades de don Luis de Góngora contra el autor del Antídoto.* Edited by Matteo Mancinelli. 2020. https://obvil.sorbonne-universite.fr/corpus/gongora/1617_examen.

Fernández de Oviedo y Valdés, Gonzalo. *La Historia General de Las Indias.* Seville: Juan Cronberger, 1535.

Fernández de Ribera, Rodrigo. *Los anteojos de mejor vista. El mesón del mundo.* Edited by Víctor Infantes de Miguel. Madrid: Legasa Literaria, 1979.

Ferrari, Giovanni Battista. *Io. Bapt. Ferrarii . . . Orationes.* Lyons: Heirs of G. Rouille, 1625.
———. *Io. Bapt. Ferrarii Senensis e Societate Iesu De florum cultura libri IV.* Rome: Stefano Paolini, 1633.
———. *Io. Bapt. Ferrarii Senensis e Societate Iesu Orationes.* Venice: Paolo Baglioni, 1644.
———. *Io. Bapt. Ferrarii Senensis e Soc. Iesu Orationes.* Rome: Francesco Corbelletti, 1627.
———. *Nomenclator Syriacus.* Rome: Stefano Paolini, 1622.
Figliucci, Flaminio. *Stanze sopra le stelle, e macchie solari scoperte col nuouo occhiale, con vna breue dichiaratione del sig. Lorenzo Salui. . . .* Rome: Giacomo Mascardi, 1615.
Fracastoro, Girolamo. *De sympathia et antipathia rerum liber unus. De contagione et contagiosis morbis et curatione libri iii.* Venice: Heirs of Lucantonio Giunti, 1546.
———. *Scritti inediti.* Edited by Francesco Pellegrini. Verona: Valdonega, 1955.
Galilei, Galileo. *Il saggiatore.* Edited by Ottavio Besomi and Mario Helbing. Rome: Antenore, 2005.
———. *Il saggiatore nel quale con bilancia esquisita e giusta siponderano le cose contenute nella Libra astronomica e filosofica.* Rome: Giacomo Mascardi, 1623.
———. *Le opere di Galileo Galilei.* Edited by Antonio Favaro. 20 vols. Florence: G. Barbèra, 1890–1909.
Garzoni, Tommaso. *La piazza universale di tutte le professioni del mondo, e nobili et ignobili.* Venice: Giovan Battista Somascho, 1585.
Gessner, Conrad. *De raris et admirandis herbis.* Zürich: Andrea and Hans Jakob Gesner, 1555.
Ghilini, Girolamo. *Teatro d'huomini letterati. . . .* 2 vols. Venice: Li Guerigli, 1633.
Glorioso, Giovanni Camillo. *De cometis dissertatio astronomico-physica publice habita in gymnasio patavino anno Domini MCDXIX.* Venice: Typographia Varisciana, 1624.
Góis, Manuel de. *Commentarij Collegij Conimbricensis Societatis Iesu in Libros Meteororum Aristotelis.* Lisbon: S. Lopes, 1593.
Gómez Miedes, Bernardino. *Comentarios sobre la sal.* Edited by Sandra Inés Ramos Maldonado. 3 vols. Alcañiz: Laberinto, 2003.
Góngora, Luis de. *Soledades.* Edited by Robert Jammes. Madrid: Castalia, 1994.
González de Salas, Jusepe Antonio. *Nueva idea de la tragedia antigua.* 2 vols. Kassel: Reichenberger, 2003.
Gracián, Baltasar. *Agudeza y arte de ingenio.* Edited by Jorge M. Ayala, Ceferino Peralta, and José María Andreu. 2 vols. Zaragoza: Prensas Universitarias de Zaragoza in collaboration with Huesca: Instituto de Estudios Altoaragoneses, 2004.
———. *Arte de ingenio, Tratado de la Agudeza.* Edited by Emilio Blanco. Madrid: Cátedra, 2010.
———. *El Criticón.* Edited by Luis Sánchez Laílla and José Enrique Laplana Gil. 2 vols. Zaragoza: Institución Fernando el Católico, 2016.
———. *El Discreto.* Edited by Aurora Egido. Madrid: Alianza editorial, 1997.
———. *Obras completas.* Edited by Santos Alonso. Madrid: Cátedra, 2011.
———. *Oráculo manual y arte de prudencia.* Edited by Emilio Blanco. Madrid: Cátedra, 2018.
Gramigna, Vincenzo. *Fantasie varie. . . .* Rome: Camera Apostolica, 1628.
Grassi, Orazio. *De tribus cometis anni M.DC.XVIII. Disputatio astronomica publice habita in Collegio Romano Societatis Iesu ab vno ex patribus eiusdem Societatis.* Rome: Giacomo Mascardi, 1619.

———. *Libra astronomica ac philosophica. . . .* Perugia: Marco Naccarini, 1619.

Guiducci, Mario, and Galileo Galilei. *Discorso delle comete*. Edited by Ottavio Besomi and Mario Helbing. Rome: Antenore, 2002.

Gutiérrez de Godoy, Juan. *Disputationes phylosophicæ ac medicæ super libros Aristotelis de Memoria et Reminiscentia. . . .* Jaén: Pedro de la Cuesta, 1629.

Halley, Edmund. "De Iride, Sive de Arcu Caelesti, Dissertatio Geometrica." *Philosophical Transactions* 22, no. 267 (1700): 714–725.

Harvey, William. *Exercitationes de generatione animalium*. Amsterdam: Ludovicus Elzevier, 1651.

Hermann, Daniel. *De rana et lacerta succino prussiaco insitis. Danielis Hermanni Prussi. Discursus Philosophicus. . . .* Cracow, 1583.

Hermogenes. *On Types of Style*. Edited by Cecil W. Wooten. Chapel Hill: University of North Carolina Press, 1987.

Herrera, Juan de. *Institución de la Academia Real Mathemática en Castellano que la Magestad del Rey Don Phelippe II N. S. mandó fundar en su corte. . . .* Madrid: Guillermo Droy, 1584.

Herrera y Tordesillas, Antonio de. *Discursos morales, políticos é históricos inéditos de don Antonio de Herrera, Cronista del rey don Felipe II*. 2 vols. Madrid: Imprenta de Ruiz, 1804.

Hesiod. *Theogony. Works and Days. Testimonia*. Translated by Glenn W. Most. Cambridge, MA: Harvard University Press, 2018.

Hodierna, Giovanni Battista. *Opvscoli del dottor Don Gio. Battista Hodierna. . . .* Palermo: Decio Cirillo, 1644.

———. *Protei Caelestis Vertigines. . . .* Palermo: Nicolò Bua, 1657.

Horace. *Satires. Epistles. The Art of Poetry*. Edited by H. Rushton Fairclough. Cambridge, MA: Harvard University Press, 1926.

Huarte de San Juan, Juan. *Examen de ingenios para las ciencias*. Edited by Guillermo Serés. Madrid: Cátedra, 1989.

Íñiguez de Lequerica, Juan, ed. *Sermones funerales, en las honras del Rey nuestro Señor don Felipe II*. Madrid: Heirs of Juan Íñiguez de Lequerica, 1599.

Jerome. *Select Letters*. Translated by F. A. Wright. Cambridge, MA: Harvard University Press, 1933.

———. *A Select Library of Nicene and Post-Nicene Fathers of the Christian Church. Second Series*. Edited by Henry Wace and Philip Schaff. Translated by W. H. Fremantle, G. Lewis, and W. G. Martley. 14 vols. New York: Charles Scribner's Sons, 1912.

Johnson, Samuel. *The Lives of the Most Eminent English Poets. . . .* 4 vols. London: C. Bathurst et al., 1781.

Kepler, Johannes. *The Harmony of the World by Johannes Kepler: Translated into English with an Introduction and Notes*. Translated by Eric J. Aiton, A. M. Duncan, and Judith Veronica Field. Philadelphia: American Philosophical Society, 1997.

———. *Ioannis Keppleri Harmonices Mundi Libri V*. Linz: Johannes Planck, 1619.

Kircher, Athanasius. *Ars magna lucis et umbrae in decem libros digesta. . . .* Rome: Ludovico Grignani, 1646.

———. *Mundus Subterraneus. . . .* 2 vols. Amsterdam: Joannes Janssonius a Waesberge and Elizaeus Weyerstraet, 1665.

———. *Musurgia universalis, sive: Ars magna consoni et dissoni in X libros digesta*. 2 vols. Rome: Heirs of Francesco Corbelletti, 1650.

Kirwitzer, Václav Pantaleon. *Observationes cometarum Anni 1616* [*sic*]. *In India Orientali factae. . . .* Oberursel: Johann Theobald Schönwetter, 1620.

Lampillas, Francisco Xavier. *Saggio storico-apologetico della letteratura spagnola. . . .* 6 vols. Genoa: Felice Repetto, 1778–1781.

Lana Terzi, Francesco. *Prodromo all'arte maestra*. Edited by Andrea Battistini. Milan: Longanesi, 1977.

Lancellotti, Secondo. *L'Hoggidì overo gl'ingegni non inferiori a' passati . . . Parte seconda. . . .* Venice: Li Guerigli, 1636.

Langbein, Lebrecht Gotthelf. *Commentatio de M. C. Sarbievii S. J. Poloni vita, studiis et scriptis*. Dresden: Georg Conrad Walther, 1753.

Leonardo de Argensola, Bartolomé. *Rimas*. Edited by José Manuel Blecua. Madrid: Espasa-Calpe, 1974.

Leonardo de Argensola, Lupercio. *Obras sueltas de Lupercio y Bartolomé Leonardo de Argensola*. Edited by Conde de la Viñaza. 2 vols. Madrid: Imprenta y Fundición de M. Tello, 1889.

Leporeo, Lodovico. *Villa Borghese di Lodouico Leporeo. . . .* Rome: Stamparìa della R. Camera Apostòlica, 1628.

Liceti, Fortunio. *De spontaneo viventium ortu: libri IV*. Vicenza: Domenico Amadei, 1618.

Lipsius, Justus, ed. C. *Cornelii Taciti Historiarum et Annalium libri qui exstant, Iusti Lipsii studio emendati & illustrati*. Antwerp: Christoph Plantin, 1574.

———. *Politicorum Sive Civilis Doctrinae Libri Sex, Qui Ad Principatum Maxime Spectant*. Leiden: Franciscus Raphaelengius, 1589.

Locher, Johann Georg, and Christoph Scheiner. *Disquisitiones mathematicae, de controversijs et nouitatibus astronomicis, quas sub praesidio Christophori Scheiner . . . publice disputandas posuit*. Ingolstadt: Elisabeth Angermaier, 1614.

Lopes da Veiga, António. *El perfeto señor. Sueño politico. Con otros varios discursos, i ultimas poesias varias. . . .* Madrid: Imprenta Real, 1652.

———. *Heraclito i Democrito de nuestro siglo. Descrivese su legitimo filosofo. Dialogos morales. . . .* Madrid: Diego Díaz de la Carrera, 1641.

López, Baltasar. *Oratio pro Instauratione Studiorum*. Mexico City: Widow of Bernardo Calderón, 1644.

López Pinciano, Alonso. *Philosophía antigua poética*. Edited by José Rico Verdú. Madrid: Fundación José Antonio de Castro, 1998.

Loyola, Ignacio de. *Exercitia Spiritualia*. Roma: Antonio Blado, 1548.

Lucian. *The Downward Journey or The Tyrant. Zeus Catechized. Zeus Rants. The Dream or The Cock. Prometheus. Icaromenippus or The Sky-Man. Timon or The Misanthrope. Charon or The Inspectors. Philosophies for Sale*. Cambridge, MA: Harvard University Press, 1915.

———. *How to Write History. The Dipsads. Saturnalia. . . .* Translated by K. Kilburn. Cambridge, MA: Harvard University Press, 1959.

Luzán, Ignacio de. *La Poetica, o Reglas de la poesia en general, y de sus principales especies. . . .* Zaragoza: Francisco Revilla, 1737.

Maelcote, Odo van. "Nuntius Sidereus Collegii Romani." In *Le opere di Galileo Galilei*, edited by Antonio Favaro, 3:291–298. Florence: G. Barbèra, 1890–1909.

Maestlin, Michael. *De Astronomiae hypothesibus sive de Circulis Sphaericis et Orbibus Theoricis, disputatio ad discutiendum proposita....* Heidelberg: Jacob Mylius, 1582.

Malapert, Charles. *Caroli Malaperti montensis e Societate Iesv Oratio habita Duaci dum lectionem mathematicam auspicaretur: in qva de nouis belgici telescopÿ phaenomenis non iniucunda quaedam academice disputantur.* Douai: Baltazar Beller, 1620.

Malpighi, Marcello. *Dissertatio epistolica de bombyce....* London: John Martyn and Jacob Allestry, 1669.

Malvezzi, Virgilio. *Discorsi sopra Cornelio Tacito.* Venice: Marco Ginami, 1622.

Manetti, Antonio. *Vita di Filippo Brunelleschi preceduta da La novella del Grasso.* Edited by Domenico de Robertis. Milan: Il Polifilo, 1976.

Manzini, Giovanni Battista. *Delle meteore rettoriche, Proginnasmi....* Bologna: Giacomo Monti, 1652.

———. *Lettera apologetica scritta dal Sig. Manzini Al Sig. Abb. N.* Bologna: Nicolò Tebaldini, 1637.

Marino, Giovan Battista. *Adone.* Edited by Emilio Russo. 2 vols. Milan: BUR, 2013.

———. *Epistolario: seguito da lettere di altri scrittori del seicento.* Edited by Angelo Borzelli and Fausto Nicolini. Bari: Giuseppe Laterza, 1912.

Martial, M. Valerius. *Epigrams.* Translated by D. R. Shackleton Bailey. 3 vols. Cambridge, MA: Harvard University Press, 1993.

Martini, Francesco di Giorgio. *Trattati di architettura ingegneria e arte militare.* Edited by Corrado Maltese. 2 vols. Milan: Il Polifilo, 1967.

Marvell, Andrew. *The Complete Poems.* Edited by Elizabeth Story Donno. London: Penguin, 1985.

Mascardi, Agostino. *Dell'arte historica d'Agostino Mascardi trattati cinque.* Rome: Giacomo Facciotti, 1636.

———. *Orationi di Agostino Mascardi....* Genoa: Giuseppe Pavoni, 1622.

———. *Prose vulgari....* Venice: Bartolomeo Fontana, 1625.

Mazzoni, Jacopo. *Della difesa della Comedia di Dante....* Cesena: Bartolommeo Raveri, 1587.

Mexía, Pedro. *Discurso sobre los dos cometas que se vieron por el mes de nouiembre del año passado de 1618.* Lisbon: Pedro Craesbeeck, 1619.

Minozzi, Pier Francesco. *La Bibliotheca Medicea, ingrandita ed illustrata dal Gran Duca ... Cosimo Terzo.* Lyons: Mathieu Libéral, 1673.

———. *Sfogamenti d'ingegno....* Venice: Turrini, 1641.

Minsheu, John. *Hēgemōn Eis Tas Glōssas = Id est, Ductor in Linguas, the Gvide into Tongves....* London: John Browne, 1617.

Monantheuil, Henri de. *Aristotelis Mechanica Graeca, emendata, Latina facta, & commentariis illustrata.* Paris: Jérémie Perier, 1599.

Monforte y Herrera (pseud.), Fernando. *Relación de las fiestas que ha hecho el colegio imperial de la compañía de Iesús de Madrid en la canonización de san Ignacio de Loyola, y S. Francisco Xavier.* Madrid: Luis Sánchez, 1622.

Muffet, Thomas. *Insectorum sive minimorum animalium theatrum....* London: Thomas Cotes, 1634.

Mulerius, Nicolaus. *Institutionum astronomicarum libri duo....* Groningen: J. Sassius, 1616.

Muñoz, Jerónimo. *Libro del nueuo cometa, y del lugar donde se hazen....* Valencia: Pedro de Huete, 1573.

Nájera, António de. *Discursos astrologicos sobre o Cometa que apareceo em 25 de nouembro de 618.* Lisbon: Pedro Craesbeeck, 1619.

Niceron, Jean-François. *La perspective curieuse, ou, Magie artificielle des effets merueilleux....* Paris: Pierre Billaine, 1638.

Nieremberg, Juan Eusebio. *Curiosa filosofia y tesoro de maravillas de la naturaleza examinadas en varias questiones naturales....* Madrid: Imprenta del Reyno, 1630.

———. *Historia naturae, maxime peregrinae....* Antwerp: Balthasar Moretus, 1635.

———. *Oculta filosofia de la sympatia, y antipatia de las cosas, artificio de la naturaleza, y noticia natural del mundo....* Madrid: Imprenta del Reyno, 1633.

———. *Practica del catecismo romano y doctrina christiana.* Madrid: Diego Díaz de la Carrera, 1640.

———. *Prolusion a la doctrina y historia natural....* Madrid: Andrés de Parra, 1629.

Niseno, Diego. *El politico del cielo, primera parte.* Barcelona: Sebastián de Cormellas, 1637.

Ormaza, José de. *Censura de la eloquencia para calificar sus obras y señaladamente las del pulpito.* Zaragoza: Hospital Real y General de Nuestra Señora de Gracia, 1648.

Ovid. *Metamorphoses.* Translated by Frank Justus Miller and G. P. Goold. 2 vols. Cambridge, MA: Harvard University Press, 1984.

Páez de Castro, Juan. *Memoria a Felipe II sobre la utilidad de juntar una buena biblioteca. Descubierta en la Real Librería de San Lorenzo por Blas Antonio Nasarre.* [Valladolid]: Consejería de Educación y Cultura, 2003.

Partenio, Bernardino. *Della imitatione poetica....* Venice: Gabriel Giolito de' Ferrari, 1560.

Pascal, Blaise. *Oeuvres complètes.* Edited by Michel Le Guern. 2 vols. Paris: Gallimard, 2000.

Patrizi, Francesco. *Della poetica.* Edited by Danilo Aguzzi Barbagli. 3 vols. Florence: Istituto di Studi sul Rinascimento, 1969.

Peletier du Mans, Jacques. *L'art poëtique.* Lyons: Jacques de Tournes & Guillaume Gazeau, 1555.

Peregrini, Matteo. *Che al savio è convenevole il corteggiare libri IIII.* Bologna: Nicolò Tebaldini, 1624.

———. *Delle acutezze, che altrimenti spiriti, viuezze, e concetti, volgarmente si appellano.* Genoa: Giovanni Maria Farroni, Nicolò Pesagni, and Pier Francesco Barbieri, 1639.

———. *I fonti dell'ingegno ridotti ad arte.* Bologna: Carlo Zenero, 1650.

Petau, Denis. *Dionysii Petavii Aurelianensis e Societ. Jesu Epistolarum libri tres.* Paris: Sébastien and Gabriel Cramoisy, 1652.

Petrarca, Francesco. *Letters on Familiar Matters: I–VIII.* Translated by Aldo S. Bernardo. New York: Italica Press, 2005.

Piccolomini, Alessandro. *Piena, et larga parafrase; di M. Alessandro Piccolomini; nel secondo libro della Retorica d'Aristotele a Theodette.* Venice: Giovanni Francesco Camotio, 1569.

———. *Piena, et larga parafrase, di M. Alessandro Piccolomini, nel terzo libro della Retorica d'Aristotele, a Theodette.* Venice: Giovanni Varisco, 1572.

Pico della Mirandola, Giovanni. *Oration on the Dignity of Man. A New Translation and Commentary.* Edited by Francesco Borghesi, Michael Papio, and Massimo Riva. New York: Cambridge University Press, 2012.

Plato. *Theaetetus. Sophist.* Translated by Harold North Fowler. Cambridge, MA: Harvard University Press, 1921.

Pliny the Elder. *Natural History.* Edited by Harris Rackham. 10 vols. Cambridge, MA: Harvard University Press, 1938–1963.

Plutarch. *Lives. Theseus and Romulus. Lycurgus and Numa. Solon and Publicola.* Translated by Bernadotte Perrin. 11 vols. Cambridge, MA: Harvard University Press, 1914–1926.

———. *Moralia. On the Malice of Herodotus. Causes of Natural Phenomena.* Translated by Lionel Pearson and F. H. Sandbach. Vol. 11, 16 vols. Cambridge, MA: Harvard University Press, 1927–2004.

———. *Moralia. Table-Talk, Books 1–6.* Translated by P. A. Clement and H. B. Hoffleit. Vol. 8, 16 vols. Cambridge, MA: Harvard University Press, 1927–2004.

Pons d'Icard, Lluis. *Libro de las grandezas y cosas memorables de la metropolitana, insigne y famosa ciudad de Tarragona. . . .* Lleida: Pedro de Robles and Juan de Villanueva, 1572.

Puteanus, Erycius. *Eryci Puteani De Laconismo Syntagma. Adjuncti Thyrsi Philotesii, Styli & Sermonis Aculei.* Leuven: Gerardus Rivius, 1609.

Quevedo, Francisco de. *La hora de todos y la fortuna con seso.* Edited by Pierre Geneste, Pierre Dupont, and Jean Bourg. Madrid: Cátedra, 1987.

Quintilian, Marcus Fabius. *The Orator's Education.* Translated by Donald Russell. 5 vols. Cambridge, MA: Harvard University Press, 2001.

Rapin, René. *Reflexions sur la Poétique d'Aristote, et sur les ouvrages des poètes anciens & modernes.* Paris: F. Muguet, 1674.

———. *Reflexions sur l'usage de l'éloquence de ce temps.* Paris: C. Barbin and F. Muguet, 1671.

Raulino, Francesco. *Pompa dell'eseqvie celebrate al serenissimo Odoardo dvca di Piacenza. . . .* Piacenza: Antonio Ardizzone, 1647.

Regio, Raffaele. *Quintilianus cum commento.* Venice: Boneto Locatelli, 1493.

Ripa, Cesare. *Iconologia . . . nella quale si descrivono diverse imagini.* Siena: Heirs of Matteo Florimi, 1613.

Robortello, Francesco. *Francisci Robortelli Utinensis in librum Aristotelis de arte poetica explicationes.* Florence: Lorenzo Torrentino, 1548.

Rocamora y Torrano, Ginés. *Sphera del vniuerso. . . .* Madrid: Juan de Herrera, 1599.

Rodríguez Mohedano, Rafael, and Pedro Rodríguez Mohedano. *Historia Literaria de España. Origen, progresos, decadencia y restauración de la Literatura Española. . . .* 9 vols. Madrid: Joaquín Ibarra, 1766–1791.

Salas Barbadillo, Alonso Jerónimo de. *La hija de Celestina.* Edited by Enrique García Santo-Tomás. Madrid: Cátedra, 2008.

Salviati, Lionardo. *Cinqve lezzioni . . . cioè due della speranza, vna della felicità, e l'altre due sopra varie materie. E tutte lette nell'Accademia Fiorentina. . . .* Florence: Giunti, 1575.

Sarbiewski, Maciej Kazimierz. *Wykłady z poetyki: praecepta poetica.* Edited by Stanisław Skimina. Wrocław: Zakład Narodowy Im. Ossolińskich, 1958.

Scaliger, Julius Caesar. *Ivlii Caesaris Scaligeri Exotericarum Exercitationum. . . .* Paris: Michel Vascosan, 1557.

———. *Ivlii Caesaris Scaligeri, Viri Clarissimi, Poetices libri septem. . . .* [Lyons]: Antoine Vincent, 1561.

Scarabelli, Pietro Francesco. *Museo, ò, Galeria adunata dal sapere e dallo studio del sig. canonico Manfredo Settala nobile milanese*. Tortona: Sons of Eliseo Viola, 1666.

Scheiner, Christoph. *Refractiones Coelestes sive Solis Elliptici Phaenomenon illustratum*. Ingolstadt: Elisabeth Angermaier, 1617.

———. *Sol ellipticus: hoc est novum & perpetuum solis contrahi soliti phaenomenon, quod noviter inventum*. Augsburg: Christoph Mang, 1615.

———. *Tres epistolae de maculis solaribus*. Augsburg: Ad insigne pinus, 1612.

Schott, Gaspar. *P. Gasparis Schotti . . . Technica Curiosa, sive Mirabilia Artis, Libris XII Comprehensa*. Würzburg: Jobst Hertz, 1664.

Seneca, Lucius Annaeus. *Epistles*. Translated by Richard M. Gummere. 3 vols. Cambridge, MA: Harvard University Press, 1917–1925.

———. *Natural Questions*. Translated by Thomas H. Corcoran. Cambridge, MA: Harvard University Press, 1971.

Sforza Pallavicino, Pietro. *Considerazioni sopra l'arte dello stile, e del dialogo. . . .* Rome: Heirs of Corbelletti, 1646.

———. *Trattato dello stile e del dialogo. . . .* Rome: Mascardi, 1662.

Silva y Figueroa, García. *Comentarios de D. García de Silva y Figueroa. . . .* Edited by Manuel Serrano y Sanz. 2 vols. Madrid: Sociedad de Bibliófilos Españoles, 1903–1905.

Snellius, Willebrord. *De maculis in sole animadversis, &, tanquam ab Apelle, in tabula spectandum in publica luce expositis, Batavi dissertatiuncula*. Leiden: Franciscus Raphelengius, 1612.

———. *Willebrordi Snellii à Royen R. F. Tiphys Batavvs, Sive Histiodromice, de navium cursibus et re navali*. Leiden: Elzevir, 1624.

Stelluti, Francesco. *Persio tradotto in verso sciolto e dichiarato da Francesco Stelluti Accad. Linceo da Fabriano. . . .* Rome: Giacomo Mascardi, 1630.

Stigliani, Tommaso. *Il canzoniero del signor caualier fra' Tomaso Stigliani. . . .* Rome: Heirs of Bartolomeo Zannetti, 1623.

Strada, Famiano. *Prolusiones Academicae*. Rome: Giacomo Mascardi, 1617.

Suárez, Francisco. *Doctoris Francisci Suarez . . . Partis secundae Summae theologiae tomus alter, complectens tractatum secundum de Opere sex dierum, ac tertium de Anima. . . .* Lyons: N.p., 1621.

Suárez de Figueroa, Cristóbal. *Plaza Universal de todas Ciencias y Artes, parte traducida de Toscano, y parte compuesta por el doctor Christoval Suarez de Figueroa*. Madrid: Luis Sánchez, 1615.

———. *Pusílipo. Ratos de conversación, en los que dura el passeo*. Naples: Lazaro Scoriggio, 1629.

Swift, Jonathan. *A Tale of a Tub, Written for the Universal Improvement of Mankind: To Which Is Added, An Account of a Battel between the Antient and Modern Books in St. James's Library*. London: John Nutt, 1704.

Tacitus, Publius Cornelius. *Agricola. Germania. Dialogue on Oratory*. Translated by M. Hutton and W. Peterson. Cambridge, MA: Harvard University Press, 1914.

Tacke, Johann. *Coeli Anomalon, Id Est, De Cometis, Sive Stellis Crinitis, Praeter Universi Ordinem in coelo multoties visis epikomma Physicum. . . .* Gießen: Chemlin, 1653.

Talentoni, Giovanni. *Discorso in forma di lezzione del signor Giovanni Talentone . . . sopra la maraviglia. . . .* Milan: Francesco Paganelli, 1597.

Tamayo de Vargas, Tomás. *Restauración de La Ciudad Del Salvador, i Baía de Todos-Sanctos.* Madrid: Widow of Alonso Martín, 1628.

Tasso, Torquato. *Discorsi del signor Torquato Tasso dell'arte poetica et in particolare del poema heroico.* . . . Venice: Giulio Vassalini, 1587.

———. *Poesie.* Edited by Francesco Flora. Milan: Ricciardi, 2004.

Tassoni, Alessandro. *Lettere.* Edited by Pietro Puliatti. 2 vols. Rome: Laterza, 1978.

———. *Varietà di pensieri.* . . . Modena: Heirs of Giovanni Maria Verdi, 1612.

Terzago, Paolo Maria. *Museum septalianum Manfredi Septalae.* . . . Tortona: Sons of Eliseo Viola, 1664.

Tesauro, Emanuele. *Idea delle perfette imprese: testo inedito.* Edited by Maria Luisa Doglio. Florence: Leo S. Olschki, 1975.

———. *Il cannocchiale aristotelico o sia idea delle argutezze heroiche vulgarmente chiamate imprese et di tutta l'arte simbolica et lapidaria.* Turin: Giovanni Sinibaldo, 1654.

———. *La filosofia morale derivata dall'alto fonte del grande Aristotele stagirita.* Turin: Bartolomeo Zavatta, 1670.

———. *Panegirici del Conte D. Emanvele Tesavro.* . . . 2 vols. Turin: Bartolomeo Zavatta, 1659.

———. *Panegirici, et Ragionamenti del Conte D. Emanvele Tesavro . . . Volume Terzo.* Turin: Bartolomeo Zavatta, 1660.

Tiraboschi, Girolamo. *Storia della letteratura italiana.* . . . 9 vols. Modena: Societá Tipografica, 1772–1782.

Toledo, Francisco de. *D. Francisci Toleti Societatis Jesu Commentaria una cum quaestionibus in tres libros Aristotelis De anima.* . . . Cologne: Birckmann, 1594.

Vasari, Giorgio. *Delle vite de' piu eccellenti pittori, scultori, e architettori.* . . . 2 vols. Florence: Giunti, 1568.

Vega, Lope de. *Arte nuevo de hacer comedias.* Edited by Evangelina Rodríguez Cuadros. Madrid: Castalia, 2011.

Vergil, Polydore. *On Discovery.* Edited by Brian P. Copenhaver. Cambridge, MA: Harvard University Press, 2002.

Vitruvius. *On Architecture.* Translated by Frank Granger. 2 vols. London: William Heinemann, 1931–1934.

Vittori, Girolamo. *Tesoro de Las Tres Lenguas: Francesa, Italiana, y Española. Thresor Des Trois Langues Francoise, Italiene et Espagnolle.* . . . Geneva: Philippe Albert and Alexandre Pernet, 1609.

Vives, Juan Luis. *Ioannis Lodouici Viuis . . . De anima et vita libri tres.* Basel: Robert Winter, 1538.

———. *On Education.* Translated by Foster Watson. Cambridge: Cambridge University Press, 1913.

Willis, Thomas. *Diatribae Duae Medico-Philosophicae: Quarum Prior Agit de Fermentatione, Sive, de Motu Intestino Particularum in Quovis Corpore.* . . . London: Thomas Roycroft, 1659.

Yrissarri, Alfonso Bernardo de. *Compendio de la vida de Manfredo Settala: con la descripcion de su celeberrima galeria.* . . . Milan: Marco Antonio Pandolfo Malatesta, 1681.

Zayas, María de. *Novelas amorosas y ejemplares.* Edited by Julián Olivares. Madrid: Cátedra, 2004.

Secondary Sources

Ago, Renata. *Gusto for Things: A History of Objects in Seventeenth-Century Rome.* Translated by Bouley Bradford, Corey Tazzara, and Paula Findlen. Chicago: University of Chicago Press, 2013.

Aït-Touati, Frédérique. *Fictions of the Cosmos: Science and Literature in the Seventeenth Century.* Translated by Susan Emanuel. Chicago: University of Chicago Press, 2011.

Albalá Pelegrín, Marta. "El Arte nuevo de Lope de Vega a la luz de la teoría dramática italiana contemporánea: Poliziano, Robortello, Guarini y el Abad de Rute." *e-Humanista* 24 (2013): 1–15.

Anctil, Michel. *Luminous Creatures: The History and Science of Light Production in Living Organisms.* Montreal: McGill-Queen's University Press, 2018.

Anderson, Christy, Anne Dunlop, and Pamela H. Smith, eds. *The Matter of Art: Materials, Practices, Cultural Logics c. 1250–1750.* Manchester: Manchester University Press, 2014.

Arato, Franco. "I gesuiti spagnoli nell'Italia del settecento." *Belfagor* 64, no. 2 (2009): 175–186.

Arena, Stefano. "Omnis in unum: l'Accademia torinese dei Solinghi nell'antiporta del Cannocchiale aristotelico." In *Le virtuose adunanze: la cultura accademica tra XVI e XVIII secolo,* edited by Ilaria Bianchi and Clizia Gurreri, 209–226. Avellino: Sinestesie, 2015.

Arnaudo, Marco. "Il microscopio aristotelico: ottica, religione e retorica nel Seicento." *Annali d'italianistica* 23 (2005): 105–120.

———. "Sul significato del giocoliere nel 'Cannocchiale aristotelico' di Emanuele Tesauro." *Studi secenteschi* 50 (2009): 3–14.

Ascoli, Albert Russell. "The History of a Story: Manetti's 'La Novella del Grasso Legnaiuolo.'" In *Rituals of Politics and Culture in Early Modern Europe: Essays in Honour of Edward Muir,* edited by Mark Jurdjevic and Rolf Strøm-Olsen, 211–234. Toronto: CRRS, 2016.

Ashworth, William B., Jr. "Natural History and the Emblematic World View." In *Reappraisals of the Scientific Revolution,* edited by David C. Lindberg and Robert S. Westman, 303–332. Cambridge: Cambridge University Press, 1990.

Asmussen, Tina. "Spirited Metals and the Oeconomy of Resources in Early Modern European Mining." *Earth Sciences History* 39, no. 2 (2020): 371–388. https://doi.org/10.17704/1944-6187-39.2.371.

Asmussen, Tina, and Pamela O. Long. "Introduction: The Cultural and Material Worlds of Mining in Early Modern Europe." *Renaissance Studies* 34, no. 1 (2020): 8–30. https://doi.org/10.1111/rest.12581.

Astrain, Antonio. *Historia de la Compañía de Jesús en la asistencia de España.* 7 vols. Madrid: Razón y Fe, 1905–1925.

Aubert, Jean-Marie. "'Subtile' et 'subtilité' chez saint Thomas d'Aquin." *Revue du Moyen Âge latin* 44, no. 1–2 (1988): 6–13.

Ayala Martínez, Jorge Manuel. "Introducción." In *Agudeza y arte de ingenio,* edited by Carmen Peralta, Jorge Manuel Ayala Martínez, and José M. Andreu, 1:xxiii–cxiii. Zaragoza: Universidad de Zaragoza, 2004.

Bailey, M. E., S. V. M. Clube, and W. M. Napier. *The Origin of Comets*. Oxford: Pergamon Press, 1990.

Baldassarri, Fabrizio, and Andreas Blank, eds. *Vegetative Powers: The Roots of Life in Ancient, Medieval and Early Modern Natural Philosophy*. Cham: Springer, 2021. https://doi.org/10.1007/978-3-030-69709-9_4.

Baldini, Ugo. *Legem impone subactis: studi su filosofia e scienza dei gesuiti in Italia, 1540–1632*. Rome: Bulzoni, 1992.

Baldwin, Martha. "Alchemy and the Society of Jesus in the Seventeenth Century: Strange Bedfellows?" *Ambix* 40, no. 2 (1993): 41–64. https://doi.org/10.1179/amb.1993.40.2.41.

Barker, Peter, and Bernard R. Goldstein. "Realism and Instrumentalism in Sixteenth Century Astronomy: A Reappraisal." *Perspectives on Science* 6, no. 3 (1998): 232–258.

Barker, Sheila. "The Universe of Giovanna Garzoni. Art, Mobility, and the Global Turn in the Geographical Imaginary." In *"The Immensity of the Universe" in the Art of Giovanna Garzoni*, edited by Sheila Barker, 17–29. Leghorn: Sillabe, 2020.

Barnett, Lydia. "Between Pleasure and Terror: Enlightenment Science and Maupertuis' Letter on the Comet." *Architectural Theory Review* 20, no. 1 (2015): 30–45. https://doi.org/10.1080/13264826.2015.1064142.

Barr, James. "St Jerome's Appreciation of Hebrew." *Bulletin of the John Rylands Library* 49, no. 2 (1967): 281–302. https://doi.org/10.7227/BJRL.49.2.2.

Barwich, Ann-Sophie. "Sleight of Hand Tricks." In *Secrets of Craft and Nature in Renaissance France. A Digital Critical Edition and English Translation of BNF Ms. Fr. 640*, edited by Pamela H. Smith. New York: Making and Knowing Project, 2020. https://doi.org/10.7916/rfq6-ok88.

Batllori, Miquel. *La cultura hispano-italiana de los jesuitas expulsos: españoles, hispanoamericanos, filipinos: 1767–1814*. Madrid: Gredos, 1966.

Beagon, Mary. "The Curious Eye of the Elder Pliny." In *Pliny the Elder: Themes and Contexts*, edited by Roy K. Gibson and Ruth Morello, 71–88. Leiden: Brill, 2011. https://doi.org/10.1163/ej.9789004202344.i-248.28.

Belfanti, Carlo M. "Guilds, Patents, and the Circulation of Technical Knowledge: Northern Italy during the Early Modern Age." *Technology and Culture* 45 (2004): 569–589.

Bellini, Eraldo. *Agostino Mascardi tra "ars poetica" e "ars historica."* Milan: Vita e pensiero, 2002.

Bensaude-Vincent, Bernadette, and William R. Newman, eds. *The Artificial and the Natural: An Evolving Polarity*. Cambridge, MA: MIT Press, 2007.

Beretta, Marco. *The Alchemy of Glass: Counterfeit, Imitation, and Transmutation in Ancient Glassmaking*. Sagamore Beach, MA: Watson, 2009.

Besomi, Ottavio, and Michele Camerota. *Galileo e il Parnaso Tychonico: un capitolo inedito del dibattito sulle comete tra finzione letteraria e trattazione scientifica*. Florence: Leo S. Olschki, 2000.

Beusterien, John. *Transoceanic Animals as Spectacle in Early Modern Spain*. Amsterdam: Amsterdam University Press, 2020.

Biagioli, Mario. *Galileo, Courtier: The Practice of Science in the Culture of Absolutism*. Chicago: University of Chicago Press, 1993.

———. *Galileo's Instruments of Credit*. Chicago: University of Chicago Press, 2006.

———. *Inventing Invention: Galileo's Telescope between Science and Craft*. La Jolla: University of California, San Diego, 2010.

Bigelow, Allison, and Pablo Cruz. "*Ingenios* and Ingenuity: Rethinking Indigenous Histories of Silver in the Colonial Andean Mining Industry." *Colonial Latin American Review* 30, no. 4 (2021): 520–544. https://doi.org/10.1080/10609164.2021.1996989.

Bisello, Linda. *Medicina della memoria: aforistica ed esemplarità nella scrittura barocca*. Florence: Leo S. Olschki, 1998.

Blanco, Emilio. "'Guzmán de Alfarache,' el canon teórico del siglo XVII y un modelo de Agudeza." *Komparatistik Online*, no. 1 (2014): 3–25.

Blanco, Mercedes. "Humanismo rezagado frente a difícil modernidad. Al margen de la polémica Ormaza-Céspedes sobre la oratoria sagrada." *Criticón* 84–85 (2002): 123–144.

———. *Les rhétoriques de la pointe. Baltasar Gracián et le Conceptisme en Europe*. Geneva: Slatkine, 1992.

Blumenberg, Hans. *Die Lesbarkeit der Welt*. Frankfurt am Main: Suhrkamp, 2018.

———. *The Legitimacy of the Modern Age*. Translated by Robert M. Wallace. Cambridge, MA: MIT Press, 1983.

Boadas, Sònia. "Lucien de Samosate et les voyages célestes dans la littérature espagnole du Siècle d'or." *Dix-septieme siecle* 286, no. 1 (2020): 135–151.

Bono, James J. "Medical Spirits and the Medieval Language of Life." *Traditio* 40 (1984): 91–130.

Bowe, G. S. "Thaumata in Aristotle's *Metaphysics* A." *Acta Classica* 60, no. 1 (2017): 50–72.

Bradford, Alan T. "Stuart Absolutism and the 'Utility' of Tacitus." *Huntington Library Quarterly* 46, no. 2 (1983): 127–155. https://doi.org/10.2307/3817324.

Bredekamp, Horst. *Galileo's Thinking Hand: Mannerism, Anti-mannerism and the Virtue of Drawing in the Foundation of Early Modern Science*. Translated by Mitch Cohen. Berlin: De Gruyter, 2019.

———. *The Lure of Antiquity and the Cult of the Machine: The Kunstkammer and the Evolution of Nature, Art and Technology*. Translated by Allison Brown. Princeton, NJ: Markus Wiener, 1995.

Brevaglieri, Sabina. *Natural desiderio di sapere. Roma barocca fra vecchi e nuovi mondi*. Rome: Viella, 2019.

Brodsky, Claudia. "Donne: The Imaging of the Logical Conceit." *English Literary History* 49, no. 4 (1982): 829–848.

Broughton, Peter. "The First Predicted Return of Comet Halley." *Journal for the History of Astronomy* 16 (1985): 123–133.

Brown, Gary I. "The Evolution of the Term 'Mixed Mathematics.'" *Journal of the History of Ideas* 52, no. 1 (1991): 81–102. https://doi.org/10.2307/2709583.

Bruni, Francesco. "Semantica della sottigliezza: note sulla distribuzione della cultura nel Basso Medioevo." *Studi medievali* 3, no. 19 (1978): 1–36.

Bundy, Murray Wright. *The Theory of Imagination in Classical and Medieval Thought*. Urbana: University of Illinois Press, 1927.

Burke, Peter. "Tacitism, Scepticism, and Reason of State." In *The Cambridge History of Political Thought 1450–1700*, edited by J. H. Burns and Mark Goldie, 477–498. Cambridge: Cambridge University Press, 1991.

Busi, Giulio. "Giovanni Pico and the Ideal of *Concordia Discors*: Disharmony as a Way to Esoteric Wisdom." In *Constructing Tradition: Means and Myths of Transmission in Western Esotericism*, edited by Andreas Kilcher, 293–302. Leiden: Brill, 2011.

Cain, Andrew. *Jerome and the Monastic Clergy: A Commentary on Letter 52 to Nepotian, with Introduction, Text, and Translation*. Leiden: Brill, 2013.

Campbell, Mary Baine. *Wonder and Science: Imagining Worlds in Early Modern Europe*. Ithaca, NY: Cornell University Press, 2016.

Carminati, Clizia. "Alcune considerazioni sulla scrittura laconica nel Seicento." *Aprosiana* 10 (2002): 91–112.

Carolino, Luis Miguel. "Cometas, estrelas novas e matéria celeste em Portugal: as diferentes faces de um debate cosmológico seiscentista." In *Novos horizontes, sécs. XV–XVII*, edited by Antonio Sánchez, Palmira Fontes da Costa, and Henrique Leitão, 347–372. Lisbon: Tinta-da-china, 2021.

———. "Cristoforo Borri and the Epistemological Status of Mathematics in Seventeenth-Century Portugal." *Historia Mathematica* 34, no. 2 (2007): 187–205.

———. "The Making of a Tychonic Cosmology: Cristoforo Borri and the Development of Tycho Brahe's Astronomical System." *Journal for the History of Astronomy* 39, no. 3 (2008): 313–344. https://doi.org/10.1177/002182860803900303.

———. "Manuel Bocarro Francês, the Comet of 1618, and the Impact of Stoic Cosmology in Portugal." In *Novas y cometas entre 1572 y 1618. Revolución cosmológica y renovación política y religiosa*, edited by Miguel Ángel Granada, 195–224. Barcelona: Universitat de Barcelona, 2012.

Carvalho, Mário Santiago de. "Cursus Conimbricensis." *Conimbricenses* (2019). http://doi.org/10.5281/zenodo.3234133.

———. "Manuel de Góis: The Coimbra Course and the Definition of an Early Jesuit Philosophy." In *Jesuit Philosophy on the Eve of Modernity*, edited by Cristiano Casalini, 374–372. Leiden: Brill, 2019.

Casalini, Cristiano. *Aristotle in Coimbra: The* Cursus Conimbricensis *and the Education at the College of Arts*. Translated by Luana Salvarani. London: Routledge, 2017.

———. "Discerning Skills: Psychological Insight at the Core of Jesuit Identity." In *Exploring Jesuit Distinctiveness: Interdisciplinary Perspectives on Ways of Proceeding within the Society of Jesus*, edited by Robert Aleksander Maryks, 189–211. Leiden: Brill, 2016. https://doi.org/10.1163/9789004313354_012.

Casini, Lorenzo. "Emotions in Renaissance Humanism: Juan Luis Vives' *De Anima Et Vita*." In *Emotions and Choice from Boethius to Descartes*, edited by Henrik Lagerlund and Mikko Yrjönsuuri, 205–228. Dordrecht: Springer, 2002. https://doi.org/10.1007/978-94-010-0506-7_9.

———. "Renaissance Theories of Internal Senses." In *Sourcebook for the History of the Philosophy of Mind*, edited by Simo Knuuttila and Juha Sihvola, 147–156. Cham: Springer, 2013.

Cassin, Barbara, Emily Apter, Jacques Lezra, and Michael Wood, eds. *Dictionary of Untranslatables*. Princeton, NJ: Princeton University Press, 2014.

Castro Ibaseta, Javier. "Mentidero de Madrid: la Corte como comedia." In *Opinión pública y espacio urbano en la Edad Moderna*, edited by Antonio Castillo Gómez and James S. Amelang, 43–58. Madrid: Trea, 2010.

Cave, Terence. *The Cornucopian Text: Problems of Writing in the French Renaissance.* Oxford: Clarendon, 1979.

Céard, Jean. *La nature et les prodiges: l'insolite au XVIe siècle, en France.* Geneva: Droz, 1977.

Chang, Ku-ming (Kevin). "Alchemy as Studies of Life and Matter: Reconsidering the Place of Vitalism in Early Modern Chymistry." *Isis* 102, no. 2 (2011): 322–329. https://doi.org/10.1086/660141.

Chartier, Roger. "El proceso civilizatorio. Elias, Gracian, Amelot." *Co-herencia* 11, no. 21 (2014): 13–24.

Christian, Kathleen Wren. "Poetry and 'Spirited' Ancient Sculpture in Renaissance Rome: Pomponio Leto's Academy to the Sixteenth-Century Sculpture Garden." In *Aeolian Winds and the Spirit in Renaissance Architecture*, edited by Barbara Kenda, 103–124. London: Routledge, 2006.

Clark, Henry C. *La Rochefoucauld and the Language of Unmasking in Seventeenth-Century France.* Geneva: Droz, 1994.

Clark, Stuart. *Thinking with Demons: The Idea of Witchcraft in Early Modern Europe.* Oxford: Oxford University Press, 1997.

———. *Vanities of the Eye: Vision in Early Modern European Culture.* Oxford: Oxford University Press, 2007.

Clericuzio, Antonio. "The Internal Laboratory. The Chemical Reinterpretation of Medical Spirits in England (1650–1680)." In *Alchemy and Chemistry in the 16th and 17th Centuries*, edited by Piyo Rattansi and Antonio Clericuzio, 51–83. Dordrecht: Springer, 1994. https://doi.org/10.1007/978-94-011-0778-5_3.

Cohen, Albert. "'La Supposition' and the Changing Concept of Dissonance in Baroque Theory." *Journal of the American Musicological Society* 24, no. 1 (1971): 63–84. https://doi.org/10.2307/830893.

Cohen, H. F. *Quantifying Music.* Dordrecht: Springer, 1984.

Cole, Michael W. *Ambitious Form: Giambologna, Ammanati, and Danti in Florence.* Princeton, NJ: Princeton University Press, 2011.

———. "The Cult of Materials." In *Revival and Invention: Sculpture through Its Material Histories*, edited by Sébastien Clerbois and Martina Droth, 1–15. Oxford: Peter Lang, 2010.

Connelly, Frances S. *The Grotesque in Western Art and Culture: The Image at Play.* Cambridge: Cambridge University Press, 2012.

Cook, Lexie. "Before the Fetish: Artifice and Trade in Early Modern Guinea." PhD diss., Columbia University, 2022. https://doi.org/10.7916/hr97-4p28.

Copenhaver, Brian P. *Magic in Western Culture from Antiquity to the Enlightenment.* Cambridge: Cambridge University Press, 2015.

Coster, Adolphe. *Baltasar Gracián, 1601–1658.* New York: Hispanic Society of America, 1913.

Courcelle, Pierre. *Connais-toi toi-même. De Socrate à saint Bernard.* 3 vols. Paris: Études augustiniennes, 1974.

Croce, Benedetto. *I trattatisti italiani del "Concettismo" e Baltasar Gracian.* Naples: Stab. Tipografico nella R. Università, 1899.

Croce, Franco. *Tre momenti del barocco letterario italiano.* Florence: Sansoni, 1966.

Cummings, Brian. "Metalepsis: The Boundaries of Metaphor." In *Renaissance Figures of Speech*, edited by S. Adamson, G. Alexander, and K. Ettenhuber, 217–234. Cambridge: Cambridge University Press, 2007. https://doi.org/10.1017/CBO9780511988806.027.

Cutrì, Maicol. "Leggere il 'libro aperto.' Un'introduzione al Cannocchiale aristotelico." PhD diss., Università Cattolica del Sacro Cuore, 2022.

———. "Storia editoriale del 'Cannocchiale aristotelico.'" *Testo: studi di teoria e storia della letteratura e della critica* 80, no. 2 (2020): 63–86.

Cypess, Rebecca. *Curious and Modern Inventions: Instrumental Music as Discovery in Galileo's Italy*. Chicago: University of Chicago Press, 2016.

Daston, Lorraine. *Against Nature*. Cambridge, MA: MIT Press, 2019.

———. "The Cold Light of Facts and the Facts of Cold Light: Luminescence and the Transformation of the Scientific Fact, 1600–1750." *EMF: Studies in Early Modern France* 3 (1997): 17–44.

———. "Curiosity in Early Modern Science." *Word and Image* 2, no. 4 (1995): 391–404.

———. "Fear and Loathing of the Imagination in Science." *Daedalus* 127, no. 1 (1998): 73–95.

———. "The Nature of Nature in Early Modern Europe." *Configurations* 6, no. 2 (1998): 149–172. https://doi.org/10.1353/con.1998.0014.

———. "What Can Be a Scientific Object? Reflections on Monsters and Meteors." *Bulletin of the American Academy of Arts and Sciences* 52, no. 2 (1988): 35–50.

Daston, Lorraine, and Katharine Park. *Wonders and the Order of Nature, 1150–1750*. New York: Zone Books, 1998.

Davies, Surekha. "The Unlucky, the Bad and the Ugly: Categories of Monstrosity from the Renaissance to the Enlightenment." In *The Ashgate Research Companion to Monsters and the Monstrous*, edited by Asa Simon Mittman and Peter J. Dendle, 49–75. London: Routledge, 2013. https://doi.org/10.4324/9781315241197.ch2.

Dear, Peter. "The Meanings of Experience." In *The Cambridge History of Science*, vol. 3: *Early Modern Science*, edited by Katharine Park and Lorraine Daston, 106–131. Cambridge: Cambridge University Press, 2006. https://doi.org/10.1017/CHOL9780521572446.

———. "Mixed Mathematics." In *Wrestling with Nature: From Omens to Science*, edited by Peter Harrison, Ronald L. Numbers, and Michael H. Shank, 149–172. Chicago: University of Chicago Press, 2011. https://doi.org/10.7208/9780226318035-008.

———. *Revolutionizing the Sciences: European Knowledge in Transition, 1500–1700*. 3rd ed. London: Red Globe Press, 2019.

De Armas, Frederick A. "The Maculate Moon: Galileo, Kepler and Pantaleón de Ribera's 'Vexamen de La Luna.'" *Calíope* 5, no. 1 (1999): 59–71.

De Boer, Sander W. "Imagination, Images, and (Im)Mortality." In *Image, Imagination, and Cognition: Medieval and Early Modern Theory and Practice*, edited by Christoph Lüthy, Claudia Swan, Paul J. J. M. Bakker, and Claus Zittel, 11–31. Leiden: Brill, 2018.

De Jong, Albert. "Zoroaster." In *Dictionary of Gnosis and Western Esotericism*, edited by W. J. Hanegraaff, Antoine Faivre, Roelof Van den Broek, and Jean-Pierre Brach, 1182–1183. Leiden: Brill, 2008.

Delage, Agnès. "Las vidas particulares bajo el reinado de Felipe IV: ¿un problema de definición genérica?" *Criticón* 97–98 (2006): 61–74.

———. "Tacitisme et hypertacitisme: les écritures du secret politique dans l'Espagne du XVIIe siècle." In *Le partage du secret. Cultures du dévoilement et de l'occultation en Europe, du Moyen Âge à l'époque moderne*, edited by Bernard Darbord and Agnès Delage, 305–328. Paris: Armand Colin, 2013.

Des Chene, Dennis. "Forms of Art in Jesuit Aristotelianism (with a Coda on Descartes)." In *The Artificial and the Natural: An Evolving Polarity*, edited by Bernadette Bensaude-Vincent and William R. Newman, 135–147. Cambridge, MA: MIT Press, 2007.

———. *Life's Form: Late Aristotelian Conceptions of the Soul*. Ithaca, NY: Cornell University Press, 2000.

———. *Physiologia: Natural Philosophy in Late Aristotelian and Cartesian Thought*. Ithaca, NY: Cornell University Press, 1996.

———. *Spirits and Clocks: Machine and Organism in Descartes*. Ithaca, NY: Cornell University Press, 2001.

Detienne, Marcel, and Jean-Pierre Vernant. *Les ruses de l'intelligence. La mètis des grecs*. Paris: Flammarion, 1974.

Dickerman, Sherwood Owen. "Some Stock Illustrations of Animal Intelligence in Greek Psychology." *Transactions and Proceedings of the American Philological Association* 42 (1911): 123–130. https://doi.org/10.2307/282577.

Didi-Huberman, Georges. *Survival of the Fireflies*. Translated by Lia Swope Mitchell. Minneapolis: University of Minnesota Press, 2018. https://doi.org/10.5749/j.ctv4rfqxs.

Dillinger, Johannes. "The Divining Rod: Origins, Explanations and Uses in the Thirteenth to Eighteenth Centuries." In *Contesting Orthodoxy in Medieval and Early Modern Europe: Heresy, Magic and Witchcraft*, edited by Louise Nyholm Kallestrup and Raisa Maria Toivo, 127–143. Cham: Springer, 2017.

Doglio, Maria Luisa. "Introduzione." In *Idea delle perfette imprese*, edited by Maria Luisa Doglio, 5–27. Florence: Leo S. Olschki, 1975.

Donato, Eugenio. "Tesauro's Poetics: Through the Looking Glass." *MLN* 78, no. 1 (1963): 15–30. https://doi.org/10.2307/3042940.

Dooley, Brendan Maurice. *The Social History of Skepticism: Experience and Doubt in Early Modern Culture*. Baltimore: Johns Hopkins University Press, 1999.

Durling, Richard J. "The Innate Heat in Galen." *Medizinhistorisches Journal* 23, no. 3–4 (1988): 210–212.

Dym, Warren Alexander. *Divining Science: Treasure Hunting and Earth Science in Early Modern Germany*. Leiden: Brill, 2011.

Eamon, William. "Medicine as a Hunt: Searching for the Secrets of the New World." In *Renaissance Futurities: Science, Art, Invention*, edited by Charlene Villaseñor Black and Mari-Tere Álvarez, 100–117. Oakland: University of California Press, 2019.

———. *Science and the Secrets of Nature: Books of Secrets in Medieval and Early Modern Culture*. Princeton, NJ: Princeton University Press, 1994.

———. "Science as a Hunt." *Physis. Rivista Internazionale di Storia della Scienza* 31 (1994): 393–432.

Eden, Kathy. *Poetic and Legal Fiction in the Aristotelian Tradition*. Princeton, NJ: Princeton University Press, 1986.

Edgerton, Samuel Y. *Pictures and Punishment: Art and Criminal Prosecution during the Florentine Renaissance*. Ithaca, NY: Cornell University Press, 1985.

Edwards, Michael. "Digressing with Aristotle: Hieronymus Dandinus' 'De Corpore Animato' (1610) and the Expansion of Late Aristotelian Philosophy." *Early Science and Medicine* 13, no. 2 (2008): 127–170.

Esteve, Cesc. "Les poètiques de la meravella a Itàlia (1550–1700)." In *La poètica barroca a Europa. Un nou sistema epistemològic i estètic*, edited by Antoni L. Moll and Josep Solervicens, 71–96. Barcelona: Punctum & Mimesi, 2009.

Fabbri, Natacha. *Cosmologia e armonia in Kepler e Mersenne. Contrappunto a due voci sul tema dell'*Harmonice Mundi. Florence: Leo S. Olschki, 2003.

Fazzari, Michela. "Redi, Buonanni e la controversia sulla generazione spontanea: una rilettura." In *Francesco Redi: un protagonista della scienza moderna: documenti, esperimenti, immagini*, edited by Walter Bernardi and Luigi Guerrini, 97–127. Florence: Leo S. Olschki, 1999.

Feingold, Mordechai. "The Grounds for Conflict: Grienberger, Grassi, Galileo, and Posterity." In *The New Science and Jesuit Science: Seventeenth Century Perspectives*, edited by Mordechai Feingold, 121–157. Dordrecht: Springer, 2011.

Fernández, Esther. *To Embody the Marvelous: The Making of Illusions in Early Modern Spain*. Nashville, TN: Vanderbilt University Press, 2021.

Ferrari, Oreste. "Bernini e i letterati del suo tempo." In *Bernini dai Borghese ai Barberini: la cultura a Roma intorno agli anni venti*, edited by Olivier Bonfait and Anna Coliva, 59–63. Rome: De Luca, 2004.

Ferro, Roberta. *Federico Borromeo ed Ericio Puteano: cultura e letteratura a Milano agli inizi del Seicento*. Milan: Biblioteca Ambrosiana, 2007.

Findlen, Paula. "Inventing Nature: Commerce, Art, and Science in the Early Modern Cabinet of Curiosities." In *Merchants and Marvels*, edited by Pamela H. Smith and Paula Findlen, 297–323. New York: Routledge, 2002.

———. "Jokes of Nature and Jokes of Knowledge: The Playfulness of Scientific Discourse in Early Modern Europe." *Renaissance Quarterly* 43, no. 2 (1990): 292–331.

———. "Ludic Postscript." In *Ludi Naturae: Spiele Der Natur in Kunst und Wissenschaft*, edited by Natascha Adamowsky, Hartmut Böhme, and Robert Felfe, 49–77. Paderborn: Wilhelm Fink, 2011.

———. *Possessing Nature: Museums, Collecting, and Scientific Culture in Early Modern Italy*. Berkeley: University of California Press, 1994.

Fisher, Philip. *Wonder, the Rainbow and the Aesthetics of Rare Experiences*. Cambridge, MA: Harvard University Press, 1998.

Fitzgerald, William. *Variety: The Life of a Roman Concept*. Chicago: University of Chicago Press, 2016.

Frare, Pierantonio. *Per istraforo di perspettiva: il* Cannocchiale aristotelico *e la poesia del Seicento*. Rome: Istituti editoriali e poligrafici, 2000.

Fratini, Giulio, and Francesco Moriconi. "Datazione e attribuzione dell'anamorfosi di San Giovanni a Pathmos presso il Convento della Trinità dei Monti a Roma." *Mélanges de l'École Française de Rome* 122 (2010): 123–135. https://doi.org/10.4000/mefrim.563.

Frede, Dorothea. "Plato on What the Body's Eye Tells the Mind's Eye." *Proceedings of the Aristotelian Society* 99 (1999): 191–209.

Freedberg, David. *The Eye of the Lynx: Galileo, His Friends, and the Beginnings of Modern Natural History*. Chicago: University of Chicago Press, 2002.

———. "From Hebrew and Gardens to Oranges and Lemons: Giovanni Battista Ferrari and Cassiano dal Pozzo." In *Cassiano dal Pozzo: Atti del Seminario Internazionale di Studi*, edited by Francesco Solinas, 37–72. Rome: De Luca, 1989.

Freudenthal, Gad. *Aristotle's Theory of Material Substance: Heat and Pneuma, Form and Soul*. Oxford: Oxford University Press, 1999.

Fried, Michael. *The Moment of Caravaggio*. Princeton, NJ: Princeton University Press, 2010.

Fruton, Joseph. *Fermentation: Vital or Chemical Process?* Leiden: Brill, 2006.

Fuchs, Barbara. "Crusoe's Absence." *Studies in Eighteenth-Century Culture* 49 (2020): 27–42.

Fuchs, Barbara, and Mercedes García-Arenal, eds. *The Quest for Certainty in Early Modern Europe: From Inquisition to Inquiry, 1550–1700*. Toronto: University of Toronto Press, 2020. https://doi.org/10.3138/9781487535506-007.

Fulco, Giorgio. "Marino, 'Flavio' e il Parnaso barocco nella corrispondenza del 'Rugginoso.'" In *Feconde venner le carte*, edited by Tatiana Crivelli, 297–331. Bellinzona: Casagrande, 1997.

Fullenwider, Henry F. "*Concors Discordia*: Sarbiewski's *De acuto et arguto* (1627) and Jean de Serres' Commentary on Plato's *Timaeus* (1587)." *Bibliothèque d'Humanisme et Renaissance* 46, no. 3 (1984): 619–624.

———. "Die Kritik der deutschen Jesuiten an dem lakonischen Stil des Justus Lipsius im Zusammenhang der jesuitischen Argutia-Bewegung." *Rhetorica* 2, no. 1 (1984): 55–62.

Fumaroli, Marc. "*Cicero Pontifex Romanus*: la tradition rhétorique du Collège romain et les principes inspirateurs du mécénat des Barberini." *Mélanges de l'École française de Rome* 90, no. 2 (1978): 797–835. https://doi.org/10.3406/mefr.1978.2466.

———. *L'âge de l'éloquence: rhétorique et "res literaria" de la Renaissance au seuil de l'époque classique*. 3rd ed. Geneva: Droz, 2009.

———. "Les abeilles et les araignées." In *La Querelle des Anciens et des Modernes*, edited by Anne-Marie Lecoq, 7–220. Paris: Gallimard, 2001.

Gabrieli, Giuseppe, ed. *Atti della Reale Accademia Nazionale dei Lincei. Anno CCCXXXV. 1938 (XVI). Memorie della Classe di scienze morali, storiche e filologiche*. Vol. 7. Rome: Giovanni Bardi, 1938.

Gal, Ofer, and Raz Chen-Morris. *Baroque Science*. Chicago: University of Chicago Press, 2013.

Galluzzi, Paolo. *The Lynx and the Telescope: The Parallel Worlds of Cesi and Galileo*. Translated by Peter Mason. Leiden: Brill, 2017.

Gambaro, Ivana. "Geo-Heliocentric Models and the Society of Jesus: From Clavius's Resistance to Dechales's *Mathesis Regia*." *Annals of Science* 78, no. 3 (2021): 265–294. https://doi.org/10.1080/00033790.2021.1919760.

García Berrio, Antonio. *España e Italia ante el Conceptismo*. Madrid: CSIC, 1968.

García Romero, Fernando. "To Have Lynceus' Eyesight." *Paremia* 29 (2019): 77–92.

García Santo-Tomás, Enrique. *The Refracted Muse: Literature and Optics in Early Modern Spain*. Translated by Vincent Barletta. Chicago: University of Chicago Press, 2017.

———. *Signos vitales. Procreación e imagen en la narrativa áurea*. Madrid: Iberoamericana/Vervuert, 2020.

Garrard, Mary D. "The Not-So-Still Lifes of Giovanna Garzoni." In *"The Immensity of the Universe" in the Art of Giovanna Garzoni*, edited by Sheila Barker, 62–77. Leghorn: Sillabe, 2020.

Garrod, Raphaële. *Cosmographical Novelties in French Renaissance Prose (1550–1630): Dialectic and Discovery*. Turnhout: Brepols, 2016.

———. "*Subtilis, Inutilis*: The Jesuit Pedagogy of Ingenuity at La Flèche in the Seventeenth Century." In *Teaching Philosophy in Early Modern Europe: Text and Image*, edited by Susanna Berger and Daniel Garber, 139–164. Cham: Springer, 2021. https://doi.org/10.1007/978-3-030-84621-3_6.

Gattei, Stefano. *On the Life of Galileo: Viviani's Historical Account and Other Early Biographies*. Princeton, NJ: Princeton University Press, 2019.

Gelber, Jessica. "Females in Aristotle's Embryology." In *Aristotle's Generation of Animals: A Critical Guide*, edited by Andrea Falcon and David Lefebvre, 171–187. Cambridge: Cambridge University Press, 2018.

Gentilcore, David. *Medical Charlatanism in Early Modern Italy*. Oxford: Oxford University Press, 2006.

Genuth, Sarah Schechner. *Comets, Popular Culture, and the Birth of Modern Cosmology*. Princeton, NJ: Princeton University Press, 1997.

Giglioni, Guido. "Philosophy According to Tacitus: Francis Bacon and the Inquiry into the Limits of Human Self-Delusion." *Perspectives in Science* 20, no. 2 (2012): 159–182.

Gillet, J. E. "A Note on the Tragic 'Admiratio.'" *Modern Language Review* 13, no. 2 (1918): 233–238.

Gilman, Ernest B. *The Curious Perspective: Literary and Pictorial Wit in the Seventeenth Century*. New Haven, CT: Yale University Press, 1978.

Ginzburg, Carlo. "High and Low: The Theme of Forbidden Knowledge in the Sixteenth and Seventeenth Centuries." *Past and Present*, no. 73 (1976): 28–41.

Gómez López, Susana. "The Encounter of the Emblematic Tradition with Optics: The Anamorphic Elephant of Simon Vouet." *Nuncius* 31 (2016): 288–331.

Goodey, C. F. *A History of Intelligence and "Intellectual Disability": The Shaping of Psychology in Early Modern Europe*. London: Routledge, 2016. https://doi.org/10.4324/9781315564838.

Gouk, Penelope. "Music and the Emergence of Experimental Science in Early Modern Europe." *Sound Effects* 2, no. 1 (2012): 6–21.

Gozza, Paolo. *Number to Sound: The Musical Way to the Scientific Revolution*. Boston: Kluwer, 2000.

Grady, Hugh H. "Rhetoric, Wit, and Art in Gracián's Agudeza." *Modern Language Quarterly* 41, no. 1 (1980): 21–37.

Grafton, Anthony. "Humanism and Science in Rudolphine Prague: Kepler in Context." In *Literary Culture of the Holy Roman Empire*, edited by James A. Parente Jr., et al., 19–45. Chapel Hill: University of North Carolina Press, 1991.

———. "Renaissance Histories of Art and Nature." In *The Artificial and the Natural: An Evolving Polarity*, edited by Bernadette Bensaude-Vincent and William R. Newman, 185–210. Cambridge, MA: MIT Press, 2007.

Grasskamp, Anna. "Unpacking Foreign Ingenuity: The German Conquest of Artful Objects with 'Indian' Provenance." In *Ingenuity in the Making: Matter and Technique in Early Modern Europe*, edited by Richard J. Oosterhoff, José Ramón Marcaida López, and Alexander Marr, 203–218. Pittsburgh, PA: University of Pittsburgh Press, 2021.

Graziosi, Elisabetta. "Cesura per il secolo dei Genovesi: Anton Giulio Brignole Sale." *Studi secenteschi* 41 (2000), 27–87.

Green, Otis H. "Sobre el significado de 'crisi(s)' antes de 'El Criticón': una nota para la historia del conceptismo." In *Homenaje a Gracián*, edited by Charles Vincent Aubrum, 99–102. Zaragoza: Institución Fernando el Católico, 1958.

Gregoriâc, Pavel, and Filip Grgiâc. "Aristotle's Notion of Experience." *Archiv für Geschichte der Philosophie* 88, no. 1 (2006): 1–30.

Gronewoller, Brian. *Rhetorical Economy in Augustine's Theology*. Oxford: Oxford University Press, 2021.

Grootenboer, Hanneke. "Thinking with Shells in Petronella Oortman's Dollhouse." In *Conchophilia: Shells, Art, and Curiosity in Early Modern Europe*, edited by Marisa Anne Bass, Anne Goldgar, Hanneke Grootenboer, and Claudia Swan, 103–123. Princeton, NJ: Princeton University Press, 2021.

Gualandi, Andrea. *Teorie delle comete. Da Galileo a Newton*. Milan: Franco Angeli, 2009.

Guasti, Niccolò. *Juan Andrés e la cultura del Settecento*. Milan: Mimesis, 2017.

———. *L'esilio italiano dei gesuiti spagnoli. Identità, controllo sociale e pratiche culturali (1767–1798)*. Rome: Edizioni di Storia e Letteratura, 2006.

Guerrini, Luigi. "Le Stanze sopra le stelle e macchie solari scoperte col nuovo occhiale di Vincenzo Figliucci: Un episodio poco noto della visita di Galileo Galilei a Roma nel 1611." *Lettere Italiane* 50 (1998): 387–415.

———. "Tradizione astronomica e cultura matematica in un'orazione gesuitica del 1619." *Giornale critico della filosofia italiana* 20, no. 2–3 (2000): 209–235.

Guest, Clare Lapraik. "Plato's Stag Goats: Sophistic Heritage in Renaissance Grotesques." In *Paradigms of Renaissance Grotesques*, edited by Damiano Acciarino, 155–200. Toronto: Centre for Reformation and Renaissance Studies, 2019.

———. *The Understanding of Ornament in the Italian Renaissance*. Leiden: Brill, 2015.

Güntert, Georges. "Filosofar con artificio: 'El Criticón' como suma del arte del ingenio." In *Los conceptos de Gracián. Tercer Coloquio Internacional sobre Baltasar Gracián . . .*, edited by Sebastian Neumeister, 217–236. Berlin: Edition Tranvía, 2010. https://doi.org/10.5167/uzh-42516.

Hadot, Pierre. *The Veil of Isis: An Essay on the History of the Idea of Nature*. Translated by Michael Chase. Cambridge, MA: Harvard University Press, 2006.

Hall, Marie Boas. "The Spirit of Innovation in the Sixteenth Century." In *The Nature of Scientific Discovery . . .*, edited by Owen Gingerich, 309–334. Washington, DC: Smithsonian Institution Press, 1975.

Hamesse, Jaqueline. "L'apport des textes philosophiques des 12e et 13e siècles à l'étude de *machina* et de *machinatio*." In *Machina: XI Colloquio internazionale, Roma, 8–10 gennaio 2004*, edited by Marco Veneziani, 159–170. Florence: Leo S. Olschki, 2005.

Hanafi, Zakiyah. *The Monster in the Machine: Explorations of Monstrosity in Early Modern Italy*. Durham, NC: Duke University Press, 2000.

Hanß, Stefan. "New World Feathers and the Matter of Early Modern Ingenuity: Digital Microscopes, Period Hands, and Period Eyes." In *Ingenuity in the Making: Materials and Technique in Early Modern Europe*, edited by Richard J. Oosterhoff, José Ramón Marcaida López, and Alexander Marr, 189–202. Pittsburgh, PA: University of Pittsburgh Press, 2021.

Harris, Henry. *Things Come to Life: Spontaneous Generation Revisited*. Oxford: Oxford University Press, 2000.

Harrison, Evelyn B. "The Composition of the Amazonomachy on the Shield of Athena Parthenos." *Hesperia* 35, no. 2 (1966): 107–133. https://doi.org/10.2307/147300.

Harrison, Peter. "Natural History." In *Wrestling with Nature: From Omens to Science*, edited by Peter Harrison, Ronald L. Numbers, and Michael H. Shank, 117–148. Chicago: University of Chicago Press, 2011.

Harvey, E. Newton. *A History of Luminescence from the Earliest Times until 1900*. Philadelphia: American Philosophical Society, 1957.

Heath, Malcolm. "The Universality of Poetry in Aristotle's *Poetics*." *Classical Quarterly* 41, no. 2 (1991): 389–402.

Heidarzadeh, Tofigh. *A History of Physical Theories of Comets, from Aristotle to Whipple*. Berlin: Springer, 2008.

Hendrickson, D. Scott. *Jesuit Polymath of Madrid: The Literary Enterprise of Juan Eusebio Nieremberg (1595–1658)*. Leiden: Brill, 2015.

Henry, Devin. *Aristotle on Matter, Form, and Moving Causes: The Hylomorphic Theory of Substantial Generation*. Cambridge: Cambridge University Press, 2019. http://10.1017/9781108646680.009.

Hidalgo-Serna, Emilio. *El pensamiento ingenioso en Baltasar Gracián: el concepto y su función lógica*. Barcelona: Anthropos, 1993.

Hirai, Hiro. "Bodies and Their Internal Powers: Natural Philosophy, Medicine and Alchemy." In *The Routledge Companion to Sixteenth Century Philosophy*, edited by Henrik Lagerlund and Benjamin Hill, 394–410. London: Routledge, 2017.

———. "Earth's Soul and Spontaneous Generation: Fortunio Liceti's Criticism of Ficino's Ideas on the Origin of Life." In *Laus Platonici Philosophi: Marsilio Ficino and His Influence*, edited by Stephen Clucas, Peter J. Forshaw, and Valery Rees, 198–273. Leiden: Brill, 2011.

———. *Le concept de semence dans les théories de la matière à la Renaissance de Marsile Ficin à Pierre Gassendi*. Turnhout: Brepols Publishers, 2005.

———. "Spirit in Renaissance Medicine." In *Encyclopedia of Renaissance Philosophy*, edited by Marco Sgarbi. Cham: Springer, 2018. https://doi.org/10.1007/978-3-319-02848-4_1107-1.

Höpfl, Harro. *Jesuit Political Thought: The Society of Jesus and the State, c. 1540–1640*. Cambridge: Cambridge University Press, 2004.

Hunt, John M. "The Conclave from the 'Outside In': Rumor, Speculation, and Disorder in Rome during Early Modern Papal Elections." *Journal of Early Modern History* 16, no. 4–5 (2012): 355–382. https://doi.org/10.1163/15700658-12342324.

Impey, Oliver R., and Arthur Macgregor, eds. *The Origins of Museums: The Cabinet of Curiosities in Sixteenth- and Seventeenth-Century Europe*. Oxford: Clarendon, 1985.

Infelise, Mario. "The History of a Word: Gazzetta / Gazette." In *News Networks in Early Modern Europe*, edited by Raymond Joad and Noah Moxham, 243–260. Leiden: Brill, 2016.

Ingham, Patricia Clare. *The Medieval New: Ambivalence in an Age of Innovation*. Philadelphia: University of Pennsylvania Press, 2015.

Jacobs, Fredrika H. *The Living Image in Renaissance Art*. Cambridge: Cambridge University Press, 2005.

Jardine, Lisa. *Worldly Goods: A New History of the Renaissance*. New York: Doubleday, 1996.

Jorink, Eric. "Insects, Philosophy and the Microscope." In *Worlds of Natural History*, edited by Helen Anne Curry, Nicholas Jardine, James Andrew Secord, and Emma C. Spary, 131–148. Cambridge: Cambridge University Press, 2018. https://doi.org/10.1017/9781108225229.009.

———. "Snakes, Fungi and Insects: Otto Marseus van Schrieck, Johannes Swammerdam and the Theory of Spontaneous Generation." In *Zoology in Early Modern Culture: Intersections of Science, Theology, Philology, and Political and Religious Education*, edited by Karl A. E. Enenkel and Paul J. Smith, 197–234. Leiden: Brill, 2014. https://doi.org/10.1163/9789004279179_007.

Jouanen, José. *Historia de la Compañía de Jesús en la antigua provincia de Quito (1570–1774)*. 2 vols. Quito: Editorial Ecuatoriana, 1941–1943.

Kallendorf, Craig W., ed. *Humanist Educational Treatises*. Cambridge, MA: Harvard University Press, 2002.

Kapoor, Ramesh. "Nūr Ud-Dīn Jahāngīr and Father Kirwitzer: The Independent Discovery of the Great Comets of November 1618 and the First Astronomical Use of the Telescope in India." *Journal of Astronomical History and Heritage* 19, no. 3 (2016): 264–297.

Keller, Evelyn Fox. *Reflections on Gender and Science*. 2nd ed. New Haven, CT: Yale University Press, 1995.

Keller, Vera. "A Charlatan's Ingenuity: Juggling, Joking, and Medical Reform in Johann Ernst Burggrav's *Lamp of Life and Death*." In *Ingenuity in the Making: Matter and Technique in Early Modern Europe*, edited by Richard J. Oosterhoff, José Ramón Marcaida López, and Alexander Marr, 174–188. Pittsburgh, PA: University of Pittsburgh Press, 2021.

Kemp, Martin. "From Mimesis to Fantasia: The Quattrocento Vocabulary of Creation, Inspiration and Genius in the Visual Arts." *Viator* 8 (1977): 347–398. https://doi.org/10.1484/J.VIATOR.2.301573.

Kenny, Neil. *The Uses of Curiosity in Early Modern France and Germany*. Oxford: Oxford University Press, 2004.

Kimmel, Seth. "Automatons and the Early Modern Drama of Skepticism." In *Science on Stage in Early Modern Spain*, edited by Enrique García Santo-Tomás, 210–230. Toronto: University of Toronto Press, 2019. https://doi.org/10.3138/9781487519346-010.

King, Rachel. "Collecting Nature Within Nature: Animal Inclusions in Amber in Early Modern Collections." In *Collecting Nature*, edited by Andrea Gáldy and Sylvia Heudecker, 1–18. Newcastle upon Tyne: Cambridge Scholars Publishing, 2014.

Klein, Ursula, and E. C. Spary, eds. *Materials and Expertise in Early Modern Europe: Between Market and Laboratory*. Chicago: University of Chicago Press, 2010.

Klibansky, Raymond, Erwin Panofsky, Fritz Saxl, and Bill Sherman. *Saturn and Melancholy: Studies in the History of Natural Philosophy, Religion, and Art*. Montreal: McGill-Queen's University Press, 2019. https://doi.org/10.2307/j.ctvscxt1r.

Kociszewska, Ewa. "Displays of Sugar Sculpture and the Collection of Antiquities in Late Renaissance Venice." *Renaissance Quarterly* 73, no. 2 (2020): 441–488. https://doi.org/10.1017/rqx.2020.2.

Koselleck, Reinhart, and Michaela W. Richter. "Crisis." *Journal of the History of Ideas* 67, no. 2 (2006): 357–400.

Kragh, Helge. "Inventive Nature: Fossils, Stones and Marvels in Seventeenth Century Danish Natural Philosophy." In *Naturspiele. Beiträge zu einem naturhistorischen Konzept der Frühen Neuzeit*, edited by Marie-Theres Federhofer, 15–24. Heidelberg: Palatina Verlag, 2006.

Kress, Emily. "Aristotle on Spontaneous Generation, Spontaneity, and Natural Processes." In *Oxford Studies in Ancient Philosophy*, edited by Victor Caston, 157–204. Oxford: Oxford University Press, 2020. https://doi.org/10.1093/oso/9780198858997.003.0005.

Lachmann, Renate. "'Problematic Similarity': Sarbiewski's Treatise *De acuto et arguto* in the Context of Concettistic Theories of the 17th Century." *Russian Literature* 27, no. 2 (1990): 239–251. https://doi.org/10.1016/0304-3479(90)90016-A.

Lao, Eugenia. "Luxury and the Creation of a Good Consumer." In *Pliny the Elder: Themes and Contexts*, edited by Roy Gibson and Ruth Morello, 35–56. Leiden: Brill, 2011.

Laurens, Pierre. *L'abeille dans l'ambre: célébration de l'épigramme de l'époque alexandrine à la fin de la Renaissance*. Paris: Les Belles Lettres, 1989.

Lehoux, Daryn. *Creatures Born of Mud and Slime: The Wonder and Complexity of Spontaneous Generation*. Baltimore: Johns Hopkins University Press, 2017.

———. "Laws of Nature and Natural Laws." *Studies in History and Philosophy of Science Part A* 37 (2006): 527–549.

Leitão, Henrique. "Manuel Bocarro Francês e o *Tratado dos cometas de 1618*." In *Manuel Bocarro Francês e o* Tratado dos cometas de 1618, edited by Henrique Leitão, 7–17. Lisbon: Biblioteca Nacional de Portugal, 2009.

Leonhard, Karin. *The Fertile Ground of Painting: Seventeenth-Century Still Lifes and Nature Pieces*. Translated by Russell Stockman. London: Harvey Miller, 2020.

Lewis, Rhodri. "Francis Bacon and Ingenuity." *Renaissance Quarterly* 67, no. 1 (2014): 113–163.

———. *Language, Mind, and Nature: Artificial Languages in England from Bacon to Locke*. Cambridge: Cambridge University Press, 2007.

Lincoln, Evelyn. *Brilliant Discourse: Pictures and Readers in Early Modern Rome*. New Haven, CT: Yale University Press, 2014.

Lipking, Lawrence. *What Galileo Saw: Imagining the Scientific Revolution*. Ithaca, NY: Cornell University Press, 2014.

Long, Pamela O. *Artisan/Practitioners and the Rise of the New Sciences, 1400–1600*. Corvallis: Oregon State University Press, 2011.

———. "The Openness of Knowledge: An Ideal and Its Context in 16th-Century Writings on Mining and Metallurgy." *Technology and Culture* 32, no. 2.1 (1991): 318–355.

———. *Openness, Secrecy, Authorship: Technical Arts and the Culture of Knowledge from Antiquity to the Renaissance*. Baltimore: Johns Hopkins University Press, 2001.

Lugli, Adalgisa. "La ricerca come collezione." In *Enciclopedismo in Roma barocca*, edited by Maristella Casciati, Maria Grazia Ianniello, and Maria Vitale, 268–281. Venice: Marsilio, 1986.

Lüthy, Christoph H. "Atomism, Lynceus, and the Fate of Seventeenth-Century Microscopy." *Early Science and Medicine* 1, no. 1 (1996): 1–27.

Lyons, John D. *Before Imagination: Embodied Thought from Montaigne to Rousseau*. Stanford, CA: Stanford University Press, 2005.

Maitte, Corinne. "Façon de Venise: Determining the Value of Glass in Early Modern Europe." In *Concepts of Value in European Material Culture, 1500–1900*, edited by Bert De Munck and Dries Lyna, 209–237. Farnham, UK: Ashgate, 2015.

Malet, Antoni. "Early Conceptualizations of the Telescope as an Optical Instrument." *Early Science and Medicine* 10, no. 2 (2005): 237–262.

Mandabach, Marisa. "Matter as an Artist: Rubens's Myths of Spontaneous Generation." *Journal of Historians of Netherlandish Art* 13, no. 2 (2021). https://doi.org/10.5092/jhna.13.2.3.

Manning, Patricia W. "Publication Protocol in the Seventeenth-Century Jesuit Province of Aragon." *Explorations in Renaissance Culture* 33, no. 1 (2007): 133–161.

Mantero, Anne. "Énigmes de la brièveté: quelques théories de l'épigramme à l'âge baroque." In *Bagatelles pour l'éternité: L'art du bref en littérature*, edited by Philippe Baron, 203–221. Besançon: Presses universitaires de Franche-Comté, 2019. http://books.openedition.org/pufc/2611.

Marcaida, José Ramón. *Arte y ciencia en el Barroco español: historia natural, coleccionismo y cultura visual*. Madrid: Marcial Pons, 2014.

———. "Juan Eusebio Nieremberg y la ciencia del Barroco. Conocimiento y representación de la naturaleza en la España del siglo XVII." PhD diss., Universidad Autónoma de Madrid, 2011. http://hdl.handle.net/10486/7757.

Marini, Quinto. "Elisabetta Graziosi. Cesura per il secolo dei genovesi: Anton Giulio Brignole Sale. 'Studi secenteschi,' XLI, 2000, pp. 27–85." *La rassegna della letteratura italiana* 105, no. 1 (2001): 240–242.

Markey, Lia, ed. *Renaissance Invention: Stradanus's* Nova Reperta. Evanston, IL: Northwestern University Press, 2020.

Marr, Alexander. *Between Raphael and Galileo: Mutio Oddi and the Mathematical Culture of Late Renaissance Italy*. Chicago: University of Chicago Press, 2011.

———. "Introduction." In *Curiosity and Wonder from the Renaissance to the Enlightenment*, edited by R. J. W. Evans and Alexander Marr, 1–20. Aldershot, UK: Ashgate, 2006.

———. "A Mirror of Wisdom: Simon Vouet's *Satyrs Admiring the Anamorphosis of an Elephant* and Its Afterlives." In *Teaching Philosophy in Early Modern Europe: Text and Image*, edited by Susanna Berger and Daniel Garber, 265–301. Cham: Springer, 2021. https://doi.org/10.1007/978-3-030-84621-3_10.

———. "Pregnant Wit: Ingegno in Renaissance England." *British Art Studies*, no. 1 (2015). https://doi.org/10.17658/issn.2058-5462/issue-01/amarr.

———. *Rubens's Spirit: From Ingenuity to Genius*. London: Reaktion Books, 2021.

Marr, Alexander, Raphaële Garrod, José Ramón Marcaida López, and Richard J. Oosterhoff. *Logodaedalus: Word Histories of Ingenuity in Early Modern Europe*. Pittsburgh, PA: University of Pittsburgh Press, 2019.

Marsh, Ben. *Unravelled Dreams: Silk and the Atlantic World, 1500–1840*. Cambridge: Cambridge University Press, 2020. https://doi.org/10.1017/9781108289672.

Martin, John. "Inventing Sincerity, Refashioning Prudence: The Discovery of the Individual in Renaissance Europe." *American Historical Review* 102, no. 5 (1997): 1309–1342.

Maus de Rolley, Thibaut. *Élévations: L'écriture du voyage aérien à la Renaissance*. Geneva: Droz, 2011.

May, Terence E. "Gracián's Idea of the Concepto." *Hispanic Review* 18, no. 1 (1950): 15–41.

Mazzeo, Joseph Anthony. "Metaphysical Poetry and the Poetic of Correspondence." *Journal of the History of Ideas* 14, no. 2 (1953): 221–234.

McLaughlin, Peter. "Spontaneous versus Equivocal Generation in Early Modern Science." *Annals of the History and Philosophy of Biology* 10 (2005): 79–88.

Meadow, Mark A. "Quiccheberg and the Copious Object: Wenzel Jamnitzer's Silver Writing Box." In *Lure of the Object*, edited by Stephen Melville, 39–58. Williamstown, MA: Sterling and Francine Clark Art Institute, 2005.

Meeusen, Michiel. "Aristotle's Second Breath: Pneumatic Processes in the *Natural Problems* (On Sexual Intercourse)." In *The Concept of Pneuma after Aristotle*, edited by Sean Coughlin, David Leith, and Orly Lewis, 63–91. Berlin: Topoi, 2020.

Menéndez y Pelayo, Marcelino. *Historia de las ideas estéticas en España*. 5 vols. Madrid: A. Pérez Dubrull, 1883–1889.

Merchant, Carolyn. *The Death of Nature: Women, Ecology, and the Scientific Revolution*. San Francisco: Harper & Row, 1980.

———. "The Scientific Revolution and The Death of Nature." *Isis* 97, no. 3 (2006): 513–533.

Merriam-Webster.com Dictionary. Springfield, MA: Merriam-Webster, 2022. https://www.merriam-webster.com/dictionary/conceit.

Michalsky, Katarzyna, and Sergiusz Michalsky. *Spider*. London: Reaktion Books, 2010.

Micheli, Gianni. *Le origini del concetto di macchina*. Florence: Leo S. Olschki, 1995.

Miller, David Marshall. "Seeing and Believing: Galileo, Aristotelians, and the Mountains on the Moon." In *The Starry Messenger*, edited by Daniel De Simone and John Hessler, 131–145. Delray Beach, FL: Levenger Press, 2013.

Miller, Patricia Cox. *In the Eye of the Animal: Zoological Imagination in Ancient Christianity*. Philadelphia: University of Pennsylvania Press, 2018.

Minsaas, Kirsti. "Poetic Marvels: Aristotelian Wonder on Renaissance Poetics and Poetry." In *Making Sense of Aristotle: Essays on Poetics*, edited by Øivind Andersen and Jon Haarberg, 145–171. London: Gerald Duckworth, 2001.

Mintz, Sidney W. *Sweetness and Power: The Place of Sugar in Modern History*. New York: Viking, 1985.

Molina, Stefania. "Matteo Peregrini poeta." *Aevum* 66, no. 3 (1992): 571–599.

Moreau, Elisabeth. "Innate Heat." In *Encyclopedia of Renaissance Philosophy*, edited by Marco Sgarbi, 1–3. Cham: Springer, 2015. https://doi.org/10.1007/978-3-319-02848-4_399-1.

Mori, Giuliano. "Mathematical Subtleties and Scientific Knowledge: Francis Bacon and Mathematics, at the Crossing of Two Traditions." *British Journal for the History of Science* 50, no. 1 (2017): 1–21.

Morpurgo Tagliabue, Guido. "Aristotelismo e Barocco." In *Retorica e Barocco. Atti del III Congresso Internazionale di Studi Umanistici, Venezia, 15–18 giugno 1954*, edited by Enrico Castelli, 119–196. Rome: Bocca, 1955.

Moss, Jean Dietz. *Novelties in the Heavens: Rhetoric and Science in the Copernican Controversy*. Chicago: University of Chicago Press, 1993.

Muir, Edward. *The Culture Wars of the Late Renaissance: Skeptics, Libertines, and Opera*. Cambridge, MA: Harvard University Press, 2007.

Mulsow, Martin. "Idolatry and Science: Against Nature Worship from Boyle to Rüdiger, 1680–1720." *Journal of the History of Ideas* 67, no. 4 (2006): 697–712.

Murphy, Hannah. "Calligraphy and Metamorphosis: Invention and Imitation in a Sixteenth-Century Craft." In *Ingenuity in the Making: Matter and Technique in Early Modern Art and Science*, edited by Richard Oosterhoff, José Ramón Marcaida, and Alexander Marr, 79–93. Pittsburgh, PA: University of Pittsburgh Press, 2021.

Nedzinskaitė, Živilė. "'Finis Epigrammatis est Anima Eius': Transformations of the Content of the Latin Epigram in the Epoch of the Baroque." *Interlitteraria* 19, no. 2 (2014): 276.

———. "M. K. Sarbievijaus *De acuto et arguto* sklaida ir interpretacijos XVII–XVIII amžiaus rankraštiniuose mokykliniuose poetikos ir retorikos kursuose." *Senoji Lietuvos literatūra* 23 (2007): 43–80.

Neri, Janice. *The Insect and the Image: Visualizing Nature in Early Modern Europe, 1500–1700*. Minneapolis: University of Minnesota Press, 2011.

Newman, William R. *Promethean Ambitions: Alchemy and the Quest to Perfect Nature*. Chicago: University of Chicago Press, 2004.

Nikulin, Dmitri. "Memory in Ancient Philosophy." In *Memory: A History*, edited by Dmitri Nikulin, 35–84. Oxford: Oxford University Press, 2015.

Nowicki, Jürgen. *Die Epigrammtheorie in Spanien vom 16. bis 18. Jahrhundert. Eine Vorarbeit zur Geschichte der Epigrammatik*. Wiesbaden: Franz Steiner Verlag, 1974.

Nutton, Vivian. "The Reception of Fracastoro's Theory of Contagion: The Seed That Fell among Thorns?" *Osiris* 6 (1990): 196–234.

———. "The Seeds of Disease: An Explanation of Contagion and Infection from the Greeks to the Renaissance." *Medical History* 27, no. 1 (January 1983): 1–34.

O'Connor, Scott. "The Subjects of Natural Generations in Aristotle's *Physics* I.7." *Apeiron* 48, no. 1 (2015): 45–75. https://doi.org/10.1515/apeiron-2014-0020.

Ogilvie, Brian W. "Order of Insects: Insect Species and Metamorphosis between Renaissance and Enlightenment." In *The Life Sciences in Early Modern Philosophy*, edited by O. Nachtomy and J. E. H. Smith, 222–245. Oxford: Oxford University Press, 2014.

Oosterhoff, Richard J. "Ingenious Materials: Salts as Material Metaphor." In *Secrets of Craft and Nature in Renaissance France. A Digital Critical Edition and English Translation of BNF Ms. Fr. 640*, edited by Pamela H. Smith. New York: Making and Knowing Project, 2020. https://www.doi.org/10.7916/4vf4-2867.

Oosterhoff, Richard J., José Ramón Marcaida López, and Alexander Marr, eds. *Ingenuity in the Making: Materials and Technique in Early Modern Art and Science*. Pittsburgh, PA: University of Pittsburgh Press, 2021.

Ortalli, Gherardo. *La pittura infamante nei secoli XIII–XVI*. Rome: Jouvence, 1979.

Ossa-Richardson, Anthony. *A History of Ambiguity*. Princeton, NJ: Princeton University Press, 2019.

Palmerino, Carla Rita. "Reading the Book of Nature: The Ontological and Epistemological Underpinnings of Galileo's Mathematical Realism." In *The Language of Nature: Reassessing the Mathematization of Natural Philosophy in the Seventeenth Century*, edited by Geoffrey Gorham, Benjamin Hill, Edward Slowik, and C. Kenneth Waters, 29–50.

Minneapolis: University of Minnesota Press, 2016. https://doi.org/10.5749/j.ctt1d39org.4.

Park, Katharine. "The Organic Soul." In *The Cambridge History of Renaissance Philosophy*, edited by C. B. Schmitt, Quentin Skinner, Eckhard Kessler, and Jill Kraye, 464–484. Cambridge: Cambridge University Press, 1988. https://doi.org/10.1017/CHOL9780521251044.016.

Park, Katharine, and Lorraine Daston. "Introduction: The Age of the New." In *The Cambridge History of Science*, vol. 3: *Early Modern Science*, edited by Katharine Park and Lorraine Daston, 1–18. Cambridge: Cambridge University Press, 2006. https://doi.org/10.1017/CHOL9780521572446.002.

———. "Unnatural Conceptions: The Study of Monsters in Sixteenth- and Seventeenth-Century France and England." *Past and Present*, no. 92 (1981): 20–54.

Parker, Alexander A. "'Concept' and 'Conceit': An Aspect of Comparative Literary History." *Modern Language Review* 77, no. 4 (1982): xxi–xxxv.

Parker, Geoffrey. *Global Crisis: War, Climate Change and Catastrophe in the Seventeenth Century*. New Haven, CT: Yale University Press, 2014.

Pasolini, Pier Paolo. *Scritti corsari*. Milan: Garzanti, 1975.

Patiño Loira, Javier. "Acuity of Wit: Wonder, Paradox and Cooperation in Early Modern Spanish and Italian Poetic Theory (1548–1648)." PhD diss., Princeton University, 2016.

———. "Policing Talent in Early Modern Jesuit Rome: Difference, Self-Knowledge, and Career Specialization." In *The Quest for Certainty in Early Modern Europe: From Inquisition to Inquiry, 1550–1700*, edited by Barbara Fuchs and Mercedes García Arenal, 132–156. Toronto: University of Toronto Press, 2020. https://doi.org/10.3138/9781487535506-007.

Pender, E. E. "Spiritual Pregnancy in Plato's Symposium." *Classical Quarterly* 42, no. 1 (1992): 72–86.

Pérez García, Pablo. "Un texto académico del ocaso del siglo XVII: el 'Discurso sobre el microscopio' de José Ortí y Moles (Estudio y edición)." *Saitabi*, no. 60–61 (2010–2011): 347–373.

Pesic, Peter. *Music and the Making of Modern Science*. Cambridge, MA: MIT Press, 2014.

———. "Wrestling with Proteus: Francis Bacon and the 'Torture' of Nature." *Isis* 90, no. 1 (1999): 81–94.

Picciotto, Joanna. *Labors of Innocence in Early Modern England*. Cambridge, MA: Harvard University Press, 2010.

Pieri, Marzio, and Diego Varini. *Il buratto ed il punto: concettismo, retorica e pittura fra Genova e Bologna: 1629–1652*. Trento: La Finestra, 2006.

Pigman, George W. "Versions of Imitation in the Renaissance." *Renaissance Quarterly* 33, no. 1 (1980): 1–32. https://doi.org/10.2307/2861533.

Pimentel, Juan. *The Rhinoceros and the Megatherium: An Essay in Natural History*. Translated by Peter Mason. Cambridge, MA: Harvard University Press, 2017.

Pinto-Correia, Clara. *The Ovary of Eve: Egg and Sperm and Preformation*. Chicago: University of Chicago Press, 1997.

Polyakova, Irina A. "Collecting Amber Naturalia in Sixteenth-Century Prussia." In *Collection in the Space of Culture*, edited by Irina A. Polyakova, Christopher J. Duffin, and Tatiana J. Suvorova, 63–81. Kaliningrad: Kaliningrad Regional Amber Museum, 2019.

Pomata, Gianna. "Innate Heat, Radical Moisture and Generation." In *Reproduction: Antiquity to the Present Day*, edited by Lauren Kassell, Nick Hopwood, and Rebecca Flemming, 195–208. Cambridge: Cambridge University Press, 2018. https://doi.org/10.1017/9781107705647.020.

Pomian, Krzysztof. *Collectors and Curiosities. Paris and Venice, 1500–1800*. Cambridge: Polity, 1990.

Pons, Alain. "Ingenium." In *Dictionary of Untranslatables: A Philosophical Lexicon*, edited by Barbara Cassin, Emily Apter, Jacques Lezra, and Michael Wood, 487–488. Princeton, NJ: Princeton University Press, 2014.

Popkin, Richard H. *The History of Scepticism from Savonarola to Bayle*. Rev. ed. Oxford: Oxford University Press, 2003.

Pozuelo Yvancos, José María. "La Agudeza y arte de ingenio, primera neorretórica." In *Baltasar Gracián. IV Centenario (1601–2001)*, edited by Aurora Egido, María del Carmen Marín, and Luis Sánchez Laílla, 133–149. Zaragoza: Instituto de Estudios Altoaragoneses, 2003.

Praz, Mario. *Studi sul concettismo*. Milan: La Cvltvra, 1934.

Preti, Cesare, and Maria Grazia Ercolino. "Grassi, Orazio." *Dizionario biografico degli italiani* 58 (2002): 669–675.

Rabaté, Philippe. "Materia y forma en *Agudeza y arte de ingenio*." *Conceptos. Revista de investigación graciana* 6 (2009): 45–66.

Raimondi, Ezio. *Anatomie secentesche*. Pisa: Nistri-Lischi, 1966.

Rampling, Jennifer M. "Spirited Matter and Ingenious Nature: Accounting for Alchemical Change." In *Ingenuity in the Making: Matter and Technique in Early Modern Europe*, edited by Richard Oosterhoff, José Ramón Marcaida, and Alexander Marr, 13–24. Pittsburgh, PA: University of Pittsburgh Press, 2021.

Reeves, Eileen. *Evening News: Optics, Astronomy, and Journalism in Early Modern Europe*. Philadelphia: University of Pennsylvania Press, 2014.

———. *Galileo's Glassworks: The Telescope and the Mirror*. Cambridge, MA: Harvard University Press, 2008.

———. *Painting the Heavens: Art and Science in the Age of Galileo*. Princeton, NJ: Princeton University Press, 1999.

Reeves, Eileen, and Albert Van Helden. *On Sunspots*. Chicago: University of Chicago Press, 2010.

Remer, Gary. "Justus Lipsius, Morally Acceptable Deceit and Prudence in the Ciceronian Tradition." *History of Political Thought* 37, no. 2 (2016): 238–270.

Reula Baquero, Pedro. *El camarín del desengaño: Juan de Espina, coleccionista y curioso del siglo XVII*. Madrid: Centro de Estudios Europa Hispánica, 2019.

Rimell, Victoria. *Martial's Rome: Empire and the Ideology of Epigram*. Cambridge: Cambridge University Press, 2009. https://doi.org/10.1017/CBO9780511575860.

Riskin, Jessica. *The Restless Clock: A History of the Centuries-Long Argument over What Makes Living Things Tick*. Chicago: University of Chicago Press, 2016.

Roberts, Lissa, and Simon Schaffer. "Preface." In *The Mindful Hand: Inquiry and Invention from the Late Renaissance to Early Industrialisation*, edited by Lissa Roberts, Simon

Schaffer, and Peter Dear, xiii–xxvii. Amsterdam: Koninklijke Nederlandse Akademie van Wetenschappen, 2007.

Rocke, Alan J. *Image and Reality: Kekulé, Kopp, and the Scientific Imagination*. Chicago: University of Chicago Press, 2010.

Rodríguez Beltrán, Joaquín. *La agudeza del ingenio en la Nueva España: La* Oratio pro instauratione studiorum *(1644) de Baltasar López*. Mexico City: Universidad Autónoma de México, 2018.

Rolfe, Kirsty. "St Paul's and London News Culture, 1618–1632." In *Old St Paul's and Culture*, edited by Shanyn Altman and Jonathan Buckner, 319–340. Cham: Springer, 2021. https://doi.org/10.1007/978-3-030-77267-3_14.

Rolfe, W. D. Ian. "Materia Medica in the Seventeenth-Century Paper Museum of Cassiano Dal Pozzo." In *A History of Geology and Medicine*, edited by C. J. Duffin, R. T. J. Moody, and C. Gardner-Thorpe, 137–156. London: Geological Society of London, 2012. http://dx.doi.org/10.1144/SP375.3.

Romano, Antonella. *La contre-réforme mathématique: constitution et diffusion d'une culture mathématique jésuite à la Renaissance (1540–1640)*. Rome: École française de Rome, 1999.

Roos, Anna Marie Eleanor. *The Salt of the Earth: Natural Philosophy, Medicine, and Chymistry in England, 1650–1750*. Leiden: Brill, 2007.

Rosen, Edward. "The Dissolution of the Solid Celestial Spheres." *Journal of the History of Ideas* 46, no. 1 (1985): 13–31. https://doi.org/10.2307/2709773.

———. "The Title of Galileo's *Sidereus nuncius*." *Isis* 41, no. 3/4 (1950): 287–289.

Rossi, Paolo. *Francis Bacon from Magic to Science*. Translated by Sacha Rabinovitch. Chicago: University of Chicago Press, 1968.

———. *La nascita della scienza moderna in Europa*. Rome: Laterza, 1997.

Rothman, Aviva. "Johannes Kepler's Pursuit of Harmony." *Physics Today* 73, no. 1 (2020): 36–42. https://doi.org/10.1063/PT.3.4388.

———. *The Pursuit of Harmony: Kepler on Cosmos, Confession, and Community*. Chicago: University of Chicago Press, 2018.

Ruan, Felipe E. *Pícaro and Cortesano: Identity and the Forms of Capital in Early Modern Spanish Picaresque Narrative and Courtesy Literature*. Lewisburg, PA: Bucknell University Press, 2011.

Rublack, Ulinka. "Matter in the Material Renaissance." *Past and Present* 219 (2013): 41–85.

Ruestow, Edward G. *The Microscope in the Dutch Republic: The Shaping of Discovery*. Cambridge: Cambridge University Press, 1996.

Russo, Alessandra. "An Artistic Humanity: New Positions on Art and Freedom in the Context of Iberian Expansion, 1500–1600." *Res: Anthropology and Aesthetics* 65–66 (2015): 352–363. https://doi.org/10.1086/691045.

Russo, Emilio. "Introduzione." In *Adone*, by Giovan Battista Marino, 5–29. Milan: BUR, 2013.

Rusu, Doina-Cristina. "Deceiving the Senses: The Role of Vapours in Francis Bacon's Natural Philosophy." In *Ingenuity in the Making: Matter and Technique in Early Modern Art and Science*, edited by Richard Oosterhoff, José Ramón Marcaida, and Alexander Marr, 25–35. Pittsburgh, PA: University of Pittsburgh Press, 2021.

———. "Imagination in Early Modern Theory of Knowledge." In *Encyclopedia of Early Modern Philosophy and the Sciences*, edited by Dana Jalobeanu and Charles T. Wolfe, 1–9. Cham: Springer, 2020. https://doi.org/10.1007/978-3-319-20791-9_403-1.

Sakamoto, Kuni. *Julius Caesar Scaliger, Renaissance Reformer of Aristotelianism: A Study of His Exotericae Exercitationes*. Leiden: Brill, 2016.

Salvatore, Filippo. *Antichi e Moderni in Italia nel Seicento*. Montreal: Guernica Editions, 1987.

Sánchez, Francisco J. "Symbolic Wealth and Theatricality in Gracian." In *Rhetoric and Politics: Baltasar Gracián and the New World Order*, edited by Nicholas Spadaccini and Jenaro Taléns, 209–229. Minneapolis: University of Minnesota Press, 1997.

Sarmiento, Edward. "Gracián's 'Agudeza y Arte de Ingenio.'" *Modern Language Review* 27, no. 3 (1932): 280–292.

Sauter, Michael J. *The Spatial Reformation: Euclid between Man, Cosmos, and God*. Philadelphia, PA: University of Pennsylvania Press, 2019.

Savoia, Paolo. "Knowing Nature by Its Surface: Butchers, Barbers, Surgeons, Gardeners, and Physicians in Early Modern Italy." *Centaurus* 64, no. 2 (2022): 399–420. https://doi.org/10.1484/j.cnt.5.129636.

Schaffer, Simon. "Regeneration." In *Science Incarnate: Historical Embodiments of Natural Knowledge*, edited by Christopher Lawrence and Steven Shapin, 83–120. Chicago: University of Chicago Press, 1998.

Schenkeveld, Dirk M. "*Ta Asteia* in Aristotle's *Rhetoric*: The Disappearance of a Category." In *Peripatetic Rhetoric after Aristotle*, edited by William F. Fortenbaugh and David C. Mirhady, 1–14. New Brunswick, NJ: Transaction, 1994.

Schiefsky, Mark John. "Art and Nature in Ancient Mechanics." In *The Artificial and the Natural: An Evolving Polarity*, edited by Bernadette Bensaude-Vincent and William R. Newman, 67–108. Cambridge, MA: MIT Press, 2007.

Scholar, Richard. *The Je-Ne-Sais-Quoi in Early Modern Europe: Encounters with a Certain Something*. Oxford: Oxford University Press, 2005. https://doi.org/10.1093/acprof:oso/9780199274406.001.0001.

Schöttler, Tobias. "From Causes to Relations: The Emergence of a Non-Aristotelian Concept of Geometrical Proof Out of the *Quaestio de Certitudine Mathematicarum*." *Society and Politics* 6, no. 2 (2012): 29–47.

Seo, J. Mira. "Plagiarism and Poetic Identity in Martial." *American Journal of Philology* 130, no. 4 (2009): 567–593.

Shapin, Steven. *The Scientific Revolution*. Chicago: University of Chicago Press, 1996.

Shea, William R. "Galileo's Copernicanism: The Science and the Rhetoric." In *The Cambridge Companion to Galileo*, edited by Peter Machamer, 211–243. Cambridge: Cambridge University Press, 1998. https://doi.org/10.1017/CCOL0521581788.007.

Shea, William R., and Tiziana Bascelli. *Galileo's Sidereus Nuncius; or, A Sidereal Message*. Sagamore Beach, MA: Science History, 2009.

Shirilan, Stephanie. "Francis Bacon, Robert Burton, and the Thick Skin of the World: Sympathy, Transmission, and the Imaginary Early Modern Skin." *ESC: English Studies in Canada* 34, no. 1 (2008): 59–83. https://doi.org/10.1353/esc.0.0104.

Sifakis, G. M. "Aristotle, 'Poetics' 17, 1455a29–34: People in Real Life, Poets, or Spectators in the Grip of Passion?" *Classical Quarterly* 59, no. 2 (2009): 486–493.

Simeoni, Luca. "*Machina* e *machinatio* in Vitruvio." In *Machina: XI Colloquio internazionale, Roma, 8–10 gennaio 2004*, edited by Marco Veneziani, 141–157. Florence: Leo S. Olschki, 2005.

Siraisi, Nancy G. *The Clock and the Mirror: Girolamo Cardano and Renaissance Medicine.* Princeton, NJ: Princeton University Press, 1997.

Smith, A. Mark. *From Sight to Light: The Passage from Ancient to Modern Optics.* Chicago: University of Chicago Press, 2015.

Smith, Albert J. *Metaphysical Wit.* Cambridge: Cambridge University Press, 1991.

Smith, Pamela H. "Between Nature and Art: Casting from Life in Sixteenth-Century Europe." In *Making and Growing: Anthropological Studies of Organisms and Artefacts*, edited by Elizabeth Hallam and Tim Ingold, 45–63. Burlington, VT: Ashgate, 2014.

———. *The Body of the Artisan: Art and Experience in the Scientific Revolution.* Chicago: University of Chicago Press, 2004.

Smith, Pamela H., and Isabella Lores-Chavez. "Counterfeiting Materials, Imitating Nature." In *The Matter of Mimesis: Studies of Mimesis and Materials in Nature, Art and Science*, edited by Marjolijn Bol and E. C. Spary, 27–53. Leiden: Brill, 2023. https://doi.org/10.1163/9789004515413_004.

Snyder, Jon R. "Art and Truth in Baroque Italy, or the Case of Emanuele Tesauro's *Il Cannocchiale Aristotelico.*" *Modern Language Notes* 131, no. 1 (2016): 74–96.

———. *Dissimulation and the Culture of Secrecy in Early Modern Europe.* Berkeley: University of California Press, 2009.

———. *L'estetica del barocco.* Bologna: Il Mulino, 2005.

———. "Strokes of Wit: Theorizing Beauty in Baroque Italy." In *The Insistence of Art: Aesthetic Philosophy After Early Modernity*, edited by Paul A. Kottman, 194–226. New York: Fordham University Press, 2017.

Solervicens, Josep, ed. *La poètica del Barroc: textos teòrics catalans.* Lleida: Punctum, 2012.

Spingarn, Joel Elias. *Critical Essays of the Seventeenth Century.* 3 vols. Oxford: Clarendon, 1908–1909.

Squire, Michael. *The Iliad in a Nutshell: Visualizing Epic on the Tabulae Iliacae.* Oxford: Oxford University Press, 2011.

———. "Making Myron's Cow Moo? Ecphrastic Epigram and the Poetics of Simulation." *American Journal of Philology* 131, no. 4 (2010): 589–634.

Stabile, Giorgio. "*Machina* e *machinatio* in età barocca." In *Machina: XI Colloquio Internazionale, Roma, 8–10 Gennaio 2004*, edited by Marco Veneziani, 321–332. Florence: Leo S. Olschki, 2005.

Steel, Karl. "Creeping Things: Spontaneous Generation and Material Creativity." In *Elemental Ecocriticism: Thinking with Earth, Air, Water, and Fire*, edited by Lowell Duckert and Jeffrey Jerome Cohen, 209–236. Minneapolis: University of Minnesota Press, 2015.

Stewart, Philip. "Science and Superstition: Comets and the French Public in the 18th Century." *American Journal of Physics* 54, no. 1 (1986): 16–24. https://doi.org/10.1119/1.14763.

Stewart, Susan. *On Longing: Narratives of the Miniature, the Gigantic, the Souvenir, the Collection.* Baltimore: Johns Hopkins University Press, 1984.

Stolberg, Michael. "Emotions and the Body in Early Modern Medicine." *Emotion Review* 11, no. 2 (2019): 113–122. https://doi.org/10.1177/1754073918765674.

———. *Uroscopy in Early Modern Europe*. Translated by Logan Kennedy and Leonhard Unglaub. Farnham, UK: Ashgate, 2015.

Stolleis, Michael. *Arcana Imperii und Ratio Status: Bemerkungen zur politischen Theorie des frühen 17. Jahrhunderts*. Göttingen: Vandenhoeck & Ruprecht, 1980.

Strehlau, Sebastian. "Fernrohr und Metapher—Die italienische Barockdichtung im Zeichen des epistemischen Umbruchs im 17. Jahrhundert." In *Transformationen. Wandel, Bewegung, Geschwindigkeit: Beiträge zum XXXIII. Forum Junge Romanistik in Göttingen (15.–17. März 2017)*, edited by Caroline Bacciu, Jaime Cárdenas Isasi, Antje Dreyer, Aenne Gottschalk, María Ximena Ordóñez, and Ana M. Troncoso Salazar, 119–132. Munich: Akademische Verlagsgemeinschaft München, 2019.

Sumrall, Laura. "Natural Magic in Renaissance Science." In *Encyclopedia of Renaissance Philosophy*, edited by Marco Sgarbi, 1–8. Cham: Springer, 2017. https://doi.org/10.1007/978-3-319-02848-4_956-1.

Suthor, Nicola. *Bravura: Virtuosity and Ambition in Early Modern European Painting*. Princeton, NJ: Princeton University Press, 2021.

Swan, Claudia. "Counterfeit Chimeras: Early Modern Theories of the Imagination and the Work of Art." In *Vision and Its Instruments: Art, Science, and Technology in Early Modern Europe*, edited by Alina Payne, 216–237. Philadelphia: University of Pennsylvania Press, 2015.

Sydor, Małgorzata Anna. "La *concordia discors* en Sarbiewski y Gracián." In *Edad de oro cantabrigense: actas del VII Congreso de la Asociación Internacional de Hispanistas del Siglo de Oro*, edited by Anthony Close, 585–590. Vigo: AISO, 2006.

Tatarkiewicz, Wladyslaw, C. Barrett, A. Czerniaskwi, D. Petsch, and J. Harrell. *History of Aesthetics*. 3 vols. The Hague: Mouton, 1970–1974.

Tenney, James. *A History of "Consonance" and "Dissonance."* New York: Excelsior, 1988.

Thompson, Katherine Anne. "Monsters in Paradise: The Representation of the Natural World in the Historias of Bartolome de Las Casas and Gonzalo Fernandez de Oviedo." PhD diss., University of Maryland, College Park, 2010. http://hdl.handle.net/1903/10264.

Thorndike, Lynn. "Newness and Craving for Novelty in Seventeenth-Century Science and Medicine." *Journal of the History of Ideas* 12, no. 4 (1951): 584–598. https://doi.org/10.2307/2707487.

Thouvenin, Pascale. "Une histoire littéraire classique: une Europe moderne des Lettres selon René Rapin." *Dix-septième siècle* 263, no. 2 (2014): 219–232. https://doi.org/10.3917/dss.142.0219.

Torre, Andrea. "Rimirandolo coll'occhialino': piaga, straforo, protratto." *Testo: studi di teoria e storia della letteratura e della critica* 58 (2009): 35–56.

Tressaud, Alain, and Michael Vickers. "Ancient Murrhine Ware and Its Glass Evocations." *Journal of Glass Studies* 49 (2007): 143–152.

Tuck, Richard. *Philosophy and Government, 1572–1651*. Cambridge: Cambridge University Press, 1993.

Tuominen, Miira. "Ancient Theories." In *Sourcebook for the History of the Philosophy of Mind*, edited by Simo Knuuttila and Juha Sihvola, 39–59. Cham: Springer, 2013.

Tutino, Stefania. *Shadows of Doubt. Language and Truth in Post-Reformation Catholic Culture*. Oxford: Oxford University Press, 2014.

———. *Uncertainty in Post-Reformation Catholicism: A History of Probabilism*. New York: Oxford University Press, 2018.

Tuve, Rosemond. *Elizabethan and Metaphysical Imagery: Renaissance Poetic and Twentieth-Century Critics*. Chicago: University of Chicago Press, 1947.

Vaíllo, Carlos. "Crisis." In *Diccionario de conceptos de Baltasar Gracián*, edited by Elena Cantarino and Emilio Blanco, 75–80. Madrid: Cátedra, 2005.

Van Gastel, Joris. "Bernini's Metamorphosis: Sculpture, Poetry, and the Embodied Beholder." *Word and Image* 28, no. 2 (2012): 193–205.

———. *Il Marmo Spirante. Sculpture and Experience in Seventeenth-Century Rome*. Berlin: Akademie Verlag, 2012.

Van Helden, Albert. "Galileo and Scheiner on Sunspots: A Case Study in the Visual Language of Astronomy." *Proceedings of the American Philosophical Society* 140, no. 3 (1996): 358–396.

———. "The Invention of the Telescope." *Transactions of the American Philosophical Society* 67, no. 4 (1977): 1–67. https://doi.org/10.2307/1006276.

Van Hook, J. W. "'Concupiscence of Witt': The Metaphysical Conceit in Baroque Poetics." *Modern Philology* 84, no. 1 (1986): 24–38.

Varo Zafra, Juan. "Grupos tacitistas españoles del siglo XVI." *Signa* 24 (2015): 537–556.

Vázquez Manassero, Margarita Ana. *El "yngenio" en palacio: arte y ciencia en la corte de los Austrias (ca. 1585–1640)*. Madrid: Fundación Juanelo Turriano, 2018.

Vega Ramos, María José. "El azar y la maravilla en la teoría neoaristotélica de la tragedia: la estatua de Mitis en el Renacimiento." *Theatralia* 3 (2000): 123–159.

Vélez-Posada, Andrés. "Handsteine: The Generative Powers of a Mineral Artifact." In *Ingenuity in the Making: Matter and Technique in Early Modern Europe*, edited by Richard J. Oosterhoff, José Ramón Marcaida, and Alexander Marr, 50–62. Pittsburgh, PA: University of Pittsburgh Press, 2021.

Vermij, Rienk. *The Calvinist Copernicans: The Reception of the New Astronomy in the Dutch Republic, 1575–1750*. Amsterdam: Koninklijke Nederlandse Akademie van Wetenschappen, 2002.

Von Staden, Heinrich. "Physis and Techne in Greek Medicine." In *The Artificial and the Natural: An Evolving Polarity*, edited by Bernadette Bensaude-Vincent and William R. Newman, 21–49. Cambridge, MA: MIT Press, 2007.

Waddell, Mark A. *Jesuit Science and the End of Nature's Secrets*. Farnham, UK: Ashgate, 2015.

Wagschal, Steven. *Minding Animals in the Old and New Worlds: A Cognitive Historical Analysis*. Toronto: University of Toronto Press, 2018.

Wallace-Hadrill, Andrew. "Pliny the Elder and Man's Unnatural History." *Greece & Rome* 37, no. 1 (1990): 80–96.

Walsh, P. G. "The Rights and Wrongs of Curiosity (Plutarch to Augustine)." *Greece & Rome* 35, no. 1 (1988): 73–85.

Walters, Lisa. "Optics and Authorship in Margaret Cavendish's *Observations* and *The Blazing World*." *Viator* 45, no. 3 (2014): 377–393. https://doi.org/10.1484/J.VIATOR.5.102933.

Waszink, Jan. "Your Tacitism or Mine? Modern and Early-Modern Conceptions of Tacitus and Tacitism." *History of European Ideas* 36, no. 4 (2010): 375–385.

Watson, Patricia A. "Martial's Snake in Amber: Ekphrasis or Poetic Fantasy?" *Latomus* 60, no. 4 (2001): 938–943.

Watson, Robert N. *Back to Nature: The Green and the Real in the Late Renaissance*. Philadelphia, PA: University of Pennsylvania Press, 2006.

Webster, Erin. *The Curious Eye: Optics and Imaginative Literature in Seventeenth-Century England*. Oxford: Oxford University Press, 2020.

Weeks, Sophie. "Francis Bacon and the Art-Nature Distinction." *Ambix* 54, no. 2 (2007): 117–145.

Werlin, Julianne. "Francis Bacon and the Art of Misinterpretation." *PMLA* 130, no. 2 (2015): 236–251.

Werrett, Simon. *Fireworks: Pyrotechnic Arts and Sciences in European History*. Chicago: University of Chicago Press, 2010.

———. "Watching the Fireworks: Early Modern Observation of Natural and Artificial Spectacles." *Science in Context*, no. 24 (2011): 167–182.

———. "Wonders Never Cease: Descartes's 'Météores' and the Rainbow Fountain." *British Journal for the History of Science* 34, no. 2 (2001): 129–147.

Weststeijn, Thijs. "'Painting's Enchanting Poison': Artistic Efficacy and the Transfer of Spirits." In *Spirits Unseen: The Representation of Subtle Bodies in Early Modern European Culture*, edited by Christine Göttler and Wolfgang Neuber, 141–178. Leiden: Brill, 2007.

White, R. S. "Language of Emotions." In *Early Modern Emotions: An Introduction*, edited by Susan Broomhall, 33–36. Leiden: Routledge, 2017.

Wilding, Nick. "Galileian Angels." In *Conversations with Angels: Essays towards a History of Spiritual Communication, 1100–1700*, edited by Joad Raymond, 67–89. London: Palgrave Macmillan, 2011. https://doi.org/10.1057/9780230316973_4.

Wilson, Malcolm. "Heat, Meteorology, and Spontaneous Generation." In *Heat, Pneuma, and Soul in Ancient Philosophy and Science*, edited by Hynek Bartoš and Colin Guthrie King, 159–181. Cambridge: Cambridge University Press, 2020. https://doi.org/10.1017/9781108651714.

Wittkower, Rudolf, and Margot Wittkower. *Born under Saturn: The Character and Conduct of Artists. A Documented History from Antiquity to the French Revolution*. New York: Random House, 1963.

Wolfe, Jessica Lynn. "Illumination. 'Living Lamps.'" In *Lesser Living Creatures of the Renaissance*, vol. 2: *Concepts*, edited by Keith Botelho and Joseph Campana, 180–196. University Park: Pennsylvania State University Press, 2023.

Woods, Michael J. *Gracián Meets Góngora: The Theory and Practice of Wit*. Warminster, UK: Aris & Phillips, 1995.

Zagoury, David. "Minerva in the Forge of Vulcan: *Ingegno*, *Fatica*, and Imagination in Early Florentine Art Theory." In *Image, Imagination, and Cognition*, edited by Cristoph Lüthy, Claudia Swan, Paul J. J. M. Bakker, and Claus Zittel, 61–93. Leiden: Brill, 2018.

Zanardi, Mario. "La metafora e la sua dinamica di significazione nel 'Cannocchiale aristotelico' di Emanuele Tesauro." *Giornale storico della letteratura italiana* 97, no. 499 (1980): 321–368.

———. "Metafora e gioco nel 'Cannocchiale aristotelico' di Emanuele Tesauro." *Studi secenteschi* 26 (1985): 25–99.

———. "Sulla genesi del 'Cannocchiale aristotelico' di Emanuele Tesauro (I)." *Studi secenteschi* 23 (1982): 3–61.

———. "Sulla Genesi Del 'Cannocchiale Aristotelico' Di Emanuele Tesauro (II)." *Studi secenteschi* 24 (1983): 3–50.

Zucchi, Enrico. "Tiraboschi, Girolamo." In *Dizionario Biografico degli Italiani*, 95 (2019): 718–723.

Index

Page numbers in italics refer to illustrations.

About the Author

JAVIER PATIÑO LOIRA is assistant professor of Spanish at the University of California, Los Angeles. He is the author of articles and book chapters on early modern Italian and Spanish rhetorical and poetic theory, the formation of libraries, as well as ideas on education and translation.

MIX
Papier | Fördert
gute Waldnutzung
FSC® C083411

Kolibri 360 GmbH
Ferdinand-Jühlke-Straße 7
99095 Erfurt, Deutschland
produktsicherheit@kolibri360.de